Water – Energy – Food Nexus
Narratives and Resource Securities

Water - Energy - Food Nexus Narratives and Resource Securities

A Global South Perspective

Edited by

Tafadzwa Mabhaudhi

Centre for Transformative Agricultural and Food Systems (CTAFS), School of Agricultural, Earth and Environmental Sciences, University of KwaZulu-Natal, Pietermaritzburg, South Africa; International Water Management Institute (IWMI-GH) — West Africa Office, Accra, Ghana

Aidan Senzanje

School of Engineering, University of KwaZulu-Natal, Pietermaritzburg, KwaZulu-Natal, South Africa; Centre for Water Resources Research, College of Agriculture, Engineering and Science, University of KwaZulu-Natal, Pietermaritzburg, South Africa

Albert Modi

Centre for Transformative Agricultural and Food Systems (CTAFS), School of Agricultural, Earth and Environmental Sciences, University of KwaZulu-Natal, Pietermaritzburg, South Africa

Graham Jewitt

IHE Delft Institute for Water Education, Delft, the Netherlands; Civil Engineering and Geosciences, Delft University of Technology, Netherlands; Centre for Water Resources Research, School of Agricultural, Earth and Environmental Sciences, University of KwaZulu-Natal (UKZN), Durban, South Africa

Festo Massawe

School of Biosciences, University of Nottingham Malaysia, Jalan Broga, Seminyih, Malaysia; Centre for Transformative Agricultural and Food Systems (CTAFS), School of Agricultural, Earth and Environmental Sciences, University of KwaZulu-Natal, Pietermaritzburg, South Africa

ELSEVIER

Elsevier
Radarweg 29, PO Box 211, 1000 AE Amsterdam, Netherlands
The Boulevard, Langford Lane, Kidlington, Oxford OX5 1GB, United Kingdom
50 Hampshire Street, 5th Floor, Cambridge, MA 02139, United States

Copyright © 2022 Elsevier Inc. All rights reserved.

No part of this publication may be reproduced or transmitted in any form or by any means, electronic or mechanical, including photocopying, recording, or any information storage and retrieval system, without permission in writing from the publisher. Details on how to seek permission, further information about the Publisher's permissions policies and our arrangements with organizations such as the Copyright Clearance Center and the Copyright Licensing Agency, can be found at our website: www.elsevier.com/permissions.

This book and the individual contributions contained in it are protected under copyright by the Publisher (other than as may be noted herein).

Notices

Knowledge and best practice in this field are constantly changing. As new research and experience broaden our understanding, changes in research methods, professional practices, or medical treatment may become necessary.

Practitioners and researchers must always rely on their own experience and knowledge in evaluating and using any information, methods, compounds, or experiments described herein. In using such information or methods they should be mindful of their own safety and the safety of others, including parties for whom they have a professional responsibility.

To the fullest extent of the law, neither the Publisher nor the authors, contributors, or editors, assume any liability for any injury and/or damage to persons or property as a matter of products liability, negligence or otherwise, or from any use or operation of any methods, products, instructions, or ideas contained in the material herein.

ISBN: 978-0-323-91223-5

> For information on all Elsevier publications visit our website at
> https://www.elsevier.com/books-and-journals

Publisher: Candice Janco
Acquisitions Editor: Gabriela D. Capille
Editorial Project Manager: Michelle Fisher
Production Project Manager: Bharatwaj Varatharajan
Cover Designer: Mark Rogers

Typeset by TNQ Technologies

Contents

CONTRIBUTORS xiii
FOREWORD xvii

CHAPTER 1   The water—energy—food nexus: its transition into a transformative approach 1
                 Sylvester Mpandeli, Luxon Nhamo, Aidan Senzanje, Graham Jewitt, Albert Modi, Festo Massawe and Tafadzwanashe Mabhaudhi
                 1. Introduction 1
                      1.1 Defining the WEF nexus and nexus planning 3
                 2. The evolution of nexus planning 5
                      2.1 Factors driving nexus planning use in resource management 5
                      2.2 Nexus planning before and after 2015 7
                 3. Benefits for adopting nexus planning 9
                 4. Concluding remarks 10
                 References 10

CHAPTER 2   Some quantitative water—energy—food nexus analysis approaches and their data requirements 15
                 Jafaru M. Egieya, Johann Görgens and Neill Goosen
                 1. Introduction 15
                 2. WEF resource management tools 18
                      2.1 CLEWs 18
                      2.2 Water, Energy, Food Nexus Tool 2.0 21
                      2.3 REMap 22
                      2.4 MuSIASEM 22
                 3. Alternative methodologies, approaches, and frameworks 24
                      3.1 Economic analysis models 24
                      3.2 Environmental impact related 26
                      3.3 Systems analysis 26
                      3.4 Statistics 27
                      3.5 Indicators and metrics 28
                 4. Data challenges of quantitative nexus research 28

	5. Conclusions 29
	References 29
CHAPTER 3	EO-WEF: a Earth Observations for Water, Energy, and Food nexus geotool for spatial data visualization and generation 33
	Zolo Kiala, Graham Jewitt, Aidan Senzanje, Onisimo Mutanga, Timothy Dube and Tafadzwanashe Mabhaudhi
	1. Introduction 33
	2. Method 35
	2.1 Predesign steps of EO-WEF 35
	2.2 Software design.. 36
	2.3 How to use the EO-WEF? 37
	3. Capability of EO-WEF for generating data for the different sectors of the Songwe nexus 39
	3.1 Water sector 39
	3.2 Climate sector 41
	3.3 Land sectors 41
	3.4 Socioeconomic, food, and energy sectors 41
	4. Further development 45
	5. Conclusion 45
	References 45
CHAPTER 4	Scales of application of the WEF nexus approach 49
	Janez Sušnik, Sara Masia and Graham Jewitt
	1. Introduction 49
	2. The local scale: household to subnational 51
	3. The national scale.. 56
	4. Higher-level nexus studies 59
	5. Spatial interactions in the nexus 61
	6. Conclusions 62
	References 63
CHAPTER 5	Tools and indices for WEF nexus analysis........................ 67
	Janez Sušnik, Sara Masia, Graham Jewitt and Gareth Simpson
	1. Introduction 67
	2. Tools and approaches to analyze the WEF nexus 68
	2.1 Conceptual maps and causal loop diagrams 68
	2.2 System dynamics modeling................................ 71
	2.3 Agent-based modeling... 74
	2.4 (Multiregion) input—output modeling 76
	2.5 Life cycle assessment 77
	2.6 Integrated assessment models 79

    3. Indices for WEF nexus performance assessment (analysis) 81
        3.1 Human development index 82
        3.2 Environmental sustainability index 82
        3.3 Sustainability development goals 83
        3.4 WEF nexus index 84
    4. Conclusions 85
    References 85

CHAPTER 6 Transboundary WEF nexus analysis: a case study of the Songwe River Basin 91
Sara Masia, Janez Sušnik, Graham Jewitt, Zolo Kiala and Tafadzwanashe Mabhaudhi
    1. Introduction 91
    2. Case study description 92
        2.1 The Songwe River Basin Development Programme 93
        2.2 WEF nexus analysis approach for the Songwe River Basin 95
        2.3 Conceptualizing the WEF nexus in the Songwe River Basin 96
    3. Conclusions 106
    References 107
    Further reading 108
    Websites 109

CHAPTER 7 Applying the WEF nexus at a local level: a focus on catchment level 111
S. Walker, I. Jacobs-Mata, B. Fakudze, M.O. Phahlane and N. Masekwana
    1. Introduction 111
    2. Methodology and data 114
    3. Progress with WEF nexus application at catchment level 117
        3.1 WEF nexus available models 117
        3.2 Model selection and description 119
        3.3 Data sources 128
        3.4 Spatial scale 135
        3.5 Time/temporal scale 135
        3.6 Application of models 136
    4. Way forward and conclusion 137
    Acknowledgments 139
    References 139

CHAPTER 8	A regional approach to implementing the WEF nexus: a case study of the Southern African Development Community	145
	Patrice Kandolo Kabeya, Dumisani Mndzebele, Moses Ntlamelle, Duncan Samikwa, Alex Simalabwi, Andrew Takawira, Kidane Jembere and Shamiso Kumbirai	
	1. Introduction	145
	1.1 Status on water, energy, and food security in the SADC region	146
	2. Fostering water, energy, and food security nexus dialogue and multi-sector investment in the SADC region project	147
	2.1 SADC WEF nexus conceptualization	148
	3. Key planning, policy, and legal documents that are relevant for water, energy, and food security in the SADC region	149
	3.1 Regional development context and sustainable development	149
	3.2 Key energy sector planning, policy, and legal documents	152
	4. Identified challenges related to the water—energy—food nexus approach in the SADC region	156
	4.1 Inadequate coordination of the three sectors at policy- and decision-making levels	157
	5. Operationalizing the WEF nexus in Southern Africa	160
	5.1 SADC WEF nexus governance framework	160
	5.2 Implementing the SADC regional WEF nexus framework	162
	5.3 SADC WEF nexus screening tool for guiding discourse in the region	164
	5.4 Capacity development and guiding discourse in the region	164
	6. Key lessons from the implementation of the SADC WEF nexus regional dialogues project	165
	7. Summary and conclusions	165
	References	166
CHAPTER 9	Exploring the contribution of Tugwi-Mukosi Dam toward water, energy, and food security	169
	Never Mujere and Nelson Chanza	
	1. Introduction	169
	2. The WEF linkage conceptual framework	171

3. Tugwi-Mukosi Dam 171
4. Contribution of Tugwi-Mukosi toward water, energy, and food security 174
   4.1 Water security 174
   4.2 Energy security 176
   4.3 Food security 176
   4.4 Promotion of tourism 176
5. Discussion 177
6. Summary and conclusion and policy implications 178
Conflict of interest 179
Funding 179
References 179

CHAPTER 10 The water—energy—food nexus as an approach for achieving sustainable development goals 2 (food), 6 (water), and 7 (energy) 181

Aidan Senzanje, M. Mudhara and L. Tirivamwe

1. Introduction 181
   1.1 The WEF nexus—past to present discourse 181
   1.2 The WEF nexus as a tool for natural resources management 183
2. The SDGs dimensions and the WEF nexus 185
   2.1 SDG 2—zero hunger 185
   2.2 SDG 6—clean water and sanitation 186
   2.3 SDG 7—affordable and clean energy 187
3. Food and nutrition security 187
4. Synergies and trade-offs in the WEF nexus 188
   4.1 Synergies and trade-offs in the WEF nexus toward achieving food and nutrition security (SDG 2) 188
   4.2 Synergies and trade-offs for achieving clean water and sanitation (SDG 6) 190
   4.3 Synergies and trade-offs for achieving affordable and clean energy (SDG 7) 190
5. Drivers of the WEF nexus toward achievement of SDGs 2, 6, and 7 191
6. Upscaling and outscaling the WEF nexus as a natural resources management tool for attaining SDGs 2, 6, and 7 192
7. Conclusion 193
References 194
Further reading 197

CHAPTER 11	Enhancing sustainable human and environmental health through nexus planning 199	

Luxon Nhamo, Sylvester Mpandeli, Shamiso P. Nhamo, Stanley Liphadzi and Tafadzwanashe Mabhaudhi

1. Introduction 199
2. Linking socioecological interactions with nexus planning 201
    2.1 Defining the water—health—environment—nutrition nexus 202
3. Understanding the risk posed by wildlife on human health 203
    3.1 The role of nexus planning in simplifying socioecological systems 203
    3.2 WHEN nexus and sustainability indicators 204
    3.3 Modeling vulnerability and resilience 206
4. Modeling multisector and complex systems 208
5. Calculating WHEN nexus indices for South Africa 211
6. Understanding the integrated health indices 214
7. Recommendations 216
8. Conclusions 217
References 217

CHAPTER 12	Financing WEF nexus projects: perspectives from interdisciplinary and multidimensional research challenges 223	

Maysoun A. Mustafa and Christoph Hinske

1. Introduction 223
2. The interlinkages within nexus research 224
3. Transboundary systems and the need for interdisciplinary spaces 225
4. Role of funding in fostering interdisciplinary dialogue 226
5. Shared value within multidimensional challenges 228
6. The challenge of goal setting 230
7. Advancing nexus research 231
8. Concluding remarks 232
References 232
Further reading 234

CHAPTER 13	The Water—Energy—Food nexus as a rallying point for sustainable development: emerging lessons from South and Southeast Asia 235	

Andrew Huey Ping Tan, Eng Hwa Yap, Yousif Abdalla Abakr, Alex M. Lechner, Maysoun A. Mustafa and Festo Massawe

1. Introduction 235
    1.1 Brief description of South and Southeast Asia 238

2. A critical review into the WEF of South and Southeast Asia 238
3. Case study: WEF in Malaysia 242
   3.1 Introduction to Malaysia and WEF conceptual framework 242
   3.2 Complex systems approach and causality 244
   3.3 Water—Energy—Food nexus in Malaysia—challenges and opportunities 250
4. Critical findings and key take-home messages 251
Nomenclature 252
References 253

CHAPTER 14 The water—energy—food nexus: an ecosystems and anthropocentric perspective 257
Sally Williams, Annette Huber-Lee, Laura Forni, Youssef Almulla, Camilo Ramirez Gomez, Brian Joyce and Francesco Fuso-Nerini
1. Introduction 257
2. Approach 259
3. WEF case studies: MENA and Latin America 261
   3.1 Case study 1: Jordan and Morocco 261
   3.2 Case study 2: Argentina and Brazil 267
4. Comparisons of the WEF nexus in MENA and Latin America 274
5. Conclusions 275
References 276
Further reading 277

CHAPTER 15 Water—energy—food nexus approaches to facilitate smallholder agricultural technology adoption in Africa 279
Michael G. Jacobson
1. Introduction 279
2. African context 280
3. Literature review 282
4. Farmer technology adoption 286
5. Research designs for incorporating a priori assessment 289
6. Conclusion 290
References 291

CHAPTER 16 Building capacity for upscaling the WEF nexus and guiding transformational change in Africa 299
Tendai P. Chibarabada, Goden Mabaya, Luxon Nhamo, Sylvester Mpandeli, Stanley Liphadzi, Krasposy K. Kujinga, Jean-Marie Kileshye-Onema, Hodson Makurira, Dhesigen Naidoo and Michael G. Jacobson
1. Introduction 299

2. Status of WEF nexus research in Africa 301
   2.1 Understanding drivers of change 301
   2.2 WEF nexus planning 303
3. Development of a conceptual framework for WEF nexus upscaling and capacity development 305
   3.1 Upscaling and uptake of WEF nexus 305
4. Capacity development for upscaling and uptake of WEF nexus 308
   4.1 Capacity needs assessment 309
   4.2 Building WEF nexus curricula for upscaling 309
   4.3 Capacity building implementation strategy 313
   4.4 Enabling environment for capacity building 314
5. Conclusion 315
References 315

CHAPTER 17 WEF nexus narratives: toward sustainable resource security 321

Tafadzwanashe Mabhaudhi, Aidan Senzanje, Albert Modi, Graham Jewitt and Festo Massawe

1. The WEF nexus 321
2. Key messages 322
3. Conclusion 325
References 325

INDEX 327

Contributors

Yousif Abdalla Abakr Department of Mechanical, Materials and Manufacturing Engineering, University of Nottingham Malaysia, Semenyih, Selangor, Malaysia

Youssef Almulla Department of Energy Technology, KTH Royal Institute of Technology, Sweden

Nelson Chanza Department of Urban and Regional Planning, University of Johannesburg, Johannesburg, South Africa

Tendai P. Chibarabada Waternet, Mt Pleasant, Harare, Zimbabwe

Timothy Dube Institute for Water Studies, Department of Earth Sciences, University of the Western Cape, Bellville, South Africa

Jafaru M. Egieya ARUA Centre of Excellence in Energy, Centre for Renewable and Sustainable Energy Studies, Faculty of Engineering, Stellenbosch University, Stellenbosch, South Africa

B. Fakudze International Water Management Institute, Pretoria, Gauteng, South Africa

Laura Forni U.S. Water Program, Stockholm Environment Institute, Davis, CA, United States

Francesco Fuso-Nerini U.S. Water Program, Stockholm Environment Institute, Somerville, MA, United States

Neill Goosen ARUA Centre of Excellence in Energy, Centre for Renewable and Sustainable Energy Studies, Faculty of Engineering, Stellenbosch University, Stellenbosch, South Africa; Department of Process Engineering, Faculty of Engineering, Stellenbosch University, Stellenbosch, South Africa

Johann Görgens Department of Process Engineering, Faculty of Engineering, Stellenbosch University, Stellenbosch, South Africa

Christoph Hinske System Leadership & Entrepreneurial Ecosystems, School of Finance and Accounting, SAXION University of Applied Sciences, Enschede, Netherlands

Annette Huber-Lee U.S. Water Program, Stockholm Environment Institute, Somerville, MA, United States

I. Jacobs-Mata International Water Management Institute, Pretoria, Gauteng, South Africa

Michael G. Jacobson School of Forest Resources, Department of Ecosystem Science and Management, Pennsylvania State University, University Park, PA, United States

Kidane Jembere Global Water Partnership Southern Africa, Pretoria, South Africa

Graham Jewitt IHE Delft Institute for Water Education, Delft, the Netherlands; Centre for Water Resources Research, School of Agricultural, Earth and Environmental Sciences, University of KwaZulu-Natal (UKZN), Pietermaritzburg, South Africa; Civil Engineering and Geosciences, Delft University of Technology, Delft, The Netherlands; Centre for Water Resources Research, College of Agriculture, Engineering and Science, University of KwaZulu-Natal, Pietermaritzburg, South Africa

Brian Joyce U.S. Water Program, Stockholm Environment Institute, Somerville, MA, United States

Patrice Kandolo Kabeya Southern African Development Community Secretariat, Gaborone, Botswana

Zolo Kiala Centre for Transformative Agricultural and Food Systems (CTAFS), School of Agricultural, Earth and Environmental Sciences, University of KwaZulu-Natal, Pietermaritzburg, South Africa; Origins Center, School: Geography, Archaeology and Environmental Studies, University of the Witwatersrand, Johannesburg, South Africa

Jean-Marie Kileshye-Onema Waternet, Mt Pleasant, Harare, Zimbabwe; School of Industrial Engineers, University of Lubumbashi, Lubumbashi, DR Congo

Krasposy K. Kujinga Waternet, Mt Pleasant, Harare, Zimbabwe

Shamiso Kumbirai Global Water Partnership Southern Africa, Pretoria, South Africa

Alex M. Lechner Lincoln Centre for Water and Planetary Health, University of Lincoln, Lincoln, United Kingdom; School of Environmental and Geographical Sciences, University of Nottingham Malaysia, Semenyih, Selangor, Malaysia; Monash University Indonesia, Tangerang Banten, Indonesia

Stanley Liphadzi Water Research Commission of South Africa (WRC), Pretoria, South Africa; School of Environmental Sciences, University of Venda, Thohoyandou, Limpopo, South Africa

Goden Mabaya Waternet, Mt Pleasant, Harare, Zimbabwe

Tafadzwanashe Mabhaudhi Centre for Transformative Agricultural and Food Systems (CTAFS), School of Agricultural, Earth and Environmental Sciences, University of KwaZulu-Natal, Pietermaritzburg, South Africa; International Water Management Institute (IWMI-GH), West Africa Office, Accra, Ghana; Centre for Water Resources Research, School of Agricultural, Earth and Environmental Sciences, University of KwaZulu-Natal (UKZN), Pietermaritzburg, South Africa; International Water Management Institute, West Africa Regional Office, Accra, Ghana; International Water Management Institute (IWMI), Accra, Ghana

Hodson Makurira Department of Construction and Civil Engineering, University of Zimbabwe, Harare Zimbabwe

N. Masekwana Agricultural Research Council — Natural Resources & Engineering, Pretoria, Gauteng, South Africa

Sara Masia IHE Delft Institute for Water Education, Delft, The Netherlands; CMCC Foundation — Euro-Mediterranean Centre on Climate Change, IAFES Division, Sassari, Italy

Festo Massawe School of Biosciences, University of Nottingham Malaysia, Semenyih, Selangor, Malaysia; Centre for Transformative Agricultural and Food Systems (CTAFS), School of Agricultural, Earth and Environmental Sciences, University of KwaZulu-Natal, Pietermaritzburg, South Africa

Dumisani Mndzebele Southern African Development Community Secretariat, Gaborone, Botswana

Albert Modi Centre for Transformative Agricultural and Food Systems (CTAFS), School of Agricultural, Earth and Environmental Sciences, University of KwaZulu-Natal, Pietermaritzburg, South Africa

Sylvester Mpandeli Water Research Commission of South Africa (WRC), Pretoria, South Africa; Faculty of Science, Engineering and Agriculture, University of Venda, Thohoyandou, South Africa; School of Environmental Sciences, University of Venda, Thohoyandou, Limpopo, South Africa

M. Mudhara School of Agriculture, Earth and Environmental Sciences, University of KwaZulu-Natal, Pietermaritzburg, South Africa

Never Mujere Department of Geography, Geospatial Science and Earth Observation, University of Zimbabwe, Mt Pleasant, Harare, Zimbabwe

Maysoun A. Mustafa School of Biosciences, University of Nottingham Malaysia, Semenyih, Selangor, Malaysia

Onisimo Mutanga Discipline of Geography, School of Agricultural, Earth and Environmental Sciences, University of KwaZulu-Natal, Pietermaritzburg, South Africa

Dhesigen Naidoo Water Research Commission of South Africa (WRC), Pretoria, South Africa

Luxon Nhamo Water Research Commission of South Africa (WRC), Pretoria, South Africa

Shamiso P. Nhamo Department of Pharmacology, Faculty of Health Sciences, University of Pretoria, Hatfield, Pretoria, South Africa

Moses Ntlamelle Southern African Development Community Secretariat, Gaborone, Botswana

M.O. Phahlane Agricultural Research Council — Natural Resources & Engineering, Pretoria, Gauteng, South Africa

Camilo Ramirez Gomez Division of Nergy Systems KTH-dEH, KTH Royal Institute of Technology, Stockholm, Switzerland

Duncan Samikwa Southern African Development Community Secretariat, Gaborone, Botswana

Aidan Senzanje Bioresources Engineering Programme, School of Engineering, University of KwaZulu-Natal (UKZN), Pietermaritzburg, South Africa; Centre for Water Resources Research, College of Agriculture, Engineering and Science, University of KwaZulu-Natal, Pietermaritzburg, South Africa

Alex Simalabwi Global Water Partnership Southern Africa, Pretoria, South Africa

Gareth Simpson Jones and Wagener Engineering Associates, Pretoria, South Africa

Janez Sušnik IHE Delft Institute for Water Education, Delft, The Netherlands

Andrew Takawira Global Water Partnership Southern Africa, Pretoria, South Africa

Andrew Huey Ping Tan School of Intelligent Manufacturing Ecosystem, XJTLU Entrepreneur College (Taicang), Xi'an Jiaotong-Liverpool University, Suzhou, Jiangsu, People's Republic of China

L. Tirivamwe School of Agriculture, Earth and Environmental Sciences, University of KwaZulu-Natal, Pietermaritzburg, South Africa

S. Walker Agricultural Research Council — Natural Resources & Engineering, Pretoria, Gauteng, South Africa; Department of Soil, Crop & Climate Sciences, University of the Free State, Bloemfontein, Free State, South Africa

Sally Williams U.S. Water Program, Stockholm Environment Institute, Somerville, MA, United States

Eng Hwa Yap School of Robotics, XJTLU Entrepreneur College (Taicang), Xi'an Jiaotong-Liverpool University, Suzhou, Jiangsu, People's Republic of China

Foreword

Dhesigen Naidoo

We live in an age of paradox. Never before have we have such a comprehensive toolbox. Science and technology have afforded us a level of local and global connectivity unimagined in history. There is hardly a village in the world, even in the most rural and extreme settings, which does not have some level of access to the information superhighway. It is estimated that 4.66 billion people have Internet access, and 4.32 billion people have access to a mobile device.

And yet, at the same time, UN-Water reminds us that 2 billion people still do not have access to safe water, and 3.66 billion need improved sanitation. Nearly half the schools in the world have no safe handwashing facilities. UNICEF estimates that 700 under-5's die each day from diarrheal diseases related to unsafe water and sanitation. In fact, in the world's major conflict zones, under-5's are 20 times more likely to die from ailments related to unclean water than the conflict itself.

In the domain of food security, we have seen the COVID-19 pandemic pushback on vital gains as all the major statistics showing increases. We now have 811m or 9.9% of the global population suffering from hunger (up from 8.4% in 2019). With the continued pandemic, this number is rising. As the UN Food Summit has now passed, we are still faced with the reality that 2.3 billion people do not have year-round access to food and are food insecure. When we delve into the domain of nutritional security and malnutritional, this number enters the realm of alarming.

Global access to energy initiatives has made great gains. However, by 2019 the IEA, IRENA and the UN System estimated that 759 million people still did not have access to reliable energy. Some 2.6 billion still did not have access to clean cooking facilities. We also know that this number will increase further on the back of the new burdens introduced by COVID-19. There are potentially further challenges to achieving the Just Transition from a fossil fuel—dominated energy system to a low-carbon global economy.

It is abundantly clear that new approaches have to dominate the recovery from the pandemic and catalyze the actions needed to get close to the Sustainable Development Goals by 2030. Key among them is the coordinated and resource optimization approach encapsulated in the water—energy—food nexus. This primer is an important global resource to assist in the thinking, planning, and implementation journey of the nexus, very importantly adopting the lens of the Global South.

The opening chapters examine the basis of the paradigm of the nexus. The nature of the interrelatedness of these very different sectors supplements the anthropocentric perspectives with an ecosystem perspective. The nexus politics and economics are unpacked practically, as are the fundamental requirements, including data and knowledge systems, capacities, and capabilities required to realize the nexus implementation. The implications of climate change and the nexus as an accelerator of the SDGs specifically and the Sustainable Agenda, in general, are explored. Case studies from across the Global South illustrate the key challenges and solutions in the nexus approach going from catchment to international river basin scales.

This book concludes by describing a potential operationalization pathway designed to stimulate more innovative thinking and actions and deepen and expand the global epistemic community of practice in this domain as we build the bridge between science and implementation in the water—energy—food nexus.

CHAPTER 1

The water—energy—food nexus: its transition into a transformative approach

Sylvester Mpandeli[1,2], Luxon Nhamo[1], Aidan Senzanje[3], Graham Jewitt[4,5], Albert Modi[6], Festo Massawe[6,7] and Tafadzwanashe Mabhaudhi[6,8,9]

[1]Water Research Commission of South Africa (WRC), Pretoria, South Africa; [2]Faculty of Science, Engineering and Agriculture, University of Venda, Thohoyandou, South Africa; [3]Centre for Water Resources Research, College of Agriculture, Engineering and Science, University of KwaZulu-Natal, Pietermaritzburg, South Africa; [4]IHE Delft Institute for Water Education, Delft, the Netherlands; [5]Centre for Water Resources Research, College of Agriculture, Engineering and Science, University of KwaZulu-Natal, Pietermaritzburg, South Africa; [6]Centre for Transformative Agricultural and Food Systems (CTAFS), School of Agricultural, Earth and Environmental Sciences, University of KwaZulu-Natal, Pietermaritzburg, South Africa; [7]School of Biosciences, University of Nottingham Malaysia, Semenyih, Selangor, Malaysia; [8]International Water Management Institute (IWMI-GH), West Africa Office, Accra, Ghana; [9]Centre for Water Resources Research, School of Agricultural, Earth and Environmental Sciences, University of KwaZulu-Natal (UKZN), Pietermaritzburg, South Africa

1. Introduction

The water—energy—food (WEF) nexus has grown into an important transformative and circular approach since 2011 when it was presented at the World Economic Forum by the Stockholm Environment Institute (SEI) in anticipation of the Sustainable Development Goals (SDGs), which came into effect in 2015 (FAO, 2014; Hoff, 2011; UNGA, 2015). This was the same period when the SDGs were being formulated in response to the continued insecurity of water, energy, and food resources (FAO, 2014; Liphadzi et al., 2021). The three resources, termed WEF resources, are vital for human well-being, poverty reduction, and sustainable development, the reason why all the 17 SDGs are developed around the three resources (Mabhaudhi et al., 2021; UNGA, 2015). The need for the formulation of the SDGs mainly around the WEF resources was also motivated by global projections indicating that the demand for the three resources will increase significantly in the coming years due to population growth, economic development, international trade, urbanization, diversifying diets, depletion of natural resources, technological advances, and climate change (FAO, 2014; Hoff, 2011). This is happening at a time when resources are depleting due to climate change (Mpandeli et al., 2018). As a result, policymakers needed an approach capable of integrating the management and governance of the three interlinked sectors (Mabhaudhi et al., 2019; Nhamo et al., 2018; Rasul and Sharma, 2016).

The intricate interlinkages between the WEF resources are demonstrated through agriculture, which accounts for 70% of total global freshwater withdrawals, making it the largest user of water (FAO, 2014; Mabhaudhi et al., 2019), and the sector also uses about 30% of the total energy that is consumed worldwide (IEA, 2016). Apart from its importance in crop production, water is also essential in energy production and transportation in different forms (Ferroukhi et al., 2015; Owusu and Asumadu-Sarkodie, 2016). Also, energy is needed for the production, transportation, and distribution of food and the extraction, storage, and treatment of water (Woods et al., 2010). Demand for resources is projected to increase in the near future as 60% more food will have to be produced to feed the growing world population by 2050 (Fróna et al., 2019), global energy consumption will increase by over 50% by 2035 (IEA, 2016), and total global freshwater withdrawals for irrigation alone will increase by 10% by 2050 (FAO, 2014). These factors and intricate relationships in the three resources facilitated the WEF nexus to grow into an important approach to guide coherent strategic policy decisions toward sustainable development.

The envisaged importance of the WEF nexus as a decision support tool and its transformational nature promoted its rise to become vital for the simultaneous management of natural resources (Mabhaudhi et al., 2019; Nhamo et al., 2020a). It has since transitioned from a simple conceptual and discourse framework to an analytical decision support tool (Nhamo and Ndlela, 2020). In recent years, its transition into an analytical tool has further enhanced its adoption, operationalization, use, and prominence. Regions such as the Southern African Development Community (SADC) have since adopted it as a governance and policy decision framework (Mpandeli et al., 2018; Nhamo et al., 2018). However, although its prominence only increased from 2011, written evidence of the three-way mutual interlinkages among the WEF resources started well before 2011 (Hellegers et al., 2008; Khan and Hanjra, 2009; McCornick et al., 2008; Mushtaq et al., 2009; Scott et al., 2003). These previous studies also highlighted the importance of nexus planning in managing resources for sustainable development. However, its adoption and operationalization were limited by a lack of empirical evidence and practical case studies on how to simultaneously assess and manage the three resources (Naidoo et al., 2021b; Nhamo et al., 2021a; Terrapon-Pfaff et al., 2018). This has since changed in recent years as more models and tools have been developed to transitions the approach into an analytical tool (Albrecht et al., 2018; Mabrey and Vittorio, 2018; Nhamo et al., 2020a).

Therefore, this chapter addresses the WEF nexus viewpoints before and after 2015, the two periods separated by the enactment of the SDGs. Apart from providing an overview of nexus planning before and after the enactment of

1. Introduction

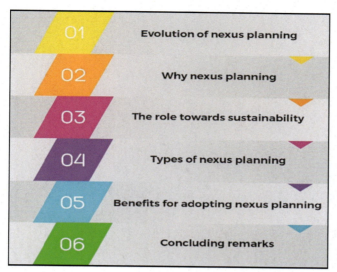

FIGURE 1.1
The main contents covered in the chapter in understanding nexus planning.

the SDGs in 2015, this chapter also discusses the definition of nexus planning, providing reasons for its rapid growth and prominence in recent years and highlighting the benefits of adopting the approach, particularly in assessing and achieving the 2030 global agenda of SDGs. This is based on the declaration of the United Nations General Assembly's (UNGA), which states that: *"The interlinkages and integrated nature of the Sustainable Development Goals are of crucial importance in ensuring that the purpose of the new Agenda is realized"* (UNGA, 2015). The chapter outline (Fig. 1.1) seeks to also identify the various types of nexus planning and the benefits of operationalizing the approach.

1.1 Defining the WEF nexus and nexus planning

WEF nexus refers to the simultaneous management of the three interconnected resources of water, energy, and food to enhance their security amid climate change, environmental degradation, and continued depletion, as any developments in any one of the three sectors will have effects in the other two sectors (FAO, 2014; Mahlknecht and González-Bravo, 2018; Nhamo et al., 2020b). The approach (Fig. 1.2) is a transformative and circular model that holistically informs the management of intricately interlinked resources and systems to achieve sustainable development (FAO, 2014; Hoff, 2011). The cross-sectoral approach is critical for identifying priority intervention areas, highlighting potential synergies, and identifying trade-offs needing immediate attention (Mahlknecht and González-Bravo, 2018; Naidoo et al., 2021b; Nhamo et al., 2020a).

Although the WEF nexus is the most known, other nexus types have been established, such as urban nexus and the water—health—environment—nutrition

4 CHAPTER 1: The water—energy—food nexus

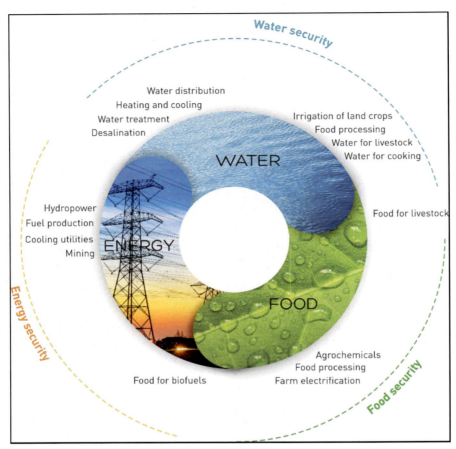

FIGURE 1.2
The three-way interconnectedness between water, energy, and food resources.

(WHEN) nexus, among others (Lehmann, 2018; Nhamo and Ndlela, 2020; Nhamo et al., 2021b; Willis, 2015). The existence of many nexuses indicates the need to assess and manage interlinked sectors holistically without prioritizing one above the others (Nhamo and Ndlela, 2020). Thus, the WEF nexus is only one of the many nexuses that have been established (Nhamo and Ndlela, 2020; Willis, 2015). As there are many nexus types, the concept has evolved from a narrow term of WEF nexus to broader nexus planning (Mabhaudhi et al., 2021; Nhamo and Ndlela, 2020).

The WEF nexus provides an opportunity for analyzing the trade-offs and synergies between various sectors to stimulate efficiency, mobilize resources, and create opportunities for policy coherent among those sectors. This development has eliminated the notion that any nexus type is always WEF nexus, which is widely known. The broader term "nexus planning" emphasizes providing integrated solutions to distinct but interlinked components (Nhamo and Ndlela, 2020). Its transformative and polycentric nature facilitates the management of trade-offs and synergies while informing coherent policy formulations and implementation to promote the transition toward sustainable and resilient socioecological systems that enhance human well-being and environmental outcomes (Mabhaudhi et al., 2019).

2. The evolution of nexus planning

As already alluded to, the World Economic Forum of 2011 marked a distinct demarcation on the knowledge, uptake, application, essence, prominence, and implementation of the WEF nexus as a holistic approach for integrated resources management (Fig. 1.3). This marked line also coincides with the SDGs' enactment by the United Nations General Assembly (FAO, 2014). Interestingly, research from both periods acknowledges the importance of nexus planning in driving the transformational change and sustainable development agenda (Liphadzi et al., 2021; McCornick et al., 2008; Naidoo et al., 2021b; Nhamo et al., 2020a). However, there are notable differences in how research approached and used the concept before and after 2015 (Fig. 1.3). However, there are marked differences where recent research has developed tools and models to inform policy decisions and integrated resource management. The initial research before 2015 informed the current progress of the approach.

2.1 Factors driving nexus planning use in resource management

The grand challenges that humankind faces today, such as climate change, resource depletion and degradation, migration, the emergence of novel pests and diseases, poverty and inequality, and unsustainable food systems, among others (Fig. 1.3), transverse all sectors and prompt a shift from the way people perceive the world from a linear view to a circular perspective, as challenges in one sector often trigger a host of other challenges in other sectors (FAO, 2014; Nhamo and Ndlela, 2020). The interconnectedness of these challenges, which cut across all sectors, require holistic and systemic interventions that promote transformational change toward greater sustainability, resilience, and equity and delivering on human health and well-being and environmental outcomes (Fogarassy and Finger, 2020; Jørgensen and Pedersen, 2018; Nhamo and Ndlela, 2020). Transformational change and related transformative and circular systems such as the WEF nexus emphasize cross-sectoral interventions to enhance socioeconomic resilience against the prevailing interconnected challenges (Jørgensen and Pedersen, 2018; Naidoo et al., 2021b). This is based

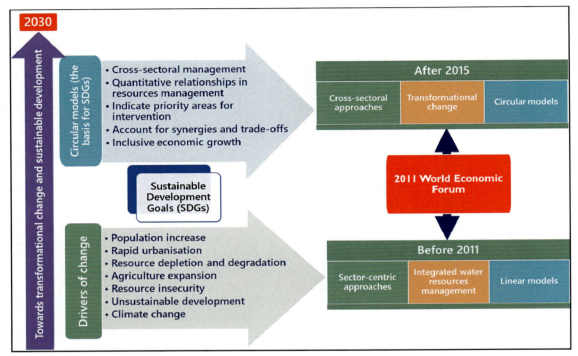

FIGURE 1.3
The progression of the WEF nexus concept before and after 2015 and its importance in achieving sustainable development by 2030.

on that current linear models forget the interconnectedness of socioeconomic and ecological systems and how their systemic properties shape their interactions, interdependencies, and interrelationships (Jørgensen and Pedersen, 2018). In contrast, nexus planning integrates, simplifies, and facilitates understanding these complex systems, indicating priority areas for intervention and reducing risk and vulnerability (FAO, 2014). This is why the WEF nexus forms the basis for SDGs (UNGA, 2015).

The 4th Industrial Revolution (4IR) relies on intricate, cross-cutting, interconnected systems to provide goods and services (Nhamo and Ndlela, 2020). Although this has brought considerable advances and opportunities for development, it has also exposed the systems to severe disruptions and shocks, as evidenced by the disruptions in global supply chains by climate change and pandemics like the COVID-19 (Madhav et al., 2017; Oppenheim et al., 2019). As in complex systems, tensions always manifest between efficiency and resilience, the ability to anticipate, absorb, recover, and adapt to unexpected disruptions (Oppenheim et al., 2019). Thus, sector-based or system-specific resilience initiatives are often associated with systemic risks, which emanate from strategies that lead to suboptimal efficiencies in one sector at the expense of others (Madhav et al., 2017; Naidoo et al., 2021a).

The COVID-19 experiences have exposed the limitations of the widely used linear models. However, circular models such as nexus planning, circular economy, scenario planning, and sustainable food systems have emerged as alternative approaches to the existing linear ones that have now been shown to have reached their physical limitations (Hofstetter et al., 2021). The disruptions caused by the COVID-19 pandemic have further exposed the fragility of linear models in addressing today's complex, cross-cutting, and interconnected challenges (Ibn-Mohammed et al., 2020; Naidoo et al., 2021a). The pandemic caused immense disruptions in all sectors of the economy while policy- and decision-makers were proving reactive measures, focusing on the health sector. The COVID-19 experience has demonstrated that focusing on a single sector during a crisis only exacerbates the stressors in other sectors as decision-makers have often viewed the world from a linear perspective, with the thought that a click of a button would get the economy and society back on track (Nhamo and Ndlela, 2020). Yet the lockdowns implemented during the various COVID-19 waves resulted in job losses, company closures, and economic recessions, demonstrating that linear approaches often overemphasize a limited set of system attributes, notably efficiency, at the expense of other aspects (Ibn-Mohammed et al., 2020; Nhamo and Ndlela, 2020). While linear approaches have been beneficial to some extent, the COVID-19 pandemic and climate change have exposed how they transfer stresses to other sectors, compromise resilience-building initiatives, weaken the adaptive capacity, and allow failure to cascade from one sector to the other.

2.2 Nexus planning before and after 2015

There is a before and an after of nexus planning, a distinction demarcated by the enactment of the SDGs in 2015. The viewpoints of the approach between the two phases differ considerably, as indicated in Figs. 1.3 and 1.4. Literature on nexus planning published before 2015 was sector-based, generally focusing on sector-specific models that were amplified, attempting to link each model around one sector (FAO, 2014; Giampietro et al., 2009; Grigg, 2019). Examples of such sector-specific models include the Integrated Water Resources Management (IWRM), Agricultural Production Systems Simulator (APSIM), Water Evaluation and Planning System (WEAP), Water Energy Food Nexus Rapid Appraisal Tool, RENA's Preliminary Nexus Assessment Tool Soil, and Water Assessment Tool (SWAT), and Physical, Economic, and Nutritional Water Productivity (Giampietro et al., 2009; Mabrey and Vittorio, 2018; Nhamo et al., 2018). This approach to nexus planning stalled its development as it remained linear, sectoral, and monocentric (Nhamo et al., 2020a). The notion at that

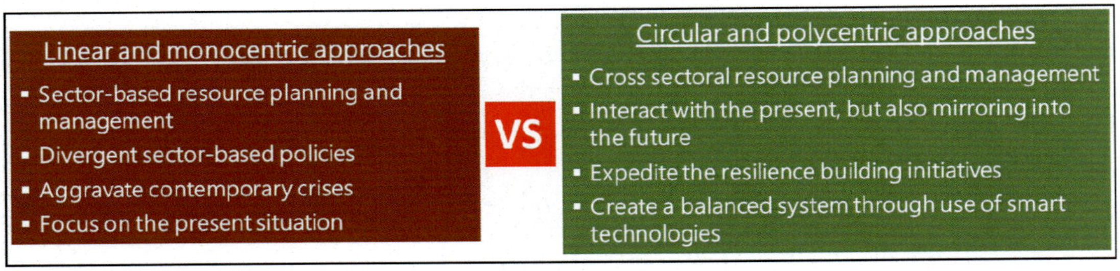

Linear and monocentric approaches
- Sector-based resource planning and management
- Divergent sector-based policies
- Aggravate contemporary crises
- Focus on the present situation

VS

Circular and polycentric approaches
- Cross sectoral resource planning and management
- Interact with the present, but also mirroring into the future
- Expedite the resilience building initiatives
- Create a balanced system through use of smart technologies

FIGURE 1.4

The advantages of adopting circular models over linear approaches in resource planning and management.

time was that other sectors should revolve around a supreme sector and should direct the other sectors. This belief was compounded by the existing sector-based policies sitting in siloed departments and ministries, a scenario that aggravates contemporary crises by the optimal development of certain sectors at the expense of other equally important sectors (Mpandeli et al., 2018; Nhamo et al., 2018; Peters, 2018). An example of such a sector-based and monocentric approach that dominated this period is the IWRM, which emphasizes that everything should evolve around water or start and end from a water perspective (Garcia, 2008).

However, this viewpoint of nexus planning as a linear approach has changed since 2015 with the promulgation of the SDGs. This period has witnessed a dynamic transitioning of the approach from only being a conceptual and discourse framework into a useful analytical decision support tool that informs strategic policy decisions for sustainable resource management (Daher and Mohtar, 2015; Mabhaudhi et al., 2019; Naidoo et al., 2021b; Nhamo et al., 2020a). To date, the approach has transitioned into a cross-sectoral, circular, and transformative model advancing the transformational change agenda and sustainable development (Mabhaudhi et al., 2021; Naidoo et al., 2021a; Nhamo and Ndlela, 2020). It unpacks the intricate interlinkages between distinct but related sectors, identifying priority areas for intervention (Mabhaudhi et al., 2019; Nhamo et al., 2020a). It now forms an integral part of initiatives aimed at achieving SDGs, mainly Goal 1 (poverty eradication), Goal 2 (zero hunger), Goal 6 (provision of water and sanitation), and Goal 7 (access to affordable and reliable energy), but with linkages to the other 13 goals (Mabhaudhi et al., 2021). It has grown into an important tool essential for assessing progress done to achieve the SDGs. Thus, nexus planning provides the following sustainable attributes: (1) the promotion and guidance of sustainable and efficient use of resources, (2) the provision of access to and equal distribution of resources to all, and (3) enhance sustainable conservation of the natural resource base (Liphadzi et al., 2021).

3. Benefits for adopting nexus planning

The WEF nexus has become valuable for understanding the intricate interlinkages and feedbacks within complex systems and for decision-making to achieve sustainable development. As the approach informed the development of the SDGs, it was factored in mainly as part of SDGs 2, 6, and 7 with linkages to the other 14 goals (Biggs et al., 2015; Ringler et al., 2013). Its strength lies in harmonizing cross-sectoral activities, synergies, trade-offs, integrated resource management, inclusive development, and sustainable development (Naidoo et al., 2021b; Nhamo et al., 2020a).

As a result, recently developed models have advanced the essence of nexus planning by indicating areas for priority interventions and guiding strategic and coherent policy formulations at all scales, leading to resilience and adaptation to climate change and sustainable development (Naidoo et al., 2021b). The benefits of the approach include addressing pertinent issues related to sustainability, such as follows:

(a) Simplifying the understanding of the intricate interconnectedness of distinct but related components of a system in space and time (Braithwaite et al., 2018). The understanding and modeling of these complex systems has facilitated enhancing the efficiency of a system as a whole and not promoting suboptimal productivity and efficiency of individual sectors (Liphadzi et al., 2021; Nhamo and Ndlela, 2020).

(b) The capability to recognize and explain intricate relationships among sectors and establish cross-sectoral planning, utilization, and management of resources informs pathways that transformational change by promoting balanced and inclusive dialogue and decision-making processes (Clarke and Crane, 2018; Liphadzi et al., 2021).

(c) Formulating cross-sectoral policy decisions that stimulate mutually beneficial responses that lead to resilience and adaptation, enhancing stakeholder cooperation through public–private partnerships at multiple scales (Naidoo et al., 2021b).

(d) The provision of decision support tools for out- and upscaling interventions enhances synergies and minimize trade-offs (Naidoo et al., 2021b; Nhamo et al., 2020a). These are anticipated to promote and catalyze the global agenda of sustainable development and eventually enhance the security of water, energy, and food resources through evidence-based and coherent policy decisions that facilitate mutual and sustainable socioecological interactions that promote environmental and human health (Boas et al., 2016; Naidoo et al., 2021a; Nhamo et al., 2021a).

4. Concluding remarks

Nexus planning has evolved into an important transformative, circular, and integrated model for informing contemporary transformative systems, including sustainable food systems, circular economy, scenario planning, SDGs progress assessment, and livelihoods transformation. Since 2015, it has transitioned into an indispensable decision support tool that provides evidence on complex sustainability challenges. These developments have allowed nexus planning to transition from a conceptual and discourse framework into an operational decision support model. It has developed into a model that provides pathways toward sustainable development and transformative solutions that lead to resilience and adaptation. As a transformative approach that assesses the spatiotemporal management of resources, it is vital to assess progress toward SDGs. These attributes are enhanced by integrating other transformative approaches such as scenario planning, circular economy, and sustainable food systems. However, it is essential to mention that nexus planning is not a panacea of humankind's many existing challenges. Still, it provides valuable information on resource management and sustainable development. It should also be noted that WEF nexus does not intend or advance creating a mega nexus ministry or department. It is a multistakeholder platform for related sectors to discuss and develop cross-sectoral strategic policies and decisions. Lastly, it is important to note that water, energy, and food production has historically been managed independently with little consideration or cognizance of cross-sectoral benefits or interactions.

References

Albrecht, T.R., Crootof, A., Scott, C.A., 2018. The Water-Energy-Food Nexus: a systematic review of methods for nexus assessment. Environ. Res. Lett. 13, 043002.

Biggs, E.M., Bruce, E., Boruff, B., Duncan, J.M., Horsley, J., Pauli, N., McNeill, K., Neef, A., Van Ogtrop, F., Curnow, J., 2015. Sustainable development and the water–energy–food nexus: a perspective on livelihoods. Environ. Sci. Pol. 54, 389–397.

Boas, I., Biermann, F., Kanie, N., 2016. Cross-sectoral strategies in global sustainability governance: towards a nexus approach. Int. Environ. Agreements Polit. Law Econ. 16, 449–464.

Braithwaite, J., Churruca, K., Long, J.C., Ellis, L.A., Herkes, J., 2018. When complexity science meets implementation science: a theoretical and empirical analysis of systems change. BMC Med. 16, 1–14.

Clarke, A., Crane, A., 2018. Cross-sector partnerships for systemic change: systematized literature review and agenda for further research. J. Bus. Ethics 150, 303–313.

Daher, B.T., Mohtar, R.H., 2015. Water–energy–food (WEF) Nexus Tool 2.0: guiding integrative resource planning and decision-making. Water Int. 40, 748–771.

FAO, 2014. The Water-energy-food Nexus: A New Approach in Support of Food Security and Sustainable Agriculture. Food and Agriculture Organization (FAO), Rome.

Ferroukhi, R., Nagpal, D., Lopez-Peña, A., Hodges, T., Mohtar, R.H., Daher, B., Mohtar, S., Keulertz, M., 2015. Renewable Energy in the Water, Energy & Food Nexus. The International Renewable Energy Agency (IRENA), Abu Dhabi.

Fogarassy, C., Finger, D., 2020. Theoretical and practical approaches of circular economy for business models and technological solutions. Resources 9 (6), 76.

Fróna, D., Szenderák, J., Harangi-Rákos, M., 2019. The challenge of feeding the world. Sustainability 11, 5816.

Garcia, L.E., 2008. Integrated water resources management: a 'small'step for conceptualists, a giant step for practitioners. Int. J. Water Resour. Dev. 24, 23—36.

Giampietro, M., Mayumi, K., Ramos-Martin, J., 2009. Multi-scale integrated analysis of societal and ecosystem metabolism (MuSIASEM): theoretical concepts and basic rationale. Energy 34, 313—322.

Grigg, N.S., 2019. IWRM and the nexus approach: versatile concepts for water resources education. J. Contemp. Water Res. Educ. 166, 24—34.

Hellegers, P., Zilberman, D., Steduto, P., McCornick, P., 2008. Interactions between water, energy, food and environment: evolving perspectives and policy issues. Water Pol. 10, 1—10.

Hoff, H., 2011. Understanding the Nexus: Background Paper for the Bonn 2011 Conference: The Water, Energy and Food Security Nexus. Stockholm Environment Institute (SEI), Stockholm.

Hofstetter, J.S., De Marchi, V., Sarkis, J., Govindan, K., Klassen, R., Ometto, A.R., Spraul, K.S., Bocken, N., Ashton, W.S., Sharma, S., 2021. From sustainable global value chains to circular economy—different silos, different perspectives, but many opportunities to build bridges. Circ. Econ. Sustain. 1, 21—47.

Ibn-Mohammed, T., Mustapha, K., Godsell, J., Adamu, Z., Babatunde, K., Akintade, D., Acquaye, A., Fujii, H., Ndiaye, M., Yamoah, F., 2020. A critical review of the impacts of COVID-19 on the global economy and ecosystems and opportunities for circular economy strategies. Resour. Conserv. Recycl. 164, 105169.

IEA, 2016. Energy and Air Pollution: World Energy Outlook Special Report 2016. International Energy Agency (IEA), Paris.

Jørgensen, S., Pedersen, L.J.T., 2018. The circular rather than the linear economy. In: Shrivastava, P., Zsolnai, L. (Eds.), RESTART Sustainable Business Model Innovation. Springer, Geneva, Switzerland, pp. 103—120.

Khan, S., Hanjra, M.A., 2009. Footprints of water and energy inputs in food production—Global perspectives. Food Pol. 34, 130—140.

Lehmann, S., 2018. Implementing the Urban Nexus approach for improved resource-efficiency of developing cities in Southeast-Asia. City Cult. Soc. 13, 46—56.

Liphadzi, S., Mpandeli, S., Mabhaudhi, T., Naidoo, D., Nhamo, L., 2021. The evolution of the water—energy—food nexus as a transformative approach for sustainable development in South Africa. In: Muthu, S. (Ed.), The Water—Energy—Food Nexus: Concept and Assessments. Springer, Kowloon, Hong Kong, pp. 35—67.

Mabhaudhi, T., Nhamo, L., Chibarabada, T.P., Mabaya, G., Mpandeli, S., Liphadzi, S., Senzanje, A., Naidoo, D., Modi, A.T., Chivenge, P.P., 2021. Assessing progress towards sustainable development goals through nexus planning. Water 13, 1321.

Mabhaudhi, T., Nhamo, L., Mpandeli, S., Nhemachena, C., Senzanje, A., Sobratee, N., Chivenge, P.P., Slotow, R., Naidoo, D., Liphadzi, S., 2019. The water—energy—food nexus as a tool to transform rural livelihoods and well-being in southern Africa. Int. J. Environ. Res. Publ. Health 16, 2970.

Mabrey, D., Vittorio, M., 2018. Moving from theory to practice in the water−energy−food nexus: an evaluation of existing models and frameworks. Water-Energy Nexus 1, 17−25.

Madhav, N., Oppenheim, B., Gallivan, M., Mulembakani, P., Rubin, E., Wolfe, N., 2017. Pandemics: risks, impacts, and mitigation. In: Jamison, D.T., Gelband, H., Horton, S. (Eds.), Disease Control Priorities: Improving Health and Reducing Poverty, third ed. The World Bank, Wasington DC, pp. 315−346.

Mahlknecht, J., González-Bravo, R., 2018. Measuring the water-energy-food nexus: the case of Latin America and the Caribbean region. Energy Proc. 153, 169−173.

McCornick, P.G., Awulachew, S.B., Abebe, M., 2008. Water−food−energy−environment synergies and tradeoffs: major issues and case studies. Water Pol. 10, 23−36.

Mpandeli, S., Naidoo, D., Mabhaudhi, T., Nhemachena, C., Nhamo, L., Liphadzi, S., Hlahla, S., Modi, A., 2018. Climate change adaptation through the water-energy-food nexus in southern Africa. Int. J. Environ. Res. Publ. Health 15, 2306.

Mushtaq, S., Maraseni, T.N., Maroulis, J., Hafeez, M., 2009. Energy and water tradeoffs in enhancing food security: a selective international assessment. Energy Pol. 37, 3635−3644.

Naidoo, D., Liphadzi, S., Mpandeli, S., Nhamo, L., Modi, A.T., Mabhaudhi, T., 2021a. Post Covid-19: a water-energy-food nexus perspective for South Africa. In: Stagner, J., Ting, D. (Eds.), Engineering for Sustainable Development and Living: Preserving a Future for the Next Generation to Cherish. Brown Walker Press, Florida, USA, p. 295.

Naidoo, D., Nhamo, L., Mpandeli, S., Sobratee, N., Senzanje, A., Liphadzi, S., Slotow, R., Jacobson, M., Modi, A., Mabhaudhi, T., 2021b. Operationalising the water-energy-food nexus through the theory of change. Renew. Sustain. Energy Rev. 149, 10.

Nhamo, L., Mabhaudhi, T., Mpandeli, S., Dickens, C., Nhemachena, C., Senzanje, A., Naidoo, D., Liphadzi, S., Modi, A.T., 2020a. An integrative analytical model for the water-energy-food nexus: South Africa case study. Environ. Sci. Pol. 109, 15−24.

Nhamo, L., Mpandeli, S., Senzanje, A., Liphadzi, S., Naidoo, D., Modi, A.T., Mabhaudhi, T., 2021a. Transitioning toward sustainable development through the water−energy−food nexus. In: Ting, D., Carriveau, R. (Eds.), Sustaining Tomorrow via Innovative Engineering. World Scientific, Singapore, pp. 311−332.

Nhamo, L., Ndlela, B., 2020. Nexus planning as a pathway towards sustainable environmental and human health post Covid-19. Environ. Res. 110376, 7.

Nhamo, L., Ndlela, B., Mpandeli, S., Mabhaudhi, T., 2020b. The water-energy-food nexus as an adaptation strategy for achieving sustainable livelihoods at a local level. Sustainability 12, 8582.

Nhamo, L., Ndlela, B., Nhemachena, C., Mabhaudhi, T., Mpandeli, S., Matchaya, G., 2018. The water-energy-food nexus: climate risks and opportunities in southern Africa. Water 10, 567.

Nhamo, L., Rwizi, L., Mpandeli, S., Botai, J., Magidi, J., Tazvinga, H., Sobratee, N., Liphadzi, S., Naidoo, D., Modi, A., Slotow, R., Mabhaudhi, T., 2021b. Urban Nexus and Transformative Pathways towards a Resilient Gauteng City-Region, South Africa. Cities 116.

Oppenheim, B., Gallivan, M., Madhav, N.K., Brown, N., Serhiyenko, V., Wolfe, N.D., Ayscue, P., 2019. Assessing global preparedness for the next pandemic: development and application of an epidemic preparedness index. BMJ Glob. Health 4, e001157.

Owusu, P.A., Asumadu-Sarkodie, S., 2016. A review of renewable energy sources, sustainability issues and climate change mitigation. Cogent Eng. 3, 1167990.

Peters, B.G., 2018. The challenge of policy coordination. Policy Design Pract. 1, 1−11.

Rasul, G., Sharma, B., 2016. The nexus approach to water−energy−food security: an option for adaptation to climate change. Clim. Pol. 16, 682−702.

References

Ringler, C., Bhaduri, A., Lawford, R., 2013. The nexus across water, energy, land and food (WELF): potential for improved resource use efficiency? Curr. Opin. Environ. Sustain. 5, 617–624.

Scott, C.A., El-Naser, H., Hagan, R.E., Hijazi, A., 2003. Facing water scarcity in Jordan: reuse, demand reduction, energy, and transboundary approaches to assure future water supplies. Water Int. 28, 209–216.

Terrapon-Pfaff, J., Ortiz, W., Dienst, C., Gröne, M.-C., 2018. Energising the WEF nexus to enhance sustainable development at local level. J. Environ. Manag. 223, 409–416.

UNGA, 2015. Transforming Our World: The 2030 Agenda for Sustainable Development, Resolution Adopted by the General Assembly (UNGA). United Nations General Assembly, New York.

Willis, C., 2015. The contribution of cultural ecosystem services to understanding the tourism—nature—wellbeing nexus. J. Outdoor Rec. Tour. 10, 38–43.

Woods, J., Williams, A., Hughes, J.K., Black, M., Murphy, R., 2010. Energy and the food system. Phil. Trans. Biol. Sci. 365, 2991–3006.

CHAPTER 2

Some quantitative water—energy—food nexus analysis approaches and their data requirements

Jafaru M. Egieya[1], Johann Görgens[2] and Neill Goosen[1,2]

[1]ARUA Centre of Excellence in Energy, Centre for Renewable and Sustainable Energy Studies, Faculty of Engineering, Stellenbosch University, Stellenbosch, South Africa; [2]Department of Process Engineering, Faculty of Engineering, Stellenbosch University, Stellenbosch, South Africa

1. Introduction

Since the year 2000, there have been evident shifts in the earth's climate, with significant negative effects on the natural environment. In the face of an increasing human population, estimated to reach approximately 9 billion by 2050 (Ezeh et al., 2012), which will require an estimated increase of 70% in food production (Zhang and Vesselinov, 2017), severe strain will likely be placed on limited water, energy, and food resources. Long-term solutions that optimally utilize these natural resources to ensure water-, energy-, and food security are complex, mainly due to interlinkages between these resources and how they are managed. As a consequence, long-term solutions need to explicitly consider the interlinkages between the supply and demand for water, energy, and food, while also taking into account policy aspects and how possible policy interventions could mitigate the situation (Hoff, 2011).

The interlinkages of water-, energy-, and food systems are referred to as the water—energy—food (WEF) nexus. The explicit realization and acknowledgment of these interlinkages and the role that they play in how these resource systems interact and are governed has given rise to several quantitative and qualitative methodologies that describe and analyze the WEF nexus. This nexus approach presents an indication of how water, energy, and food security can be sustainably enhanced by improving efficiency, forming synergies, reducing trade-offs, and improving governance across disparate sectors. Two categories of the definition of the WEF nexus have been put forward (Zhang et al., 2018): (1) the WEF nexus may be defined as an interaction between subsystems or sectors within the nexus system, while (2) the more prevalent definition views the WEF nexus as an analytical approach used to quantify the interlinkages between each of the nexus nodes (water, energy, and food), which is also the approach that will be followed in this chapter. Developing any WEF nexus analysis, be it quantitative or qualitative, should be able to address two questions: "What are the available optimal strategies to sustainably manage the water, energy, and food resources?" and "How will the effects of the policies and management strategies interact within and outside a sector?" These questions can also be asked at different scales, from, e.g., large regional scale, down to smaller scales where technologies are integrated and implemented at local level. The latter seems to be somewhat neglected in WEF nexus literature.

Developing a thorough understanding of the WEF nexus is important for long-term planning purposes. If applied, a WEF nexus approach should aid implementation of policies or projects within WEF sectors to enhance the overall system efficiency, without causing unwanted consequences within the other sectors. It is also seen to directly foster achieving the Sustainable Development Goals (SDGs) of the United Nations (Mpandeli et al., 2018). For instance, water security, food security, and energy security directly impact reducing poverty (SDG 1), ending hunger (SDG 2), and providing clean water and sanitation (SDG 6), which by extension enables good health and well-being (SDG 3). However, to sustainably supply water and food to meet these goals, energy needs to be provided affordably and cleanly (SDG 7), which by extrapolation tends to alleviate climate change (SDG 13). Since food resources are derived from land (SDG 15) and water (SDG 14) bodies, a useful WEF nexus assessment inevitably offers guidance on policy frameworks that must be developed to exploit available resources more sustainably. In effect, the importance of assessing the interrelations of water, energy, and food has ramifications for present and future generations.

Qualitative WEF nexus assessments are mostly performed by either applying existing tools and frameworks that employ preselected underlying methodologies, or by taking a more fundamental approach and building analyses "from scratch" by using existing modeling approaches, or developing new ones. Using existing ready-made tools has advantages in terms of convenience and prior validation by other others, whereas self-building models allow a certain degree of flexibility to customize the analysis (particularly when small-scale situations are assessed), something that more rigid predefined tools are not necessarily able to achieve. Several tools and frameworks are available for the quantitative assessment of the WEF nexus and can be delineated according to the inputs required, outputs provided, and their general characteristics. When inputs to quantitative assessments of the WEF nexus are considered, two categories of WEF nexus approaches are encountered, namely a fully integrated approach or an entry point approach (IRENA, 2015) as shown in Fig. 2.1. While the fully integrated approach to assessing the WEF nexus considers a situation whereby a relationship occurs between each of the nodes, the entry-point approach implies a situation whereby a (policy) change in one node (e.g., energy) causes a domino effect on the other two nodes.

1. Introduction   17

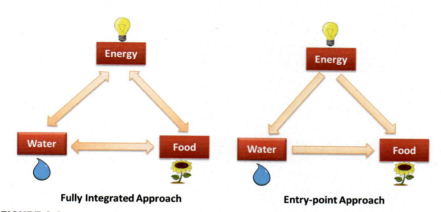

FIGURE 2.1
Nexus approaches.

Alternatively, the WEF tools or framework may be selected based on the outputs derived from perceived future policy implications of varying the inputs. Examples of such outputs are quantifying greenhouse gas (GHG) emissions, cost implications of changing technology, the effects on water resources or the energy burden, or a change in agricultural practices.

Whether existing tools or custom models are employed for WEF nexus analyses, quantitative WEF nexus analyses rely heavily on the availability of good quality data. The exact nature of the data required to perform these analyses will be dependent on the framework or methodology chosen. Additionally, issues of scale need to be considered, especially whether there are data available for the specific scale of interest, or whether such data can be derived (i.e., upscaled or downscaled) from available data sets. Conversely, the lack of data availability might constrain the possible types of WEF nexus analyses, which can be performed in a particular case, which in turn might limit the utility of the analysis performed.

To understand the data requirements of typical quantitative WEF nexus analyses, a number of existing tools and modeling approaches are described in the following, with brief discussions on what types of data are required for these analyses, and what the outputs of these tools and techniques are. It is not the aim of this chapter to provide a comprehensive review of WEF nexus methods as such reviews already exist in literature. Rather, this chapter aims to provide a brief overview of a sufficient number of often-employed methods to guide the reader on the types of data that will be required before quantitative analyses can be performed. This is done both for some existing WEF nexus tools, and for some more fundamental modeling approaches. This chapter further highlights specific data challenges to the reader and contextualizes it for the African continent.

2. WEF resource management tools

A range of WEF nexus assessment tools are available for use. Some of the more commonly utilized tools include the Climate, Land, Energy Use, and Water Strategies (CLEWs) (Howells et al., 2013), Water, Energy, Food Nexus Tool 2.0 (Daher and Mohtar, 2015), REMap (IRENA, 2015), Multi-Scale Integrated Analysis of Societal and Ecosystem Metabolism (MuSIASEM) (Giampietro et al., 2013), International Institute for Sustainable development (IISD) Water-Energy-Food Security Analysis Framework (Bizikova et al., 2013), Transboundary Basin Nexus Assessment (TBNA) Methodology (Roidt and Strasser, 2015), and iSDG planning Model (Millennium Institute, 2021). Generally, the tools were developed to assess WEF nexus scenarios on fairly large scales (e.g., national level), although if sufficient data are available some tools can be used for smaller scale analyses. Four tools (CLEWs, Water, Energy, Food Nexus Tool 2.0, REMap, and MuSIASEM) are selected and discussed in the following, and summarized in Table 2.1.

2.1 CLEWs

The CLEWs tool (Howells et al., 2013) is a fully integrated framework composed of three models including LEAP (Long-range Energy Alternatives Planning tool) developed by the Stockholm Energy Institute (SEI), WEAP (Water Evaluation and Planning tool, also by SEI), and AEZ (Agro-Ecological Zoning) by the International Institute for Applied Systems Analysis (IIASA) and Food and Agriculture Organization (FAO), interacting with climate change scenarios. The framework assesses binary relationships. For the water–energy binary relationship, the required energy is assessed for water processing and treatment, water pumping, and desalination while simultaneously determining the volume of water required for hydropower, power plant cooling, and biofuel processing. For the binary energy–food relationship, energy needed for fertilizer production, field preparation, and harvesting and conversely, determining the type and quantity of biomass feedstock required for biofuel production and other energy uses, is assessed. For the binary water–food relationship, the quantity of water required to grow biofuel crops (rain-fed and irrigation), water needs for food, fiber, and fodder crops (rain-fed and irrigated), is assessed.

A module-based approach is used to integrate the three models (i.e., LEAP, WEAP, and AEZ) whereby the required data are iteratively passed between each model (see Fig. 2.2). Hence, the output of one model becomes the input to the other two models and vice versa. The models are then solved sequentially until a convergent solution is determined.

For the CLEWs framework, the key data inputs are (1) the water requirements for land-use and energy systems, (2) land requirements for energy and water infrastructure, and (3) energy needs for the supply of water and the particular

Table 2.1 Summary of advantages and disadvantages of the tools.

S/No.	Tool	Scale	Advantages	Disadvantages
1.	CLEWs (Howells et al., 2013)	City National Global	• Less time and cost needed to develop new methodologies from the scratch because of previously existing modeling methodologies. • It makes it easy to bring in experts from different fields to work together. • It allows users to compare results in an integrated mode compared to its stand-alone models.	• It does not show the quantitative effects of changes to ecosystem services such as the impact of change in cropping practice on loss of biodiversity. • The iteration process and time required to check results may be too long for a policymaking process. • The accuracy of the climate models is very uncertain. • Access to the individual models has to be requested.
2.	Water, Energy, Food Nexus Tool 2.0 (Daher and Mohtar, 2015)	National	• It is a preliminary assessment tool that highlights potential resource trends and alarms, which may serve as input to a more detailed methodology. • The tool is publicly available on www.wefnexustool.org	• The tool can only be tailored to a national context and falls short of assessing the effects on a global scale. • It does not capture the effects of future projections on process, population growth, and increased resource demand. • The relationship between each node is empirical instead of process-based. • It does not assess the financial implication of using different water and energy sources.
3.	REMap (IRENA, 2015)	National	• It is analytically simple use. • The output obtained could potentially serve as an input to a more comprehensively formulated. • It provides information about the nodes' interaction and the energy policy.	• It uses a linear once-through framework that does not show nodes continuously feeding into each other. • It is static tool.
4.	MuSIASEM (Giampietro et al., 2013)	National	• It is able to link GIS-based data used in spatial analysis with socioeconomic sciences, thus establishing a bridge	• The approach generates large chunks of information, which is hard to decipher and summarize.

Continued

CHAPTER 2: Quantitative WEF nexus analysis approaches

Table 2.1 Summary of advantages and disadvantages of the tools. *Continued*

S/No.	Tool	Scale	Advantages	Disadvantages
			between ecological and socioeconomic dimensions. • Its complexity mirrors reality since it recognizes fuzziness and uncertainties because there are no optimal solutions. • It is able to infer missing information or data gaps since the quantitative information used is based on the rules of inference exploited in the Sudoku games. • It can check the data quality by assessing top-down assessments that use aggregate statistical data with bottom-up assessments that use empirical data for local plants and processes technical coefficients.	• The accounting scheme can use official statistics, but using this may require some form of preliminary data sets processing, which can be time-consuming and expensive to undertake. • The transdisciplinary character of the approach presents a major challenge to the team members who not only contribute their expertise but must also strive to understand the whole system. • It is critical to have a good environmental economist (which is specialist role) in the team to assist in linking biophysical and economic flows, which is a complex task.

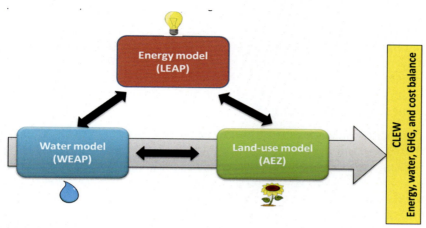

FIGURE 2.2

The CLEWs framework. *Adapted from Howells, M., Hermann, S., Welsch, M., Bazilian, M., Segerström, R., Alfstad, T., Gielen, D., Rogner, H., Fischer, G., Van Velthuizen, H., Wiberg, D., Young, C., Alexander Roehrl, R., Mueller, A., Steduto, P., Ramma, I., 2013. Integrated analysis of climate change, land-use, energy and water strategies. Nat. Clim. Change 3 (7), 621–626. https://doi.org/10.1038/nclimate1789.*

land use. As CLEWs is developed for national scale studies, users should be aware of national resource plans, and how these may change over time (Bazilian et al., 2011). For CLEWs, future climate scenarios are employed to formulate future rainfall (and potentially other parameters), and it is therefore a powerful tool to assess how the WEF nexus status might change for a particular country under different climate change scenarios.

2.2 Water, Energy, Food Nexus Tool 2.0

The Water, Energy, Food Nexus Tool 2.0 (Daher and Mohtar, 2015) is a nexus tool developed at Texas A&M University, United States. It uses a scenario-based framework to assess the interrelationships between WEF nodes and can be seen as an initial screening tool to highlight particular areas of concern. The data requirements for each of the water, energy, and food nodes are quite specific. For the water node, data inputs include water utilization per unit of production (m^3/ton) for agriculture, municipal and industrial applications, and water sources on annual basis (i.e., rainfall, surface water, groundwater, treated water, desalinated water, and treated wastewater). Furthermore, there is a need to quantify the available conventional and nonconventional water sources, and to take into account the agricultural practices and technology employed during production, e.g., irrigation method. Inputs for food need to be specified in units of weight and constitute consumption (imports and exports). Energy inputs related to water are specified in kJ/m^3 and include energy needs for water transport and treatment (surface and groundwater pumping, water and wastewater treatment, desalination, and other potential uses), whereas energy requirements for food production are specified in kJ/ton production (including energy needs for tillage, fertilizer production, harvest, and local transport). Additionally, the greenhouse gas emissions need to be specified in ton CO_2 emitted/ton production. Energy requirements are based on technologies implemented and their efficiencies.

Additional considerations that need to be addressed include the local climate and land availability, and financial implications of the WEF system, and any modifications brought about. Local climatic conditions, coupled with the biophysical environment, govern the types of agriculture that can be engaged in and what types of crops can be produced, whereas land availability will determine whether certain thresholds (e.g., total food self-sufficiency) are achievable. Financial considerations include local costs of production in US$/ton (capital, operating, and maintenance costs), energy costs in US$/kJ, and the environmental cost impact on air, water, and soil. Carbon footprints are further also taken into account measured in ton $CO_2/kJ/m^3$ and ton $CO_2/kJ/ha$.

Based on user input for different scenarios, the tool calculates a sustainability index for each scenario. For each scenario, the output of the tool summarizes total land requirements (ha), total water requirements (m^3), emissions (ton CO_2), and energy consumption (kJ) during import of produce, local energy requirement-based water supply needs and food production systems (kJ), financial costs (US$), and total carbon footprint (ton CO_2).

2.3 REMap

The REMap tool is used for the preliminary assessment of the nexus nodes as a result of changing energy policies (IRENA, 2015). Hence, the REMap is an entry-point tool. REMap is heavily focused on energy scenarios and requires substantial knowledge of the local energy landscape. The first step of the procedure is the provision of an energy balance (current or future) that serves as a base case, after which alternative energy balances (normally based on energy modeling from local agencies) are compared against the base case, using a WEF nexus outlook. The analysis allows the evaluation of the change in water, land, emissions, and costs between the initial base case and alternative scenarios. These comparisons provide insights on the basic resources, cost, and emissions implications of the assessed change in energy balances, with the assumption that the changes are based on changes in energy policy.

2.4 MuSIASEM

MuSIASEM (Giampietro et al., 2013) is an integrated approach that exploits the characterization of flows of different systems within a society to analyze the nexus between water, energy, and food security. The model also considers the dynamics of population, GHG emissions, and land-use changes at both the national and subnational levels. It can therefore be used as a diagnostic tool and for simulation purposes.

As a diagnostic tool, the model can characterize existing socioeconomic system patterns (i.e., population, workforce, capital, total land, etc.) and flows of water, energy, food, and money (referred to as flow elements). However, as a simulation tool, the model's viability, desirability, and feasibility check of scenarios proposed can be validated.

MuSIASEM as a framework comprises a flow-fund model and three conceptual tools: multiscale accounting, multipurpose grammar, and impredicative loop analysis. While flow elements in the model refer to those elements that appear and disappear throughout the analysis such as the consumption and production of water, energy, food, and other key materials, the fund elements refer to what the system is made of or production factors such as managed land uses, rivers, human beings, and technological capital.

The procedure to implement the MuSIASEM model is involved and requires a number of steps:

- **Step 1**—Define the socioeconomic system as a set of functional compartments essential to guarantee its survival, reproducibility, and adaptability. In essence, the boundary conditions should be defined using a nested hierarchy structure of a socioeconomic system.
- **Step 2**—Select relevant fund elements and quantify them across the various compartments defined in Step 1.
- **Step 3**—Define and quantify the flow elements (water, energy, food, and money) associated with the fund elements (in step 2) and functional compartments (in step 1).
- **Step 4**—Develop a multilevel, multidimensional matrix of the flow elements (in step 3) and the fund elements (in step 2).
- **Step 5**—Check the viability and desirability of the metabolic pattern. The viability of the model is checked by cross-checking the stability of the dynamic budgets of the individual flows (water, energy, food, and money), whereas the desirability is the ratio of production factors and flows allocated to the total amount of funds and flow elements.
- **Step 6**—Check the feasibility of the results obtained of the metabolic pattern concerning the environmental loading (demand side) and resource supplies.

Each of the discussed tools has advantages and disadvantages, and tool selection should explicitly consider whether an initial or comprehensive analysis is required, the availability of the particular data needed to run the tool, and the scale at which the analysis needs to be performed. Some of the most common advantages and disadvantages of the tools discussed are given in Table 2.1.

3. Alternative methodologies, approaches, and frameworks

Apart from the specific modeling tools discussed in Section 2, other authors have proposed several quantitative modeling frameworks, approaches, or methodologies to assess the WEF nexus from more conceptual and systematic perspectives. Holistically, these modeling approaches fall under five categories including economic related, environmental assessment, systems analysis, statistical models, and indicators/metrics (Albrecht et al., 2018; Zhang et al., 2018). Again, it is not the aim of this chapter to provide a comprehensive overview of all possible modeling approaches, but rather to provide the reader with sufficient information to identify the type of approach suitable for their particular needs.

3.1 Economic analysis models

Economic models and frameworks are developed to gain a clearer picture of the existing economic climate of a country, region, or company. Concerning WEF nexus research and policy assessment, some of the most widely employed economic models include input—output analysis frameworks (Bellezoni et al., 2018; Li et al., 2016), cost—benefit analysis (Endo et al., 2015), profit maximization (Egieya et al., 2018), and trade-off analysis (Siderius et al., 2020; Wu et al., 2021). The scales at which economic analyses can be applied vary from very large (e.g., regional or continental) to very small scale (e.g., household level).

Models that incorporate economic and/or cost considerations are powerful and flexible, as they allow assessment of customized scenarios and they can be combined with other analysis approaches. Numerous examples of such tailored approaches exist, which examine particular cases or scenarios that may be unlikely to be examined within the predefined frameworks of existing tools. Although a comprehensive review of all the types of economic WEF nexus analyses are outside the scope of this text, a selection of recent studies is summarized in Table 2.2.

The studies highlighted in Table 2.2 show some of the different analyses that are available through models that consider economic and cost impacts. Although many of the analyses that are currently being done using these methods are still geographically bound, the study of White et al. (2018) is a good example of how such analyses can be expanded beyond national borders

3. Alternative methodologies, approaches, and frameworks   25

Table 2.2 Selected economic analysis studies.

Analysis type	Question examined	Geographic region	Main findings	Citations
Hybrid economic–ecological input–output analysis	Scenario-based investigation of the impact of expanding sugarcane biofuel production in Paranaíba basin in Brazil, on local resource availability	Goiás State, Brazil	Controlled and well-planned sugarcane expansion will cause little negative impact to local resource availability	Bellezoni et al. (2018)
Data envelopment analysis (DEAs)	Comparative analysis on WEF nexus input–output efficiency in 30 different states in China, over time period 2005–2014	China	The DEA method quantified WEF nexus efficiency successfully. It was able to give current status of WEF nexus efficiency and also showed trends over time. The method therefore allows identification of regions requiring intervention	Li et al. (2016)
Interval-based multiobjective programming, incorporating sensitivity analysis	Optimization of sustainable bioenergy production in agricultural systems	Northeast China	The method could successfully identify scenarios where bioenergy production is maximized for under conditions of minimized costs and environmental impact	Li et al. (2020)
Input-output analysis supplemented with trade flows	Quantification of oil-dominated WEF nexus across scales and over time	Kuwait, Qatar and Saudi Arabia	Successfully showed domestic and international interactions within the WEF nexus, and particularly how local policies can impact resource depletion in other countries through using virtual water as metric	Siderius et al. (2020)
Multiperiod socioeconomic model	Development of a methodology that satisfies future socioeconomic WEF demands to ensure sustainable development	Not region specific	Successfully quantified trade-offs between resource constraints, environmental impacts, and economic objectives	Zhang and Vesselinov (2017)
Transnational interregional input–output analysis, including spatial linkages between local consumption and environmental impacts over long distances	How trade and global value chains impact local WEF nexus resources	Local (China) and transnational (China, South Korea, Japan)	The analysis shows how China's export-oriented economic growth strategy leads to unsustainable levels of WEF resource exports, with accompanying environmental impacts, to the benefit of trading partners	White et al. (2018)

and therefore highlight cross-boundary challenges. In terms of data requirements, these analyses are data intensive and require significant data collection and data validation efforts. Data also need to be collected from a variety of government agencies, industry associations, national and regional resource governance bodies, and other available sources. As an illustration of the effort required, readers are encouraged to view the data descriptions in the papers cited in Table 2.2.

3.2 Environmental impact related

Environmental impact-based WEF nexus analyses allow the assessment of how activities (e.g., a production process) impact natural resource systems, among which are land, water, mineral, and fossil resources. The most widely used environmental impact assessment tool is the life cycle analysis (LCA). The method has the advantage of being able to accurately measure any unit during its life cycle while simultaneously identifying and quantifying all inputs and outputs that may pose a significant impact on the environment (Zhang et al., 2018). Various investigations employed LCA within a WEF nexus context. Al-Ansari et al. (2015) carried out an LCA to assess the impact of different energy sources on the WEF nexus terrestrial and marine ecosystems in Qatar. In another study, a novel environmental input—output LCA model was applied to assess the interactions of urban food—energy—water on 382 metropolitan statistical areas within the United States (Sherwood et al., 2017). Li and Ma (2020) assessed the embodied resource consumption of the WEF nexus while distinguishing the direct and indirect consumption of resource flows in Taiwan, according to an LCA approach. Chen et al. (2020) employed a sequential approach, using material flow analysis followed by LCA to evaluate the WEF system's environmental impact based on 15 impact categories in Taiwan.

Performing environmental impact assessments of actions (including LCA) requires data on resource utilization linked to the investigated activity, the inputs required to make such resources available, along with the products of the activity and the possible waste generated therefrom. Although LCA is a powerful method that can assess a wide range of scenarios, it yields a static output that only provides a "snapshot" of environmental impacts at any one time under a single set of assumptions. Each additional scenario requires the method to be reprogrammed and rerun.

3.3 Systems analysis

Systems analysis studies usually involve the use of mathematical representations to define how systems operate and to propose procedures or make recommendations for the systems to operate more efficiently. Systems analysis is an integrated approach often applied in the form of systems dynamics modeling (SDM). SDM unlike other models is employed to assess complex feedback-driven systems by emulating any system to the level of detail required (Sušnik et al., 2012) and has been utilized by policymakers as a hands-on tool to solve organizational challenges (Sterman, 2000). However, unlike the economic-related and environmental assessment methods for quantifying the WEF nexus, systems analysis studies have been less often employed for WEF nexus work. The low utilization of system analysis for WEF nexus studies may be due to the perceived risk of too much analyzing that may be time-consuming and expensive (Sterman, 2000).

Despite its relatively lower frequency to describe WEF nexus situations, SDM is a very powerful methodology that can incorporate significant complexity, as is an inherent requirement in WEF nexus work. It allows investigation of WEF nexus questions on micro- and macrolevels and is an especially appropriate methodology to describe multiactor and multidisciplinary questions (Zhang et al., 2018). However, to ensure that the modeling procedure represents reality, significant stakeholder involvement is required during the setup of models (Sušnik et al., 2021), meaning that the model is only as good as the input received from the stakeholders.

SDM can generate a wide range of outputs, depending on the exact research question. Among others, a well setup model can be combined with qualitative work (Huey et al., 2020; Sušnik et al., 2021), it can be employed to elucidate interactions and trade-offs within the WEF nexus (Wu et al., 2021), it can estimate resource fluxes (Walker et al., 2014), and it can examine economic or financial aspects (White et al., 2018). Overall, SDM is powerful to elucidate how proposed policy changes could impact the overall WEF system and how certain unintended consequences may render proposed policies ineffective. However, there is a challenge in communicating such a complex method and its results to policymakers, who in themselves may not have modeling expertise.

3.4 Statistics

The collection and interpretation of statistics forms an important part of WEF nexus assessments. These statistics can be in the form of field surveys, questionnaires, expert panels, and data collection from public government institutions or local agencies (Zhang et al., 2018). Apart from the collection of data and information, the method relies heavily on human input to identify links between the different subsystems in the WEF nexus, e.g., by utilizing expert inputs. However, it remains a very useful approach to combine qualitative and quantitative approaches and to preliminarily identify important metrics and interlinkages to take forward in more comprehensive analyses. The method is, however, severely limited in that it fails to illuminate any of the mechanisms causing interrelations between different statistics and is therefore not greatly useful in evaluating or making policy recommendations (Zhang et al., 2018).

3.5 Indicators and metrics

Indicators and metrics are useful methods to indicate the status of the WEF nexus at any one point, and they indicate the properties of such a system (Endo et al., 2015; Zhang et al., 2018). These indicators tend to be most valuable in providing preliminary assessment results for more detailed tools (Dargin et al., 2019), and as comparative indicators to assess and benchmark different systems (Yigitvanlar et al., 2015). These indices and metrics are underpinned by data collection and reporting initiatives, and rely heavily on good databases (Schlör et al., 2018). Despite their utility, there remain concerns in how a good index system is put together for these metrics and indicators, and how to remove bias when weighted indicators are employed (Yigitvanlar et al., 2015).

4. Data challenges of quantitative nexus research

Quantitative WEF nexus research is data-driven, and the accuracy of the overall analysis is highly dependent on the quality, completeness, and accuracy of the input data. However, seeing that by its nature, research on the WEF nexus requires data from different sectors, over different scales and in some instances over time, there are a number of challenges associated with this kind of research. Some general challenges will be highlighted, and some challenges are particular to the Global South.

In general, data availability remains the most extensive limiting factor to carrying out nexus assessment studies (Kaddoura and El Khatib, 2017). There is also often significant disparity in the basis or form in which data from different sectors are presented, which requires a significant time investment at project initiation to achieve alignment. For example, food and energy consumption might be given in annual amounts, but water requirements are often given in hourly values. Furthermore, different sectors might report data for different geographic regions (e.g., country-based data vs. transboundary watershed-based data), which may also result in significant effort to ensure alignment.

Even if data are available, there are instances where data or software for WEF nexus analysis are not accessible. Examples would be where specific data sets are hidden behind paywalls, or held at institutions or departments that are unwilling to grant access to the data, and where expensive specialist software is required for a specific type of analysis. In other instances, the data or software may be publicly or freely available but would contain many assumptions that make it imperative to get an expert to validate and explain the data quality.

The issue of system boundaries remains ever present in WEF nexus research and has significant data implications. The globalization of trade in water, food, and energy across national and regional boundaries presents a significant barrier to carrying out nexus research. Sometimes, authors are constrained to make assumptions on data outside the system boundary or study area, with the implication that the results generated do not necessarily reflect reality. Even if researchers attempt to obtain data across the system boundary, it may once again prove costly, laborious, and time-consuming (Simpson and Jewitt, 2019).

In the Global South, there may be some additional challenges in obtaining the data required for quantitative WEF nexus research. Although it is taking shape, the policy environment making data openly available to researchers is still lagging that of the Global North. An implication of this is that data sets collected and curated by one research project are not necessarily accessible to a following project, as the open access and data availability policies of funding bodies are still developing. Additionally, it might be more difficult to obtain regional data in less well-integrated regions of the Global South, than, for example, the European Union where substantial transboundary integration has already been established. Finally, the different levels of data digitization between different entities (Government Departments, Ministries, countries, etc.) may render some data inaccessible to researchers, even though these data exist.

5. Conclusions

Quantitative WEF nexus analyses are very valuable tools for describing the state of WEF nexus at a particular time and for future planning purposes. There are two main approaches to performing these quantitative analyses: utilization of existing tools such as CLEWs, or employing more fundamental modeling methodologies. Irrespective of whether existing WEF nexus tools or more fundamental modeling frameworks are employed, quantitative WEF nexus research remains data-driven. However, there are significant challenges to obtaining good quality, coherent, and complete data sets for WEF nexus research, and data availability remains one of the largest constraints to research in this field.

References

Al-ansari, T., Korre, A., Nie, Z., Shah, N., March 2015. Development of a life cycle assessment tool for the assessment of food production systems within the energy, water and food nexus. Sustain. Prod. Consum. 2, 52–66. https://doi.org/10.1016/j.spc.2015.07.005.

Albrecht, T.R., Crootof, A., Scott, C.A., 2018. The Water-Energy-Food Nexus: a systematic review of methods for nexus assessment. Environ. Res. Lett. 13 (4). https://doi.org/10.1088/1748-9326/aaa9c6.

Bazilian, M., Rogner, H., Howells, M., Hermann, S., Arent, D., Gielen, D., Steduto, P., Mueller, A., Komor, P., Tol, R.S.J., Yumkella, K.K., 2011. Considering the energy, water and food nexus: towards an integrated modelling approach. Energy Pol. 39 (12), 7896–7906. https://doi.org/10.1016/j.enpol.2011.09.039.

Bellezoni, R.A., Sharma, D., Villela, A.A., Pereira Junior, A.O., April 2018. Water-energy-food nexus of sugarcane ethanol production in the state of Goiás, Brazil: an analysis with regional input-output matrix. Biomass Bioenergy 115, 108–119. https://doi.org/10.1016/j.biombioe.2018.04.017.

Bizikova, L., Roy, D., Swanson, D., Venema, H.D., McCandless, M., February 28, 2013. The water-energy-food security nexus: towards a practical planning and decision-support framework for landscape investment and risk management. Int. Insti. Sustain. Develop. http://www.iisd.org/sites/default/files/pdf/2013/wef_nexus_2013.pdf.

Chen, C.F., Feng, K.L., Wen Ma, H., May 2020. Uncover the interdependent environmental impacts associated with the water-energy-food nexus under resource management strategies. Resour. Conserv. Recycl. 160, 104909. https://doi.org/10.1016/j.resconrec.2020.104909.

Daher, B.T., Mohtar, R.H., 2015. Water–energy–food (WEF) Nexus Tool 2.0: guiding integrative resource planning and decision-making. Water Int. 40 (5–6), 748–771. https://doi.org/10.1080/02508060.2015.1074148.

Dargin, J., Daher, B., Mohtar, R.H., 2019. Complexity versus simplicity in water energy food nexus (WEF) assessment tools. Sci. Total Environ. 650, 1566–1575. https://doi.org/10.1016/j.scitotenv.2018.09.080.

Egieya, J., Čuček, L., Zirngast, K., Isafiade, A., Pahor, B., Kravanja, Z., 2018. Synthesis of Biogas supply networks using various biomass and manure types. Comput. Chem. Eng. https://doi.org/10.1016/j.compchemeng.2018.06.022.

Endo, A., Burnett, K., Orencio, P.M., Kumazawa, T., Wada, C.A., Ishii, A., Tsurita, I., Taniguchi, M., 2015. Methods of the water-energy-food nexus. Water (Switzerland) 7 (10), 5806–5830. https://doi.org/10.3390/w7105806.

Ezeh, A.C., Bongaarts, J., Mberu, B., 2012. Global population trends and policy options. Lancet 380 (9837), 142–148. https://doi.org/10.1016/S0140-6736(12)60696-5.

Giampietro, M., Aspinall, R.J., Bukkens, S.G.F., Benalcazar, J.C., Diaz-Maurin, F., Flammini, A., Gomiero, T., Kovacic, Z., Madrid, C., Ramos-Martin, J., Serrano-Tovar, T., 2013. Application of the MuSIASEM Approach to Three Case Studies an Innovative Accounting Framework for the Food-Energy-Water Nexus. http://www.fao.org/docrep/019/i3468e/i3468e.pdf.

Hoff, H., November, 2011. Understanding the Nexus. Background Paper for the Bonn2011 Nexus Conference: Stockholm Environment Institute, pp. 1–52.

Howells, M., Hermann, S., Welsch, M., Bazilian, M., Segerström, R., Alfstad, T., Gielen, D., Rogner, H., Fischer, G., Van Velthuizen, H., Wiberg, D., Young, C., Alexander Roehrl, R., Mueller, A., Steduto, P., Ramma, I., 2013. Integrated analysis of climate change, land-use, energy and water strategies. Nat. Clim. Change 3 (7), 621–626. https://doi.org/10.1038/nclimate1789.

Huey, A., Tan, P., Yap, E.H., 2020. A Complex Systems Analysis of the Water-Energy Nexus in Malaysia, pp. 1–21.

IRENA, January 2015. Renewable Energy in the Water, Energy and Food Nexus. International Renewable Energy Agency, pp. 1–125.

Kaddoura, S., El Khatib, S., May 2017. Review of water-energy-food Nexus tools to improve the Nexus modelling approach for integrated policy making. Environ. Sci. Pol. 77, 114–121. https://doi.org/10.1016/j.envsci.2017.07.007.

Li, G., Huang, D., Li, Y., 2016. China's input-output efficiency of water-energy-food nexus based on the data envelopment analysis (DEA) model. Sustainability 8 (9). https://doi.org/10.3390/su8090927.

Li, P., Ma, H., March 2020. Evaluating the environmental impacts of the water-energy-food nexus with a life-cycle approach. Resour. Conserv. Recycl. 157, 104789. https://doi.org/10.1016/j.resconrec.2020.104789.

Millennium 2021. https://www.millennium-institute.org/isdg. (Accessed 7 August 2021).

Mpandeli, S., Naidoo, D., Mabhaudhi, T., Nhemachena, C., Nhamo, L., Liphadzi, S., Hlahla, S., Modi, A.T., 2018. Climate change adaptation through the water-energy-food nexus in Southern Africa. Int. J. Environ. Res. Publ. Health 15 (10), 1–19. https://doi.org/10.3390/ijerph15102306.

Roidt, M., Strasser, L., 2015. Methodology for Assessing the Water-Food-Energy-Ecosystems Nexus in Transboundary Basins and Experiences from its Application: Synthesis.

Schlör, H., Venghaus, S., Jürgen-Friedrich, H., 2018. The FEW-Nexus city index – Measuring urban resilience. Appl. Energy 210, 382–392. https://doi.org/10.1016/j.apenergy.2017.02.026.

Sherwood, J., Cqlabeaux, R., Carbajales-Dale, M., 2017. An extended environmental input – output lifecycle assessment model to study the urban food – energy – water nexus an extended environmental input – output lifecycle assessment model to study the urban food – energy – water nexus. Environ. Res. Lett. 12, 1–13.

Siderius, C., Conway, D., Yassine, M., Murken, L., Lostis, P.L., Dalin, C., 2020. Multi-scale analysis of the water-energy-food nexus in the Gulf region. Environ. Res. Lett. 15 (9). https://doi.org/10.1088/1748-9326/ab8a86.

Simpson, G.B., Jewitt, G.P.W., February 2019. The development of the water-energy-food nexus as a framework for achieving resource security: a review. Front. Environ. Sci. 7. https://doi.org/10.3389/fenvs.2019.00008.

Sterman, J.D., 2000. Business Dynamics: Systems Thinking.

Sušnik, J., Masia, S., Indriksone, D., Brēmere, I., Vamvakeridou-Lydroudia, L., 2021. System dynamics modelling to explore the impacts of policies on the water-energy-food-land-climate nexus in Latvia. Sci. Total Environ. 1–16. https://doi.org/10.1016/j.scitotenv.2021.145827.

Sušnik, J., Vamvakeridou-lyroudia, L.S., Savi, D.A., Kapelan, Z., 2012. Integrated system dynamics modelling for water scarcity assessment: case study of the Kairouan region. Sci. Total Environ. 440, 290–306. https://doi.org/10.1016/j.scitotenv.2012.05.085.

Walker, R.V., Beck, M.B., Hall, J.W., Dawson, R.J., Heidrich, O., 2014. The energy-water-food nexus : strategic analysis of technologies for transforming the urban metabolism. J. Environ. Manag. 141, 104–115. https://doi.org/10.1016/j.jenvman.2014.01.054.

White, D.J., Hubacek, K., Feng, K., Sun, L., Meng, B.O., 2018. The Water-Energy-Food Nexus in East Asia: a tele-connected value chain analysis using inter-regional input-output analysis. Appl. Energy 550–567. https://doi.org/10.1016/j.apenergy.2017.05.159.

Wu, L., Elshorbagy, A., Pande, S., Zhuo, L., 2021. Trade-offs and synergies in the water-energy-food nexus: the case of Saskatchewan, Canada. Resour. Conserv. Recycl. 164, 105192. https://doi.org/10.1016/j.resconrec.2020.105192 (September 2020).

Yigitvanlar, T., Dur, F., Dizdaroglu, D., 2015. Towards prosperous sustainable cities: a multiscalar urban sustainability assessment approach. Habitat Int 45 (1), 36–46. https://doi.org/10.1016/j.habitatint.2014.06.033.

Zhang, C., Chen, X., Li, Y., Ding, W., Fu, G., 2018. Water-energy-food nexus: concepts, questions and methodologies. J. Clean. Prod. 195, 625–639. https://doi.org/10.1016/j.jclepro.2018.05.194.

Zhang, X., Vesselinov, V.V., 2017. Integrated modeling approach for optimal management of water, energy and food security nexus. Adv. Water Resour. 101, 1–10. https://doi.org/10.1016/j.advwatres.2016.12.017.

CHAPTER 3

EO-WEF: a Earth Observations for Water, Energy, and Food nexus geotool for spatial data visualization and generation

Zolo Kiala[6], Graham Jewitt[1,2], Aidan Senzanje[7], Onisimo Mutanga[5], Timothy Dube[3] and Tafadzwanashe Mabhaudhi[4,6]

[1]*IHE Delft Institute for Water Education, Delft, the Netherlands;* [2]*Centre for Water Resources Research, School of Agricultural, Earth and Environmental Sciences, University of KwaZulu-Natal (UKZN), Pietermaritzburg, South Africa;* [3]*Institute for Water Studies, Department of Earth Sciences, University of the Western Cape, Bellville, South Africa;* [4]*International Water Management Institute (IWMI), Accra, Ghana;* [5]*Discipline of Geography, School of Agricultural, Earth and Environmental Sciences, University of KwaZulu-Natal, Pietermaritzburg, South Africa;* [6]*Centre for Transformative Agricultural and Food Systems (CTAFS), School of Agricultural, Earth and Environmental Sciences, University of KwaZulu-Natal, Pietermaritzburg, South Africa;* [7]*Bioresources Engineering Programme, School of Engineering, University of KwaZulu-Natal (UKZN), Pietermaritzburg, South Africa*

1. Introduction

The energy, water, and food resources are highly connected and interdependent. Globally, food production is strongly dependent on both water availability and energy production, with the agricultural sector using 71% of the former and 30% of the latter (Sadegh et al., 2020; Simpson and Jewitt, 2019b). So, if the security of one of the resources is threatened, the security of the two resources will also be jeopardized. Improving our understanding of the complex interrelationships of these resources is therefore critical for sustainable development, policymaking, and human well-being (McCarl et al., 2017). Decisions based on a nexus consideration rather than individual sectors of a nexus are likely to yield better, if not more informed, outcomes. Since the Bonn Conference on Water Energy and Food Security Nexus—Solutions for the Green Economy, a paradigm shift, dubbed "nexus approach" has been occurring where the energy, water, and food sectors are assessed as an interlinked system by considering their trade-offs and synergies (Simpson and Jewitt, 2019a). When compared with other integrated approaches, stronger demand for operationalization and solution-orientation of the water—energy—food (WEF) nexus approach has been made by resource managers, policymakers, and other stakeholders (Liu et al., 2018).

Although the approach has evolved from a nexus thinking to a nexus action, existing implementations have failed to address complex interlinkages. Liu et al. (2018) noted that most of nexus studies have focused on two sectors (e.g., as energy and water, food, and energy). One of the main reasons of this failure is the lack of data sharing and availability for mining available observations that can uncover complex interactions of the different components of a system (Purwanto et al., 2021). Some of these data may already be available in public databases, while others will require intensive engineering efforts. Even when data exist, data holdings due to reasons such as preservation of confidentiality, security, etc., may hamper their accessibility. Moreover, data assembly for a WEF nexus analysis is also challenged by the questions under investigation and its boundary. As an illustration, aggregate data may be more acceptable and easier to find at a national scale than at a local level where more detailed studies are required (McCarl et al., 2017). Although it is practically impossible to cater data at different scales to all the components of a nexus system, there is a need to develop multisectoral information systems that consolidate data from different sources and provide the basis for insights on the WEF nexus.

Earth observations (EOs) and geospatial data sets have the potential of addressing the aforementioned challenges. Many key water, energy, and food sector variables can be reliably estimated from space at a relatively low cost and various temporal and spatial resolutions. Standard products such as precipitation, evapotranspiration, soil moisture, and so on are regularly generated from algorithms, which correlate reflectance of objects on the ground to physical atmospheric and surface variables. In combination with ancillary data, such as census, remotely sensed data have been used to produce socioeconomic data (e.g., population) (Lawford, 2019). Furthermore, the advent of cloud computing platforms has made it possible to process massive information from numerous remotely sensed data (Wang et al., 2018; Amani et al., 2020).

Over the past decades, several attempts have been made to come up with data systems that target nexus issues, using EO data (Huntington et al., 2017; Berrick et al., 2008; Eugene et al., 2017; Murray et al., 2018; Deng et al., 2013). However, they are not always easy to use and are not mobile responsive, making them less user-friendly. Moreover, some of them are sectoral (Eugene et al., 2017; Murray et al., 2018; Deng et al., 2013). In this chapter, we introduce Earth Observation for WEF nexus, in short EO-WEF, a multisectoral information system that generates data for the different components of a generic nexus system. The application is powered by Google Earth Engine, a cloud computing platform that includes data archives of regularly updated EO and scientific data sets for more than 40 years. EO-WEF is developed with nexus thinking and aims at providing data in a user-friendly manner.

2. Method
2.1 Predesign steps of EO-WEF

The EO-WEF app is inspired by the system-thinking approach of the WEF nexus. The first step of developing the EO-WEF app was to map the conceptual model of a nexus system, where the trade-offs and synergies between the different components of the system are identified. As Google Earth Engine powers EO-WEF, various data sets contained in its data catalog were then added to the conceptual model to determine how they can provide data for the different components of the nexus. Finally, a database was designed based on the main sectors of the nexus and the variables of different components. EO-WEF displays the image data of each variable for time series retrieval.

Fig. 3.1 illustrates the conceptual model of the nexus system of the Songwe River Basin (SRB). The SRB is a transboundary basin that partly borders Malawi

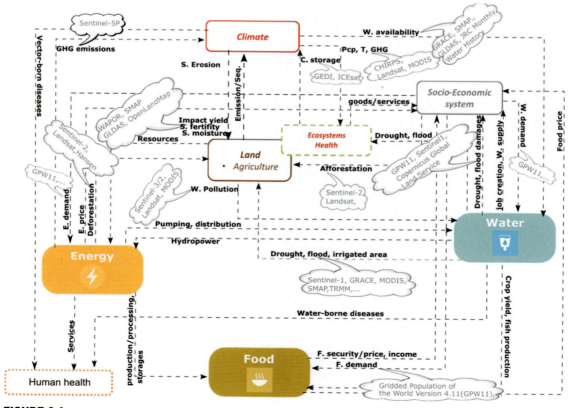

FIGURE 3.1
Conceptual model of the Songwe River Basin with examples of remotely sensed data that can be provided for the different components of the nexus.

and Tanzania. It covers an area of about 4214 km^2 (54.6% in Tanzania and 45.4% in Malawi) (Kalisa et al., 2010). The basin is home to mainly rural communities with low access to electricity, water, and sanitation and who depend largely on agriculture as a source of livelihood. Over the past decades, population growth and climate change have stressed land resources, hence exacerbating food and energy insecurity and poverty (Gwambene, 2017). To address these issues, the Songwe River Basin Development Programme has proposed the construction of multipurpose dams for ensuring flood control, and water and energy supply. The development of the conceptual model of the Songwe WEF nexus is crucial for identifying trade-offs and synergies between the different components of the system. More detail of this conceptual model is summarized in Chapter 6.

2.2 Software design

The EO-WEF app was developed using the Tethys platform. The Tethys platform is an open-source app Web development framework for EOs. It leverages recent advances in cloud computing for easy handling of large EO data sets. It was by Nathan and his team at Brigham Young University, Provo, Utah in 2015 (Swain, 2015). Its architecture is subdivided into three major components: a Python software development kit (SDK), a portal, and a software suite. The SDK provides Python modules that link various software packages supported by the Tethys platform to apps, making their functionality easy to incorporate on Web applications. Tethys portal is an application library page where users can access installed applications. Other tools and functionalities, such as user permissions (e.g., login to an app), portal design, and portal settings, are included on the page. The software suite contains commonly used Web programming frameworks (e.g., Django, Plotly Python Library, OpenLayers, etc.), which allows users to create Web apps for visualizing, analyzing, and modeling EO data (Nelson et al., 2019; Markert et al., 2019).

The EO-WEF app follows the development pattern of the Tethys platform, which has three components: Model, View and Controller (Swain, 2015). The Model is responsible to initialize and manage databases. As the EO-WEF app is powered by Google Earth Engine, it does not use any database. Instead, a package called gee in the main app directory was created to house modules that contain functions and variables for Google Earth Engine—related logic. For instance, the function "get_image_collection_asset" in the method module builds the map tile service URL from the platform, sensor, and product and the date range and reducer method input by a user. The variable EE_Product in the product module is a python dictionary that lists all the data sets of Google Earth Engine that are supported by the EO-WEF with some metadata. The View represents the front end of the EO-WEF with all the HTML pages, CSS, and JavaScript logic. HTML pages of the EO-WEF mainly include a homepage that can display maps, data controllers, etc., and a plot page for displaying graphs. The Controller holds the logic needed for connecting the data from the Model (from various functions and variables in this case) to the front end. For example, the home controller of the EO-WEF app connects the different values of the EE_Products to the front end as per user's input. The get_image_collection controller creates Google Earth Engine XYZ tile layer endpoints that can be used to call the get_image_collection_asset function from the front end.

2.3 How to use the EO-WEF?

The EO-WEF app can be subdivided into two parts: the search and map interactive sections:

In the Search section, the following steps can be used (Fig. 3.2):

1. Use the search bar to look up in the EO-WEF database the variable of interest and its sector, satellite (or model) and product, and the availability dates of its data.
2. Click the product link(s) to learn about the data associated to the selected variable.
3. Record the information related to the variable. This will help to visualize its data and time series on the map area.
   The map section (Figs. 3.3 and 3.4) of the application consists of visualizing the data of the selected variable and retrieving the time series. The steps are as follows:
4. From the data controller panel of the app, enter the information that was captured in step 3, choose the reducer method (e.g., median) to

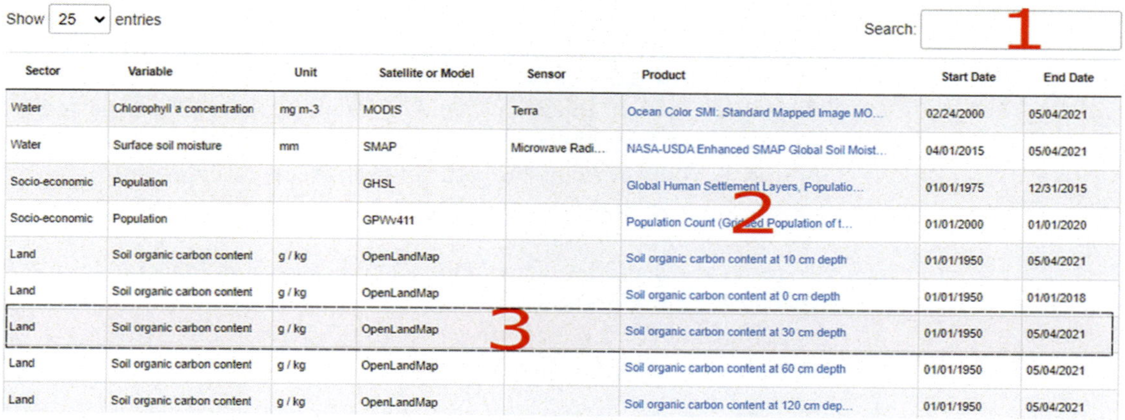

Sector	Variable	Unit	Satellite or Model	Sensor	Product	Start Date	End Date
Water	Chlorophyll a concentration	mg m-3	MODIS	Terra	Ocean Color SMI: Standard Mapped Image MO...	02/24/2000	05/04/2021
Water	Surface soil moisture	mm	SMAP	Microwave Radi...	NASA-USDA Enhanced SMAP Global Soil Moist...	04/01/2015	05/04/2021
Socio-economic	Population		GHSL		Global Human Settlement Layers, Populatio...	01/01/1975	12/31/2015
Socio-economic	Population		GPWv411		Population Count (Gridded Population of t...	01/01/2000	01/01/2020
Land	Soil organic carbon content	g / kg	OpenLandMap		Soil organic carbon content at 10 cm depth	01/01/1950	05/04/2021
Land	Soil organic carbon content	g / kg	OpenLandMap		Soil organic carbon content at 0 cm depth	01/01/1950	01/01/2018
Land	Soil organic carbon content	g / kg	OpenLandMap		Soil organic carbon content at 30 cm depth	01/01/1950	05/04/2021
Land	Soil organic carbon content	g / kg	OpenLandMap		Soil organic carbon content at 60 cm depth	01/01/1950	05/04/2021
Land	Soil organic carbon content	g / kg	OpenLandMap		Soil organic carbon content at 120 cm dep...	01/01/1950	05/04/2021

FIGURE 3.2
Search table of the EO-WEF app.

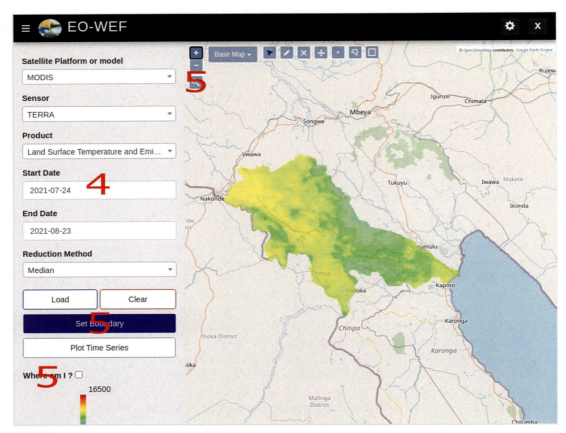

FIGURE 3.3
User interface of the EO-WEF.

display the image collection for a date range, which should be within the availability dates of the variable data, and click the load button to display the reduced image. A legend of your data should be displayed.

5. To customize the visualization of images to a specific area, use the "set boundary" button to upload your own shapefile or enter your location in the location search bar or click the geolocation button or use the geolocation button.

6. To plot the time series of a variable data for a selected date range at a specific location, use the draw tool menu, which is located on the upper-left side of the map view area, to drop a point or draw a polygon and click the "Plot time Series" button.

The following link can be visited to view the application: http://eo-wef.com/.

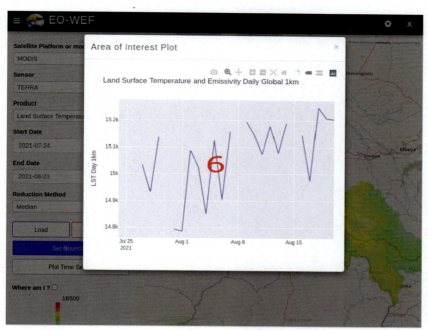

FIGURE 3.4
Piloting of time series in EO-WEF.

3. Capability of EO-WEF for generating data for the different sectors of the Songwe nexus

In this section, the capability of the EO-WEF in delivering reliable data for the different components of the Songwe River Basin nexus system is evaluated.

3.1 Water sector

Table 3.1 displays the variables that can be derived from different components of the water sector using the EO-WEF. For example, the NASA Global Land Data Assimilation System Version 2 (GLDAS 2) can be reliably used to derive data on water available in soil. Bi et al. (2016) compared in situ soil moisture data and GLDAS in the Tibetan Plateau (China) and found that correlation between data sets was above 0.5. Considering water pollution, the Sentinel-3 data may be suitable to monitor water pollution at a temporal resolution of 2 days. Sentinel-3 has several products (e.g., water temperature, chlorophyll) that generate data for evaluating water quality. Cherif et al. (2020) confirmed the usefulness of the water surface temperature product of Sentinel-3 to estimate the bathing water quality of the west coast of Tangier, Morocco.

CHAPTER 3: EO-WEF: a Earth Observations for Water, Energy, and Food nexus geotool

Table 3.1 Examples of satellite and model data sets for different variables in the water sector offered by EO-WEF.

Variables	Satellites plattform/Models	Spatial resolution	Temporal resolution	Data set availability	References
Surface soil moisture	Soil Moisture Active Passive (SMAP)	0.25°	Daily	January 2015 –present	O'Neill et al. (2019)
Groundwater	GLDAS-2	0.1°	Daily	May 2018 –present	Rodell et al. (2004)
Water pollution (chlorophyll content, etc.)	Sentinel-3	300 m	2 days	October 2016 –present	Donlon et al. (2012)

3.2 Climate sector

Table 3.2 illustrates the data that can be derived from different components of the climate sector using the EO-WEF. The climate change is one of the chief causes of flood events and drought recorded in the Songwe River Basin (Ipyana and Mikova, 2019). Greenhouse gas (GHG) emission activities from other sectors, such as deforestation or afforestation, also impact the climate sector. Variable data on the different components of this sector (e.g., precipitation, land surface, evapotranspiration, GHG emissions) can be reliable derived from satellite imagery or models utilized by EO-WEF. For instance, Katsanos et al. (2016) found a good correlation between the Climate Hazards Group Infrared Precipitation with Station data (CHIRPS) and in situ measurements of rainfall from dense and reliable network of rain gauges in the island of Cyprus over a period of 30 years. Shikwambana et al. (2020) used Sentinel-5 P data to observe the sulfur dioxide (SO_2) and nitrogen dioxide (NO_2) over the provinces of Gauteng, Mpumalanga, Limpopo, in South Africa, and found out increasing trends of SO_2 on account of the emissions from coal-fired power stations.

3.3 Land sectors

Scarcity of arable land and declining soil fertility are among the factors that have caused a rapid land conversion in the Songwe River Basin (Kalisa et al., 2013). Climate change has also caused drought in the area. As summarized in Table 3.3, EO-WEF can provide data for these components of the Songwe nexus using satellites and scientific data sets contained in the data catalog of the Google Earth Engine. The Palmer Drought Severity Index (PDSI), for example, is one of the widely used indices for quantifying drought and monitoring its development (Dai, 2011). It was used to assess the variability and intensity of droughts in Iran during the period 1951 to 2005. It was found that intense droughts occurred 1999–2002 (Zoljoodi and Didevarasl, 2013). The Copernicus Global Land Cover Layers: CGLS-LC100 Collection 3 can provide a valuable information on land conversion in the basin. Xu et al. (2019) achieved an accuracy of more 60% with validation data sets in Africa. They also found areas covered by different land-covers to be consistent to FAO statistics. However, overestimations of CGLS-LC100 Collection 3 were found for some of the regions such as Eastern Africa Plateau, due probably to elevation changes, among others.

3.4 Socioeconomic, food, and energy sectors

EO-WEF provides the least data sets for the socioeconomic, food, and energy sectors. As depicted in Fig. 3.1, population growth influences the WEF sectors in terms of resources demand. Over the past decades, population growth has exerted considerable pressure on land resources, exacerbating food and energy insecurity and poverty in Songwe. Gridded Population of the World Version 4.11 (GPW411) (Table 3.4), developed by the Columbia University, may

Table 3.2 Examples of satellite and model data sets for different variables in the climate sector provided by EO-WEF.

Variables	Satellites platform/Models	Spatial resolution	Temporal resolution	Data set availability	References
Precipitation	CHIRPS	5 days	0.05°	January 1981–present	Funk et al. (2015)
Land surface temperature	MODIS	8 days	1 km	March 2000–present	Wan and Hook (2015)
Evapotranspiration	WaPOR (Water Productivity)	10 days	0.00223°	January 2009–present	FAO (2018)
GHG emissions	Sentinel-5 Precursor	Near real time	0.01°	July 2018–Present	Sentinel-5P (2018)

Table 3.3 Examples of satellite and model data sets for different variables in the climate sector provided by EO-WEF.

Variables	Satellites platform/ Models	Spatial resolution	Temporal resolution	Data set availability	References
Soil organic carbon content	OpenLandMap	250 m	year	January 1950–January 2018	Hengl and Wheeler (2018)
Afforestation	MODIS	250 m	16 days	July 2002–present	Didan (2015)
Drought	TerraClimate	2.5 arc minutes	5 days	January 1958–December 2020	Abatzoglou et al. (2018)
Land-cover	CGLS-LC100 Collection 3	100 m	1 year	2015–2020	Buchhorn et al. (2020)

Table 3.4 Examples of satellite and model data sets for different variables in the socioeconomic sector provided by EO-WEF.

Variables	Satellites platform/Models	Spatial resolution	Temporal resolution	Data set availability	References
Population	GPWv411	~1 km	5 years	January 2000–January 2020	Center for International Earth Science Information Network - CIESIN - Columbia University (2018)

play an important role in addressing those issues. Holtedahl and Joutz (2004) successfully modeled the urban electricity demand in Taiwan from 1950 to 1996 as a function of population growth, among others. It is worth noting that the latest ingestion of other kind of population data such as the "WorldPop Global Project Population Data: Constrained Estimated Age and Sex Structures of Residential Population per 100 × 100 m Grid Square" (Sorichetta et al., 2015) in Google Earth Engine will further improve the modeling of energy demand. Kim et al. (2019) proved that residential electricity demand in the short- and long-run and increase in the youth and people aged 65 population are strongly correlated in Korea.

4. Further development

EO-WEF is developed to provide spatial data for different sectors of a nexus system. Further development will include the following:

- Adding more data sets for continuous (e.g., climate projections) and categorical variables (e.g., land cover, forest change, etc.).
- Allowing users to download layers of displayed images and data drawn on time series graphs in a csv format for further analyses by users.
- Allowing users to compare two variables on graphs. This will give insight on the interactions between different components of nexus system.

5. Conclusion

In this chapter, we introduced a newly developed application, EO-WEF. The application is a multisectoral information system that uses remotely sensed and geospatial data sets, contained in the data catalog of Google Earth Engine, to produce data for different components of a nexus system. Specifically, it addresses the issues of data holdings, data variation over time, and customization of data to fit a given scope of analysis. We also showed that data sets used by EO-WEF were also reliably used in previous studies with good correlation to in situ measurements. The tool will be beneficial to nexus analysts during the process of data assembly for modeling the nexus. However, it is worth noting that EO-WEF is not a one-stop shop for all the data needed for a nexus analysis but can supplement other data sources such national and regional statistics. Furthermore, accuracies of data sets used by EO-WEF depend on the area under investigation. They should therefore always be validated with in situ data.

References

Abatzoglou, J.T., Dobrowski, S.Z., ParkS, S.A., Hegewisch, K.C., 2018. TerraClimate, a high-resolution global dataset of monthly climate and climatic water balance from 1958–2015. Sci. Data 5, 1–12.

Amani, M., Ghorbanian, A., Ahmadi, S.A., Kakooei, M., Moghimi, A., Mirmazloumi, S.M., Moghaddam, S.H.A., Mahdavi, S., Ghahremanloo, M., Parsian, S., 2020. Google earth engine cloud computing platform for remote sensing big data applications: a comprehensive review. IEEE J. Sel. Top. Appl. Earth Obs. Rem. Sens. 13, 5326–5350.

Berrick, S.W., Leptoukh, G., Farley, J.D., Rui, H., 2008. Giovanni: a web service workflow-based data visualization and analysis system. IEEE Trans. Geosci. Rem. Sens. 47, 106–113.

Bi, H., Ma, J., Zheng, W., Zeng, J., 2016. Comparison of soil moisture in GLDAS model simulations and in situ observations over the Tibetan Plateau. J. Geophys. Res. Atmos. 121, 2658–2678.

Buchhorn, M., Lesiv, M., Tsendbazar, N.-E., Herold, M., Bertels, L., Smets, B., 2020. Copernicus global land cover layers—collection 2. Rem. Sens. 12, 1044.

Center for International Earth Science Information Network - CIESIN - Columbia University, 2018. Gridded Population of the World, Version 4 (GPWv4): Basic Demographic Characteristics, Revision 11. NASA Socioeconomic Data and Applications Center (SEDAC), Palisades, NY.

Cherif, E.K., Vodopivec, M., Mejjad, N., Esteves da Silva, J.C., Simonovič, S., Boulaassal, H., 2020. COVID-19 pandemic consequences on coastal water quality using WST sentinel-3 data: case of tangier, Morocco. Water 12, 2638.

Dai, A., 2011. Characteristics and trends in various forms of the palmer drought severity index during 1900–2008. J. Geophys. Res. Atmos. 116, D12.

Deng, M., Di, L., Han, W., Yagci, A.L., Peng, C., Heo, G., 2013. Web-service-based monitoring and analysis of global agricultural drought. Photogramm. Eng. Rem. Sens. 79, 929–943.

Didan, K., 2015. MOD13Q1 MODIS/Terra Vegetation Indices 16-Day L3 Global 250m SIN Grid V006. NASA EOSDIS LP DAAC. <Website> (accessed ?????date.month.year).

Donlon, C., BerrutI, B., Buongiorno, A., Ferreira, M.-H., Féménias, P., Frerick, J., Goryl, P., Klein, U., Laur, H., Mavrocordatos, C., 2012. The global monitoring for environment and security (GMES) sentinel-3 mission. Remote Sens. Environ. 120, 37–57.

Eugene, G.Y., Di, L., Rahman, M.S., Lin, L., Zhang, C., HU, L., Shrestha, R., Kang, L., Tang, J., Yang, G., 2017. Performance improvement on a Web Geospatial service for the remote sensing flood-induced crop loss assessment web application using vector tiling. In: 2017 6th International Conference on Agro-Geoinformatics, vol. 2. IEEE, pp. 1–6.

FAO, 2018. WaPOR Database Methodology: Level 1. Remote Sensing for Water Productivity Technical Report: Methodology Series. FAO, Rome.

Funk, C., Peterson, P., Landsfeld, M., Pedreros, D., Verdin, J., Shukla, S., Husak, G., Rowland, J., Harrison, L., Hoell, A., 2015. The climate hazards infrared precipitation with stations—a new environmental record for monitoring extremes. Sci. Data 2, 1–21.

Gwambene, B., 2017. Potential corollaries of land degradation on rural livelihoods in upper Songwe transboundary river catchment, Tanzania. J. Agric. Ext. Rural Dev. 3 (1), 139–148.

Hengl, T., Wheeler, I., 2018. Soil organic carbon content in x 5 g/kg at 6 standard depths (0, 10, 30, 60, 100 and 200 cm) at 250 m resolution (Version v0. 2)[Data set]. Zenodo. https://zenodo.org/record/2525553#.YhzHwegzaUk.

Holtedahl, P., Joutz, F.L., 2004. Residential electricity demand in Taiwan. Energy Econ. 26, 201–224.

Huntington, J.L., Hegewisch, K.C., Daudert, B., Morton, C.G., Abatzoglou, J.T., Mcevoy, D.J., Erickson, T., 2017. Climate engine: cloud computing and visualization of climate and remote sensing data for advanced natural resource monitoring and process understanding. Bull. Am. Meteorol. Soc. 98, 2397–2410.

Ipyana, M., Mikova, K., 2019. Flood analysis and short-term prediction of water stages in river Songwe catchment. In: IOP Conference Series: Earth and Environmental Science, vol. 321. IOP Publishing, Bristol, United Kingdom, p. 012034, 1.

Kalisa, D., Lyimo, J., Majule, A., 2010. Role of Wetlands Resource Utilisation on Community Livelihoods: The Case of Songwe River Basin, Tanzania.

Kalisa, D., Majule, A., Lyimo, J., 2013. Role of wetlands resource utilisation on community livelihoods: the case of Songwe River Basin, Tanzania. Afr. J. Agric. Res. 8, 6457–6467.

Katsanos, D., Retalis, A., Michaelides, S., 2016. Validation of a high-resolution precipitation database (CHIRPS) over Cyprus for a 30-year period. Atmos. Res. 169, 459–464.

Kim, J., Jang, M., Shin, D., 2019. Examining the role of population age structure upon residential electricity demand: a case from Korea. Sustainability 11, 3914.

Lawford, R.G., 2019. A design for a data and information service to address the knowledge needs of the Water-Energy-Food (WEF) Nexus and strategies to facilitate its implementation. Front. Environ. Sci. 7, 56.

Liu, J., Hull, V., Godfray, H.C.J., Tilman, D., Gleick, P., Hoff, H., Pahl-Wostl, C., Xu, Z., Chung, M.G., Sun, J., 2018. Nexus approaches to global sustainable development. Nat. Sustain. 1, 466–476.

Markert, K.N., Pulla, S.T., Lee, H., Markert, A.M., Anderson, E.R., Okeowo, M.A., Limaye, A.S., 2019. AltEx: an open source web application and toolkit for accessing and exploring altimetry datasets. Environ. Model. Software 117, 164–175.

Mccarl, B.A., Yang, Y., Srinivasan, R., Pistikopoulos, E.N., Mohtar, R.H., 2017. Data for WEF nexus analysis: a review of issues. Curr. Sustain. Renew. Energy Rep. 4, 137–143.

Murray, N.J., Keith, D.A., Simpson, D., Wilshire, J.H., Lucas, R.M., 2018. Remap: an online remote sensing application for land cover classification and monitoring. Methods Ecol. Evol. 9, 2019–2027.

Nelson, E.J., Pulla, S.T., Matin, M.A., Shakya, K., Jones, N., Ames, D.P., Ellenburg, W.L., Markert, K.N., David, C.H., Zaitchik, B.F., 2019. Enabling stakeholder decision-making with earth observation and modeling data using Tethys platform. Front. Environ. Sci. 7, 148.

O'Neill, P., Chan, S., Njoku, E., Jackson, T., Bindlish, R., 2019. SMAP L3 Radiometer Global Daily 36 Km EASE-Grid Soil Moisture, Version 6.[Indicate Subset Used]. NASA National Snow and Ice Data Center Distributed Active Archive Center, Colorado.

Purwanto, A., Sušnik, J., Suryadi, F.X., De Fraiture, C., 2021. Water-energy-food nexus: critical review, practical applications, and prospects for future research. Sustainability 13 (4), 1919.

Rodell, M., Houser, P., Jambor, U., Gottschalck, J., Mitchell, K., Meng, C.-J., Arsenault, K., Cosgrove, B., Radakovich, J., Bosilovich, M., 2004. The global land data assimilation system. Bull. Am. Meteorol. Soc. 85, 381–394.

Sadegh, M., Aghakouchak, A., Mallakpour, I., Huning, L.S., Mazdiyasni, O., Niknejad, M., Foufoula-Georgiou, E., Moore, F.C., Brouwer, J., Farid, A., 2020. Data and analysis toolbox for modeling the nexus of food, energy, and water. Sustain. Cities Soc. 61, 102281.

SENTINEL-5P, C., 2018. TROPOMI Level 2 Ozone Total Column Products, Version 01. European Space Agency. Accessed 01.09.21. https://sentinels.copernicus.eu/web/sentinel/data-products/-/asset_publisher/fp37fc19FN8F/content/tropomi-level-2-tropospheric-ozone.

Shikwambana, L., Mhangara, P., Mbatha, N., 2020. Trend analysis and first time observations of sulphur dioxide and nitrogen dioxide in South Africa using TROPOMI/Sentinel-5 P data. Int. J. Appl. Earth Obs. Geoinf. 91, 102130.

Simpson, G.B., Jewitt, G.P., 2019a. The development of the water-energy-food nexus as a framework for achieving resource security: a review. Front. Environ. Sci. 7, 8.

Simpson, G.B., Jewitt, G.P., 2019b. The water-energy-food nexus in the anthropocene: moving from 'nexus thinking' to 'nexus action. Curr. Opin. Environ. Sustain. 117–123.

Sorichetta, A., Hornby, G.M., Stevens, F.R., Gaughan, A.E., Linard, C., Tatem, A.J., 2015. High-resolution gridded population datasets for Latin America and the Caribbean in 2010, 2015, and 2020. Sci. Data 2, 1–12.

Swain, N.R., 2015. Tethys Platform: A Development and Hosting Platform for Water Resources Web Apps. Brigham Young University, Utah.

Wan, Z., Hook, G.H., 2015. MOD11A2 MODIS/Terra Land Surface Temperature/Emissivity 8-Day L3 Global 1km SIN Grid V006. https://ladsweb.modaps.eosdis.nasa.gov/missions-and-measurements/products/MOD11A2/. NASA EOSDIS Land Processes DAAC. (accessed 09.07.21).

Wang, L., Ma, Y., Yan, J., Chang, V., Zomaya, A.Y., 2018. pipsCloud: high performance cloud computing for remote sensing big data management and processing. Future Generat. Comput. Syst. 78, 353–368.

Xu, Y., Yu, L., Feng, D., Peng, D., Li, C., Huang, X., Lu, H., Gong, P., 2019. Comparisons of three recent moderate resolution African land cover datasets: CGLS-LC100, ESA-S2-LC20, and FROM-GLC-Africa30. Int. J. Rem. Sens. 40, 6185–6202.

Zoljoodi, M., Didevarasl, A., 2013. Evaluation of spatial-temporal variability of drought events in Iran using palmer drought severity index and its principal factors (through 1951-2005). Atmos. Clim. Sci. 3 (2), 193–207.

CHAPTER 4

Scales of application of the WEF nexus approach

Janez Sušnik[1], Sara Masia[1,2] and Graham Jewitt[1,3,4]

[1]*IHE Delft Institute for Water Education, Delft, The Netherlands;* [2]*CMCC Foundation — Euro-Mediterranean Centre on Climate Change, IAFES Division, Sassari, Italy;* [3]*Centre for Water Resources Research, School of Agricultural, Earth and Environmental Sciences, University of KwaZulu-Natal (UKZN), Pietermaritzburg, South Africa;* [4]*Civil Engineering and Geosciences, Delft University of Technology, Delft, The Netherlands*

1. Introduction

Water, energy, and food (WEF) form a coherent system, commonly referred to as the WEF nexus (Hoff, 2011), which exists as a "hyperconnected" system akin to ecological systems governed by complexity and feedback (WEF, 2016). The WEF nexus, including its management, is operational from global to local scales, where local level impacts and measures (e.g., climate change mitigation and adaptation measures) can add up to have larger-scale consequences. Similarly, global-level system behavior modulates the local level response regarding nexus resources and management (Fig. 4.1). The nexus resource base, and its effective functioning, is essential for human well-being at all scales (e.g., human development demands abundant, high-quality, easily accessible resources). Despite this, about 1 billion people lack access to clean water, 2.5 billion people lack basic sanitation, 1.4 billion have no electricity, and over 850 million are chronically malnourished while global food waste is about 30% of production (Moe and Rheingans, 2006; IMechE, 2013; World Bank, 2013a,b; World Hunger, 2013). In addition, because of the connected nature of WEF resources, and their dependence on climate and socioeconomic pathways, the interconnected nature and scale of global risks resulting from climate change and socioeconomic development (Cramer et al., 2018; Byers et al., 2018) may feedback to impact the functioning of the WEF resource nexus.

Since about 2010 when research into the WEF nexus started in earnest, researchers on the WEF nexus have recognized the multiscalar facet of the WEF nexus. As such, studies have been conducted on the nexus, both qualitative and quantitative, at household, local, regional, national, and global scales. These studies demonstrate the potential of the nexus approach when adapted

50 CHAPTER 4: Scales of application of the WEF nexus approach

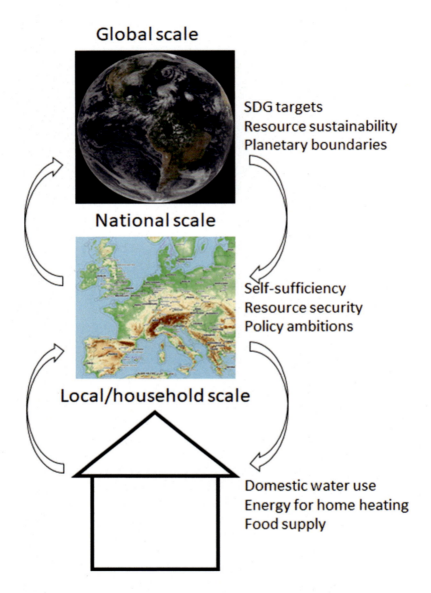

FIGURE 4.1
Schematic showing the mutual relationships between the WEF nexus at scales from global, to national, and down to household. At each scale, some key issues of consideration are highlighted. "Planetary boundaries" refers to theoretical boundaries on various global metrics proposed by Steffan et al. (2015). Resource security includes quantity, quality, and accessibility. *SDG*, Sustainable Development Goal; *WEF*, water—energy—food.

appropriately to studying WEF nexus issues at a range of scales, tailoring the methodology, focus, and research objectives to meet the scale under consideration.

In this chapter, case studies from the literature at spatial scales from household to global are presented so as to offer an overview of nexus issues being considered and studied, the range of applicability of the nexus approach, and to suggest potential options for improved policy and WEF resources decision-making relevance. These cases are mostly quantitative in nature, relying on models to assess the nexus at different scales. Yet almost all the studies presented use qualitative means at some point in the modeling process to "map" the systems under study, demonstrating that nexus modeling approaches cannot easily be split into qualitative or quantitative, but rather blending aspects from both (see Chapters 6 and 7). This chapter therefore aims to showcase the vast range in spatial scales and WEF nexus issues to which a nexus approach has been applied, demonstrating its flexibility to address a number of pertinent topics. At the same time, there remain several shortcomings, which will also be discussed. This chapter does not detail different WEF nexus methodologies, as these are covered elsewhere in this book.

2. The local scale: household to subnational

At the household level, Hussein et al. (2017) developed an integrated system dynamics model (SDM; Ford, 1999, and see Chapter 6) to capture the interactions between water, energy, and food resource at the household end-use level. The model and data were developed from a survey of over 400 local households in Duhok, Iraq. Hussein et al. (2017) tested the impacts on the WEF resource usage deriving from changes in user behavior, diets, income, family size, and climate variables. The model developed is one of the first dynamic models accounting for the interactions between water, energy, and food at the household level. Energy consumption was related to fuel type, the duration of usage of appliances and their wattage, and the desired water temperature. On the water side, the ownership of appliances, their flow rates, the duration of usage, and the frequency of usage were considered. For food, the model considered the consumption of different food commodities, the number of "cooking sessions" and their duration, the fuel or electricity usage of the cooker, the water consumption during cooking, and the amount of waste. As input to the model, family size, income, and seasonal variability could be altered, and the model outputs water, energy, and food demands, wastewater, and an amount of food waste. Water demand was shown to be most sensitive to changes in the duration of using water appliances in the garden and the number of garden watering sessions, while energy demand was most sensitive to the use of air conditioning units in the home. The study by Hussein et al. (2017) is perhaps the only one comprehensively modeling all three sectors' interactions, and assessing the impact of different scenarios, at the household level. This is important as the aggregated effects of household resource demand, and the implications of that demand on other nexus resources, may sum up over large spatial areas (e.g., a city), to have considerable regional resource implications, potentially linked to resource exploitation or carbon emissions, for example. Other studies modeling household scale resource demand either focus on single resources, or only consider interactions between two resources such as energy and water (e.g., Cheng, 2002; Kadian et al., 2007; Kenway et al., 2013; Cominola et al., 2016). This demonstrates that studies considering all three WEF nexus elements, their interactions, and their connection with wider resources availability and sustainability are rare. Such deficiencies should be addressed to improve understanding of how household-level resource use interacts both as a discrete system and as part of the wider (i.e., national or regional level) resources systems, and how policy may impact on both local-level resources use and higher-level resource sustainability.

Bahri (2020) developed "system archetypes" for the water, energy, food, and land sectors surrounding the Jatiluhur reservoir, West Java, Indonesia. Each archetype is represented as a causal loop diagram (CLD, cf. Chapter 6) describing the critical causal relationships between elements within each sector of the WEF system (e.g., water level in the reservoir, fish production, turbine flow for power generation, and the link to industrial and economic development). Each sectoral archetype was combined to form a complete nexus model of the region. While no modeling is conducted per se, the relational diagrams developed can aid nonexperts and policymakers "trace" causal relations through the whole nexus (Purwanto et al., 2019) in a qualitative way, leading to a better appreciation of potential whole-nexus response to (policy) interventions, and potentially contributing to more efficient nexus-wide policymaking.

At the city scale, Valek et al. (2017) assess the water—energy nexus in the Mexico City water system (supply and wastewater). While it can be argued that the water—energy nexus is "obvious," it is often overlooked as to just how closely connected these two resources are, especially in an urban context where water and energy may almost be seen as two sides of the same coin. This study helps to elucidate the urban water—energy link. This study is also interesting because while in most places it is the groundwater supply that consumes most energy resources (in a relative sense compared to surface water supplies), here it is the surface water supply that consumes most energy due to the topographical peculiarities of the city. At the time of the study, system water losses where c. 40%, and wastewater treatment was minimal. Specific features of the Mexico City water supply system were that surface water was pumped over large distances and a topographic barrier exceeding 1000 m elevation, and that local groundwater sources are overexploited, leading to city center subsidence. The supply from surface and groundwater sources was split about 50:50, each contributing about half to the water supply (Fig. 4.2A). On the energy side,

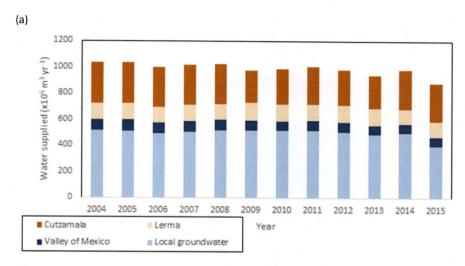

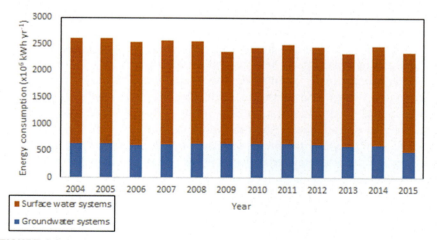

FIGURE 4.2
(A) The contribution to water supply in Mexico City from surface water (Cutzamala and Lerma) and groundwater (Valley of Mexico and Local groundwater) systems. The split is almost 50:50; (B) the energy attributed to the water supply of the water and groundwater sources in Mexico City. Surface water sources consume the vast majority of energy for water supply. *Figures adapted from Valek, A.M., Sušnik, J., Grafakos, S., 2017. Quantification of the urban water-energy nexus in México City, México, with an assessment of water-system related carbon emissions. Sci. Total Environ. 590–591, 258–268.*

the vast majority of water system energy consumption was related to surface water supply, and not to groundwater as is often the case (Fig. 4.2B), due to the vast topographic barrier that needs to be overcome, which requires considerable energy. Of the small amount of energy associated with wastewater treatment at the time of the study, most was attributed to treatment itself, and only a small fraction was due to pumping as the wastewater system was largely gravity-fed. Since the study, a large wastewater treatment facility has been built for the city, meaning that more wastewater is now treated and to a higher degree, but also that the water—energy nexus relationships and divisions in the city water system have changed and therefore need to be updated.

Using a multiregional input—output (MRIO; Chapter 6) analysis, Chen et al. (2018a) show how much water and energy resources of the hinterland of a city contribute to the consumption within that city by considering the resources "embedded" within the products consumed in the city. This is important as many cities globally rely significantly on their hinterlands and beyond to provide the resources required to allow for optimal functioning of city services. Likewise, the characteristics of a city's resources demands have a profound impact on resources exploitation and sustainability in locations that may be distant to the city itself. It is therefore important to better understand the resource demands of a city, and where these resources are sources from to better mitigate and adapt to potential constraints in the future. The study is focused on Hong Kong and its dependency on the Guangdong hinterland. It is shown that 79% of freshwater in Hong Kong was imported from Gaungdong, placing a large resources stress on that region. Note that this demand might place the Hong Kong water demand in conflict with water demands within the Gaungdong region itself. It is also shown that the energy "embedded" in this water supply from Guangdong was higher than the local Hong Kong water-related energy consumption, again leading to Hong Kong placing a high energy demand on a distant region. It is also shown that wastewater treatment consumes more energy than the water supply sector, mainly due to the high level of wastewater treatment standards in the city. While the local water consumption in the energy sectors is 10 times that of residential consumption in Hong Kong, this water is sourced from the sea, and much is returned after use (i.e., a low consumptive fraction). Expected growth in population in the city will likely lead to an 8%—9% increase in resource demands, both within the city and from the hinterland, potentially leading to resource-related constraints in the future.

Moving up to the regional scale, Purwanto et al. (2019, 2021) conducted a WEF resource security analysis (i.e., including aspects of resource quantity, quality, and accessibility) in Karawang Regency, Indonesia. In Indonesia, much resource-related decision-making is devolved to the Regency level but is guided by nationally determined plans and objectives. In Purwanto et al. (2019), a

2. The local scale: household to subnational   55

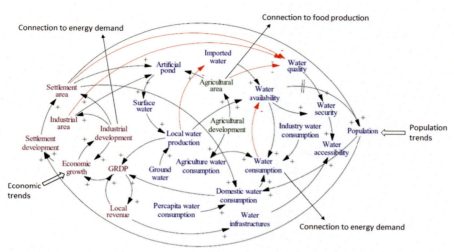

FIGURE 4.3
The CLD of the water sector in Karawang Regency, Indonesia, with links to food production, energy demand, population, and economic trends indicated. *CLD, causal loop diagram. Modified from Purwanto, A., Sušnik, J., Suryadi, F.X., de Fraiture, C., 2019. The use of a group model building approach to develop causal loop diagrams of the WEF security nexus in a local context: a case study in Karawang Regency, Indonesia. J. Clean. Prod. 240, 118170.*

detailed qualitative CLD of the WEF nexus (the so-called K-WEFS model; Fig. 4.3) was developed in a collaborative group model building exercise with local expert stakeholders from all the WEF sectors and from planning agencies. The model connects the water, energy, and food sectors with developments in local population and economic trends (Fig. 4.3). Without the development of a quantitative model, the significant added value of a qualitative approach including stakeholders is demonstrated. Local policymakers have a better appreciation of the complexity of this local WEF nexus and are better placed to assess the wider impacts of implementing sectoral-specific policy measures, and to develop policies that address many nexus issues simultaneously while identifying and attempting to minimize detrimental trade-offs and negative impacts across sectors.

In a follow-up study, Purwanto et al. (2021) develop and demonstrate a quantitative system dynamics model (SDM; Chapter 6) using the causal loop diagram (Fig. 4.3) as a guide. The SDM tries to accurately recreate the CLD within the constraints of data availability. Once the model was validated against historical observations, it was used to assess the impact of proposed interventions in the Regency. From these analyses, unanticipated synergies and trade-offs were identified and quantified. The CLD, together with the SDM, has resulted in a series of practical recommendations for local policymakers, therefore making this study a good example of moving from nexus thinking to policy-relevant nexus implementation, something that is urgently called for (Brouwer et al., 2018).

Bakhshianlamouki et al. (2020) conducted a WEF nexus analysis in the Urmia Lake Basin, Iran, in a basin-scale study. A conceptual mapping of the WEF nexus system is developed with input from local experts from which a quantitative SDM is developed. As Urmia Lake is undergoing declining lake levels due to water overuse, the Urmia Lake Restoration Programme (ULRP) has proposed a series of restoration measures in an attempt to halt and even reverse this decline. The study tested these measures for their efficacy and impact across the wider nexus in the basin. While some measures are broadly beneficial, some unanticipated negative impacts such as increasing fuel demand for water supply in agriculture were uncovered. Through such an analysis and reporting back to the ULRP, measures can be revised and adjusted to minimize such adverse impacts, leading to win-win situations in the efforts to restore Urmia Lake water level. This study better "grounds" nexus research in real issues and could serve as an example for other studies where real policy decisions are becoming critically important. It shows that modeling nexus issues can lead to suggestions as to potential ways forward and can act as a point of discussion in future policy talks.

3. The national scale

Moving up to national scale nexus analyses, Wang et al. (2018) used MRIO and ecological network analysis to assess the water—energy nexus in China. It was found that major cities consume significant "embedded" water and energy resources and that water resources are generally transferred from west to east and north to south across China. While this analysis is detailed, using considerable national-level trade data to demonstrate the flow of resources between cities and regions in China, little is suggested with regard to practical policy changes that could be implemented. Nor is the methodology capable of assessing dynamic interactions or system changes over time. However, as with the local-level study of Chen et al. (2018a), this study helps to place cities and their resource consumption within wider environmental and resource sustainability contexts, especially as, alluded to aforementioned in Valek et al. (2017), water and energy are especially interconnected in urban areas. As cities continue to grow, more emphasis will have to be given as to where cities get their resources from, and how resource interconnectedness may have implications for resource sustainability in the long run.

In South Africa, Nhamo et al. (2020) use an analytical hierarchical process (AHP) methodology and WEF nexus indicators suggested by the World Bank to assess the performance of the nexus in South Africa. Food self-sufficiency and water productivity targets are well achieved, but apparently at the cost of progress in other nexus goals such as energy accessibility and water availability. In addition, the high dependence of South Africa on coal for energy generation means that while energy productivity targets are all met, the climate impacts are scored poorly. Nhamo et al. (2020) go further, by demonstrating how South Africa has performed in achieving national-level SDGs related to the nexus, and how progress has been made between 2015 and 2018. While water and energy productivity have improved, efforts in cereal production and energy accessibility have got worse. This analysis appears to suggest national-level conflicts in South Africa in its ambitions to achieve multiple SDG targets simultaneously. Through such a nexus analysis, these trade-offs can be identified and addressed, potentially leading to more holistic policy being developed that attempts to harness synergies and avoid trade-offs.

Laspidou et al. (2020) conduct a highly detailed nexus analysis for Greece, using dedicated sectoral thematic models and data combined in an SDM modeling framework and disaggregated into 14 interacting regions in Greece. The study uses extensive data from several thematic models, EUROSTAT data (ec.europa.eu/eurostat/home), as well as data from national Greek statistics to develop a highly complex, interacting system dynamics model. The model is disaggregated to represent 14 regions in Greece, which interact and "trade" resources with each other (Fig. 4.4). It is shown that in Greece, about half of water resources come from surface water. Over 80% of water demand in Greece is consumed by agriculture and livestock for food production, demonstrating a strong and critical nexus connection between these sectors. The direct climate impact of the energy sector to climate emissions is also shown, with those regions having large fossil-based energy industries contributing significantly to national greenhouse gas (GHG) emissions totals. Most energy consumption in Greece is in the "built environment."

Through such a detailed analysis of resource demand and of resource requirements in different sectors and their impacts (e.g., the amount of energy used in the water sector, and the concomitant climate impact of that water-related energy demand) at the national and regional levels in Greece, policy- and decision-makers are better placed to consider policy design that results in lower overall resource consumption along with lower environmental impact. In addition, cross-sectoral synergies can be better identified, and suggestions can be made as to where best to apply a policy within the country, as national blanket policy implementation may not be most efficient. Although such suggestions are not made in the study, follow-up work could use these findings as a starting point for such policy-relevant advice.

58 CHAPTER 4: Scales of application of the WEF nexus approach

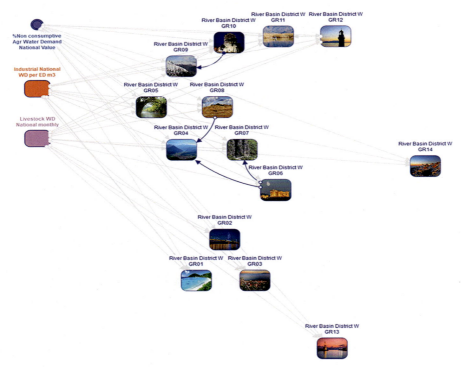

FIGURE 4.4
The Greek national-level SDM disaggregated into 14 interacting regions. Regional totals are summed to national values. *SDM*, system dynamics model.

As a final example of national-scale WEF nexus analysis, Janssen et al. (2020) investigate the nexus of water, energy, food, land, and climate in the Netherlands (Fig. 4.5) using the drivers—pressures—states—impact—response (DPSIR) methodology combined with systems thinking. The study also assesses the impacts of various resource-related innovations on the nexus.

A wide suite of innovations across the sectors is first identified, and then both the effort to implement the innovation and its expected impact (assessed on a relative scale from 1 to 10) are assessed for each. For example, the wider implementation of district heating scores relatively low on effort, but highly on impact. This is because the impact to the energy sector is high, and at the same time, district heating is already partly implemented in the Netherlands, making it easy to extend to new areas. From the work, concrete suggestions for policy are made, bringing traditionally theoretical nexus modeling to real-world application, a feature lacking in many nexus studies.

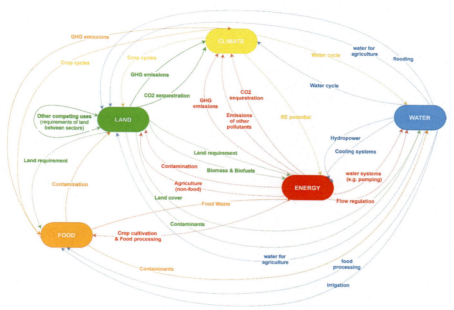

FIGURE 4.5
Interactions between the water, energy, food, land, and climate sectors in the Netherlands. *From Janssen, D.N.G., Ramos, E.P., Linderhof, V., Polman, N., Laspidou, C., Fokkinga, D., de Mesquita e Sousa, D., 2020. The climate, land, energy, water, and food nexus challenge in a land scarce country: innovations in The Netherlands. Sustainability 12, 10491.*

4. Higher-level nexus studies

At the highest spatial scales, studies tend to be conducted at the global level, deriving broad conclusions about the nature of the global WEF system. Sometimes, such studies are disaggregated into smaller global regions to analysis variability and particularities in results.

Perhaps one of, if not the first global-scale nexus study was conducted by Meadows et al. (1972) in the classic "Limits to Growth" study. The high-level global scale dynamics of the population–pollution–capital–resource nexus was simulated using an early application of the system dynamics paradigm. In this study, common global finite resources such as cropland, fossil fuel stocks, and metals contribute to development, population growth, and pollution of the environment. Simulations suggest that as resources are depleted and pollution crosses thresholds, the global system is unable to support further development, leading to collapses in food and industrial output, and ultimately in population. While criticized at the time for being too speculative, recent reanalysis has shown that the *trends* predicted in the 1972 study are reflected in observed data over the intervening 30 years (Turner, 2008), leading to concerns about the future of resources, production, and ecosystem and human health.

More recently, Chen et al. (2018b) carry out a global-scale MRIO analysis to examine agricultural land and freshwater use that is embodied in global supply chains. It is shown that globally, developed and major developing economies such as China are major drivers in land and freshwater use due to their large and increasing product demand. This is restated as a "transfer" of water and land resources from resource rich but less-developed economies to resource poor but economically highly developed nations. For example, Africa as a continent is shown to contribute significant water and land resources to the production of goods destined for Europe. It is expected that the intensification of globalization will only increase this disparity and drive land and water "displacement" from poor to rich nations. Through this analysis, suggestions can be made on how to optimize supply chains and improve efficiencies to reduce these inequities.

Focusing largely on global water system dynamics, Simonovic (2002) uses a system dynamics modeling approach to investigate the behavior of the global water system and its response to changes in arable land, industrial capital and production, and population, somewhat similar to the study of Meadows et al. (1972), albeit with a different focus. Indeed, the standard runs in the Simonovic model derived from Meadows et al. (1972). The intimate relationship between population growth and water abstractions is clearly demonstrated in the simulations. The importance of water quality is also highlighted as being critically important for continued development, an issue that appears to have been underappreciated until recently. It is also shown that water must be considered as one of the most important factors for continued human development globally, again a fact that is only recently starting to get more attention.

As final examples, Sušnik (2015, 2018) analyses the global-scale WEF nexus using a system dynamics approach in combination with data-based correlative and causal statistical analysis between the WEF sectors and gross domestic product (GDP) as a proxy for development. In the earlier work, Sušnik (2015) demonstrated temporally robust correlative relationships between WEF parameters and GDP. From this, using global GDP projections, global water withdrawals, food production, and energy (electricity) generation were projected to 2100. It was shown that the trajectory of growth in these three WEF sectors was strongly related to the GDP growth scenario. The stronger the GDP growth, the more resources were expected to be exploited. This is a crucial finding, as GDP is still the "benchmark" by which economic "performance" and "development" are measured. If resource use is indeed closely connected to GDP change, then constantly increasing global GDP implies an ever-growing demand for, and exploitation of, water, energy, and food (land) resources, something that is clearly not possible on a finite planet. A suggestion from this work is that GDP and resource use must be "decoupled" as soon as possible. This early work did not consider causal relations between the WEF sectors, nor did it consider dynamic feedbacks.

To address these gaps, a follow-up study (Sušnik, 2018) was carried out, where the WEF—GDP nexus was further explored (Fig. 4.6). In this study, apart from including feedback between the sectors (Fig. 4.6), the strength of causal relationships was quantitatively analyzed. This allowed the dominant causal direction between two variables to be assessed, and the relative strength of linking equations forming feedback loops was scaled to represent causal asymmetries. In addition, uncertainty in future projections was accounted for using a Monte Carlo modeling procedure. While water and food historical values were well captured by the model, energy was overestimated. For the future projections, as in the earlier study, the trend of resource use depends largely on the GDP projection. Stronger GDP growth implies more resource use. There is also considerable bandwidth in projections. The estimates in Sušnik (2018) agree well with independent projections from the literature for water and food resources, but overestimate energy production projections. While useful regarding global scale trends, these two studies do not capture national-level variability, highlighting a need for future research.

5. Spatial interactions in the nexus

The studies presented deal with specific cases at specific and fixed spatial scales (e.g., household, national). However, it must be acknowledged that actions within the nexus take place between these scales, and the different scales interact with each other (cf. Fig. 4.1), affecting the processes at the different

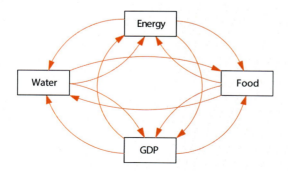

FIGURE 4.6

Schematic representation of the WEF—GDP nexus analyzed in Sušnik (2018). *GDP*, gross domestic product; *WEF*, water—energy—food.

scales. For example, national policies might be informed by higher-level policy goals. Within Europe, many water and agricultural policies at the national level are guided and shaped by the EU-level Water Framework Directive (WFD; ec.europa.eu/environment/water/water-framework/index_en.html) and Common Agricultural Policy (CAP; ec.europa.eu/info/food-farming-fisheries/key-policies/common-agricultural-policy/cap-glance_en) respectively. However, there is sufficient flexibility in the EU Directives to allow for "fine-tuning" or interpreting these policies according to national-level characteristics and priorities, giving rise to considerable heterogeneity in their implementation. Likewise, national-level policies may be reflected differently in local-level policymaking, having an influence on local nexus-related decisions and resource management. It is less clear how actions at the local level may feed up to influence policy decisions at higher levels. One recent example from the United Kingdom is that of household wood-burning stoves that have recently undergone a popular resurgence. Due to the level of uptake and the resultant particulate pollution levels being recorded, it is possible that changes in national level energy and clean air policy will be implemented in an attempt to mitigate this effect, though this is yet to be seen.

Considering the nexus in terms of resources, Bijl et al. (2018) show that the WEF resources vary greatly in terms of their locations of production and the extent to which they are traded. They also show that the spatial scales of trade are related to the physical characteristics of the resources and that global- and continental-scale trade characteristics are important when considering local and national solutions to nexus issues. Abulibdeh and Zaiden (2020) present a framework where different scales are nested within each other. For example, the national scale may be concerned with households and urban area dynamics. This is nested within the regional scale, concerned with wider water, energy, and food resources production. Nexus risks such as population growth, sectoral coupling, and energy prices are considered, as are the impact of policies, which are seen to influence all scales and can mitigate risks, and thus form the highest-level nest.

6. Conclusions

This chapter has shown that the WEF nexus is operational at scales from household to global, with numerous studies being applied at all these scales. The WEF nexus approach, comprising a number of methodological approaches dealt with in other chapters in this book, is shown to be highly flexible and has been adapted to many geographical and socioeconomic conditions, as well as to many nexus issues specific to each study or location. This flexibility is arguably the greatest asset of the nexus approach, making it suitable to study many pertinent issues at a range of spatial and temporal scales. At the same time, there are interactions between these spatial scales, which have not been addressed in nexus studies. Global-, regional-, and national-level policies and ambitions can have significant implications for local-level resource demand, use, and sustainability. Likewise, local-level resource exploitation can aggregate up to have national- and global-level implications, thus impacting on policy formulation and implementation. This interacting multiscale aspect of the WEF nexus has not yet been explored and represents a major challenge for future research efforts. What is much less studied is differences in the temporal scales of nexus sectors, impacts, and feedbacks, and how these temporal scales differ at different spatial scales. This is a research gap that represents a major challenge for future nexus research. It is likely that local-level change occurs relatively quickly when compared with national and international scales; however, this is yet to be robustly tested.

References

Abulibdeh, A., Zaiden, E., 2020. Managing the water-energy-food nexus on an integrated geographical scale. Environ. Devel. 33, 100498.

Bahri, M., 2020. Analysis of the water, energy, food, and land nexus using the system archetypes: a case study in the Jatiluhur Reservoir, West Java, Indonesia. Sci. Total Environ. 71, 137025.

Bakhshianlamouki, E., Masia, S., Karimi, P., van der Zaag, P., Sušnik, J., 2020. A system dynamics model to quantify the impacts of restoration measures on the water-energy-food nexus in the Urmia Lake Basin, Iran. Sci. Total Environ. 708, 134874.

Bijl, D.L., Bogaart, P.W., Dekker, S.C., van Vuuren, D.P., 2018. Unpacking the nexus: different spatial scales for water, food, and energy. Glob. Environ. Chang. 48, 22−31.

Brouwer, F., Anzaldi, G., Laspidou, C., Munaretto, S., Schmidt, G., Strosser, P., Sušnik, J., Vamvakeridou-Lyroudia, L.S., 2018. Commentary to SEI Report 'Where Is the Added Value? A Review of the Water-Energy-Food Nexus Literature'. www.sim4nexus.eu/page.php?wert=Publications (Accessed June 2021).

Byers, E., Gidden, M., Leclere, D., Balkovic, J., Burek, P., Ebi, K., Greve, P., Grey, D., Havlik, P., Hillers, A., Johnson, N., Kahil, T., Krey, V., Langan, S., Nakicenovic, N., Novak, R., Obersteiner, M., Pachauri, S., Palazzo, A., Parkinson, S., Rao, N., Rogelj, J., Satoh, Y., Wada, Y., Willaarts, B., Riahi, K., 2018. Global exposure and vulnerability to multi-sector development and climate change hotspots. Environ. Res. Lett. 13, 055012.

Chen, P.-C., Alvarado, V., Hsu, S.-C., 2018a. Water energy nexus in city and hinterlands: multiregional physical input-output analysis for Hong Kong and South China. Appl. Energy 225, 986−997.

Chen, B., Han, M.Y., Peng, K., Zhou, S.L., Shoa, L., Wu, X.F., Wei, W.D., Liu, S.Y., Li, Z., Li, J.S., Chen, G.Q., 2018b. Global land-water nexus: agricultural land and freshwater use embodied in worldwide supply chains. Sci. Total Environ. 613−614, 931−943.

Cheng, C.L., 2002. Study of the inter-relationship between water use and energy conservation for a building. Energy Build. 34, 261−266.

Cominola, A., Giuliani, M., Castelletti, A., Abdallah, A.M., Rosenberg, D.E., 2016. Developing a stochastic simulation model for the generation of residential water end-use demand time series. In: Proceedings of the 8th International Congress on Environmental Modelling and Software (IEMSs 2016). 10-14 July 2016, Toulouse, France.

Cramer, W., Guiot, J., Fader, M., Garrabou, J., Gattuso, J.-P., Iglesias, A., Lange, M.A., Lionello, P., Llasat, M.C., Paz, S., Penuelas, M., Snoussi, M., Toreti, A., Tsimplis, M.N., Xoplaki, E., 2018. Climate change and interconnected risks to sustainable development in the Mediterranean. Nat. Clim. Change 8, 972−980. https://doi.org/10.1038/s41558-018-0299-2.

Ford, A., 1999. Modelling the Environment: An Introduction to System Dynamics Modeling of Environmental Systems. Island Press, Washington, D.C.

Hoff, H., 2011. Understanding the Nexus: Background Paper for the Bonn2011 Nexus Conference: The Water, Energy and Food Security Nexus. Stockholm Environment Institute (SEI), Stockholm.

Hussein, W.A., Memon, F.A., Savić, D.A., 2017. An integrated model to evaluate water-energy-food nexus at a household scale. Environ. Model. Software 93, 366−380.

IMechE, 2013. Global Food: Waste Not, Want Not. Institute of Mechanical Engineers (IMechE), London.

Janssen, D.N.G., Ramos, E.P., Linderhof, V., Polman, N., Laspidou, C., Fokkinga, D., de Mesquita e Sousa, D., 2020. The climate, land, energy, water, and food nexus challenge in a land scarce country: innovations in The Netherlands. Sustainability 12, 10491.

Kadian, R., Dahiya, R.P., Garg, H.P., 2007. Energy-related emissions and mitigation opportunities from the household sector in Delhi. Energy Pol. 35, 6195−6211.

Kenway, S.J., Scheidegger, R., Larsen, T.A., Lant, P., Bader, H., 2013. Water-related energy in households: a model designed to understand the current state and simulate possible measures. Energy Build. 58, 378−389.

Laspidou, C.S., Mellios, N.K., Spyopoulou, A.E., Kofinas, D.T., Papadopoulou, M.P., 2020. Systems thinking on the resource nexus: modeling and visualization tools to identify critical interlinkages for resilient and sustainable societies and institutions. Sci. Total Environ. 717, 137264.

Meadows, D.H., Meadows, D.L., Randers, J., Behrens, W.W., 1972. The Limits to Growth. Universal Books.

Moe, C.L., Rheingans, R.D., 2006. Global challenges in water, sanitation and health. J. Water Health 4, 41−57.

Nhamo, L., Mabhaudi, T., Mpandeli, S., Dickens, C., Nhemachena, C., Senzanje, A., Naidoo, D., Liphadzi, S., Modi, A.T., 2020. An integrative analytical model for the water-energy-food nexus: South Africa case study. Environ. Sci. Pol. 109, 15−24.

Purwanto, A., Sušnik, J., Suryadi, F.X., de Fraiture, C., 2019. The use of a group model building approach to develop causal loop diagrams of the WEF security nexus in a local context: a case study in Karawang Regency, Indonesia. J. Clean. Prod. 240, 118170.

Purwanto, A., Sušnik, J., Suryadi, F.X., de Fraiture, C., 2021. Quantitative simulation of the water-energy-food (WEF) security nexus in a local planning context in Indonesia. Sustain. Prod. Consum. 25, 198−216.

Simonovic, S.P., 2002. World water dynamics: global modelling of water resources. J. Environ. Manag. 66, 249−267.

Steffan, W., Richardson, K., Rockstrom, J., Cornell, S.E., Fetzer, I., Bennett, E.M., Biggs, R., Carpenter, S.R., de Vries, W., de Wit, C.A., Folke, C., Gerten, D., Heinke, J., Mace, G.M., Persson, L.M., Ramanathan, V., Reyers, B., Sorlin, S., 2015. Planetary Boundaries: Guiding Human Development on a Changing Planet. Science.

Sušnik, J., 2015. Economic metrics to estimate current and future resource use, with a focus on water withdrawals. Sustain. Prod. Consum. 2, 109−127.

Sušnik, J., 2018. Data-driven quantification of the global water-energy-food system. Resour. Recyc. & Conser. 133, 179−190.

Turner, G.M., 2008. A comparison of *the Limits to Growth* with 30 years of reality. Glob. Environ. Chang. 18, 397–411.

Valek, A.M., Sušnik, J., Grafakos, S., 2017. Quantification of the urban water-energy nexus in México City, México, with an assessment of water-system related carbon emissions. Sci. Total Environ. 590–591, 258–268.

Wang, S., Liu, Y., Chen, B., 2018. Multiregional input–output and ecological network analyses for regional energy–water nexus within China. Appl. Energy 227, 353–364.

WEF (World Economic Forum), 2016. Global Risks Report 2016, eleventh ed. Available at: http://wef.ch/risks2016 (Accessed June 2021).

World Bank, 2013a. Energy Fact-File. http://www.worldbank.org/ (Accessed June 2021).

World Bank, 2013b. Water Papers: Thirst Energy (No. 78923). World Bank, Washington, D.C.

World Hunger, 2013. http://www.wfp.org/hunger (Accessed June 2021).

CHAPTER 5

Tools and indices for WEF nexus analysis

Janez Sušnik[1], Sara Masia[1,2], Graham Jewitt[1,3,4] and Gareth Simpson[5]

[1]*IHE Delft Institute for Water Education, Delft, The Netherlands;* [2]*CMCC Foundation — Euro-Mediterranean Centre on Climate Change, IAFES Division, Sassari, Italy;* [3]*Centre for Water Resources Research, School of Agricultural, Earth and Environmental Sciences, University of KwaZulu-Natal (UKZN), Pietermaritzburg, South Africa;* [4]*Civil Engineering and Geosciences, Delft University of Technology, Delft, The Netherlands;* [5]*Jones and Wagener Engineering Associates, Pretoria, South Africa*

1. Introduction

George Box poignantly observed that "All models are wrong; some are useful" (Collins, 2009). This is particularly true of complex systems where the constituent parameters, aside from being interlinked with each other, fluctuate spatially and temporally while varying in the units that they are measured with. Many decision- and policymakers, academics, and private and public practitioners desire a universally applicable model to inform their context. Yet there is seldom a one-size-fits-all solution. Rather, actors entrusted with responsibility within a multifaceted environment must carefully consider the relative strengths and weaknesses associated with a specific model or method to inform that situation.

Within the discipline of sustainability science, the number of components, variables, unknowns (both known-unknowns and unknown-unknowns) and indicators are myriad. Practitioners should be aware of what tools are available to them. They must also garner sufficient information to weigh these methods against one another and then select the optimal tool for the task at hand. While doing this, they must be cognizant that any model is an approximation of reality based on assumptions, and constrained by several factors (e.g., data availability for complex system replicability, knowledge of governing equations of the system, placement of system boundaries, etc.). The assumptions and limitations associated with the selected method must be both understood and clearly communicated.

The WEF nexus is the context under analysis, and offers a framework that provides a perspective on integrated resource management and security. It also provides an integrated perspective on the performance of SDGs 2, 6, 7, and 13. Reasons for assessing the "nexus" include the following:

1. a desire to have a multicentric approach, seeking to prevent a "silo" approach to resource management, and
2. to seek to exploit potential synergies and avoid tradeoffs associated with the implementation of resource-based policies.

Meadows et al. (1972) cautioned almost half a century ago, before the word "sustainability" became a buzzword, "If the present growth trends in world population, industrialisation, pollution, food production and resource depletion continue unchanged, the limits to growth on this planet will be reached sometime within the next one hundred years." Approximately 30 years later, it has been reported that "the human economy is exceeding important limits now and that this overshoot will intensify greatly over the coming decades" (Meadows et al., 2004). The goal of this chapter is to introduce some prominent tools and/or approaches to studying the WEF nexus from different perspectives and to highlight indices by which to analyze the anthropogenic effects on earth and efforts toward reversing detrimental trends.

2. Tools and approaches to analyze the WEF nexus
2.1 Conceptual maps and causal loop diagrams

Conceptual maps and causal loop diagrams are closely related approaches and/or tools that are discussed together. Conceptual maps can be thought of as a mapping of the most important connections within a system at an abstract level that is usually accessible for nonexperts to understand. Conceptual maps can help define the system boundary as well as identify the main issues under investigation, including the connections between those issues. They can start to elucidate the mechanisms of the interactions. Conceptual maps should be developed as much as possible with the involvement of local experts and with a wide group of interested stakeholders. This will help ensure that the developed map is representative of the case study and as accurate in reflecting WEF nexus issues as possible. Conceptual maps are usually developed iteratively over a series of meetings or workshops, with details being gradually added to the level desired of the study and to refine ideas. Conceptual maps can be "high level conceptual maps" where the main sectors and major links are highlighted without details (Fig. 5.1), or "extended conceptual maps" where details with specific nexus sectors and the links among its subsectors and all the other sectors of the system are shown (Fig. 5.2). In Fig. 5.1, a high-level map between the water, energy, food, land, and climate sectors in the Netherlands is shown, indicating the connections between sectors and potential mechanisms. For example, a connection is shown between the climate and water sectors, indicating that climate change may impact on future water availability in the Netherlands. In this example, each sector has its own separate

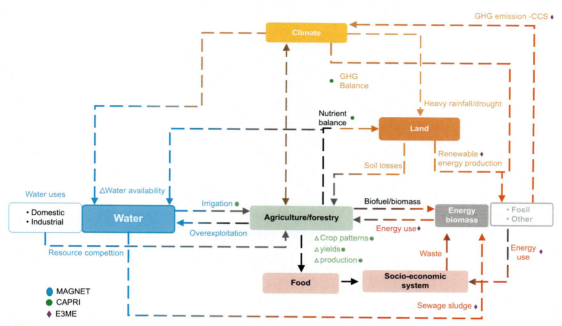

FIGURE 5.1
"High-level" nexus conceptual map for the Netherlands case study in SIM4NEXUS (www.sim4nexus.eu; Vamvakeridou-Lyroudia et al., 2019).

conceptual map developed (e.g., food sector in Fig. 5.2), thus forming a "Russian doll" of nested conceptual maps. A detailed example of conceptual model development for the Songwe River Basin, located in the border between Malawi and Tanzania, is shown in Chapter 7. Within each sector, more detail is added on how that sector behaves and the detailed connections to the other nexus sectors. Through such high-level understanding, communicating complex nexus issues to nonexpert stakeholders becomes considerably easier than when trying to communicate model output. Although these maps seem simple, their development and refinement may take weeks to months, especially when developed in a collaborative setting with expert advisors. Their importance should not be underestimated, as they play critical roles in data mapping and quantitative model development and in communication regarding complex nexus issues in an accessible way.

Causal loop diagrams (CLDs; Ford, 2010) are a mapping of interconnections between system elements to better understand causal connections between those elements. They go beyond conceptual maps (but are complementary to them) by introducing the concept of causality between elements, allowing

70 CHAPTER 5: Tools and indices for WEF nexus analysis

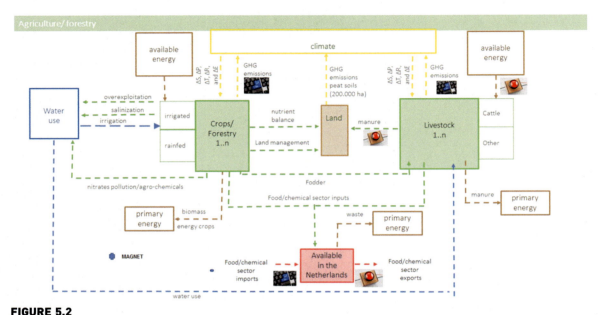

FIGURE 5.2

"Extended" nexus conceptual map for the Netherlands case study in SIM4NEXUS (www.sim4nexus.eu; Vamvakeridou-Lyroudia et al., 2019).

one to define reinforcing and balancing feedback connections and loops. They offer different information than conceptual maps. CLD is an approach that can be applied in the process toward developing quantitative systems models (Binder et al., 2004) and are helpful in assisting nonexpert stakeholders in developing a better understanding of the main interconnections in a complex system, such as the WEF nexus. Wolstenholme (1999) explains that CLDs are able to be developed and applied independently of any quantitative modeling exercise. Through this mapping, complex feedback loops through a system can be explored. CLDs assign "polarity" to connections between variables (Sterman, 2000). Connections with positive polarity (indicated with a "+" next to the arrowhead) indicate that variable "A" changes *with the same direction* as variable "B" (e.g., if "A" increases, "B" also increases). Connections with negative polarity (indicated with a "−" next to the arrowhead) mean the opposite (i.e., if "A" increases, "B" decreases). Tracing polarities around loops allows one to assign a "type" to a complete feedback loop. Reinforcing feedback loops suggest runaway behavior, potentially leading to exponential growth in a system. This is the situation when the values of a system double in the same period of time. For example, if it takes 10 years to go from 10,000 to 20,000 people, it would also take 10 years to go from 1,000,000 to 2,000,000 people. Fig. 5.3 shows an example of exponential growth, using world population over time

2. Tools and approaches to analyze the WEF nexus

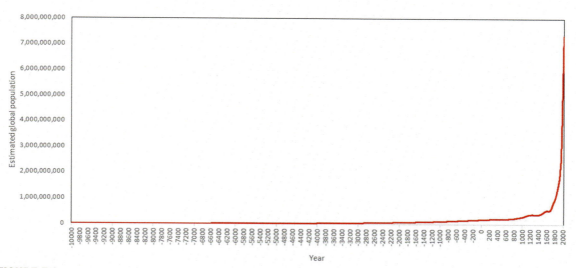

FIGURE 5.3
Example of exponential growth using world population as an example. *Data source: Our World in Data (ourworldindata.org).*

as an example. In a CLD, two, four, six, or any even number of positive connections through an entire loop mean that loop has a reinforcing character. Balancing feedback loops suggest "goal-seeking" behavior resulting in dampened growth. A good example is that of an ecological predator–prey dynamic between two species. Rises in prey populations are followed by temporally delayed rises in predator numbers, who consume members of the prey population, which subsequently starts to decline. This is followed by a decline in the predator population, which allows the prey population to rebound again. This behavior can continue over time, resulting in wave-like patterns of population numbers, oscillating around an approximate mean value. While simplistic, this gives an idea of the notion of dampening of system behavior and oscillatory behavior. Interactions between reinforcing and balancing loops can lead to oscillatory system behavior as the system transitions between dominant behavior modes. Table 5.1 summarizes visual representation of CLD notation. CLDs can be very useful in guiding the development of system dynamics models, especially when developed with (local) experts in the fields of water, food, energy, and systems analysis (Purwanto et al., 2019, 2021).

2.2 System dynamics modeling

System dynamics models (SDMs; see Ford, 2010 for a comprehensive introduction) may be thought of as the "next step" from CLDs, with CLDs guiding the development of SDMs. The concept of SDM was developed in the 1960s by Jay

CHAPTER 5: Tools and indices for WEF nexus analysis

Table 5.1 Basic elements in causal loop diagrams.

Notation	Description	Example
A —(+)→ B Connector	Change in A causes change in B in the same direction. If A increases/decreases, B also increases/decreases	Temperature —(+)→ Evaporation Cultivated land —(+)→ Water demand
A —(−)→ B Connector	Change in A causes change in B in the opposite direction. If A increases/decreases, B also increases/decreases	Infiltration —(−)→ Run-off Groundwater table —(−)→ Pumping cost
(R) or ◂(+)▸	Reinforcing or positive feedback loop, if it contains an even number of negative causal links	Birth Rate ⇌ Population (R), both +
(B) or ◂(−)▸ or ⫮	Balancing or negative feedback loop, if it contains an odd number of negative causal links. Delay, the situation when the systems respond slowly in certain condition	Population ⇌ Death Rate (B), +/+ Number of Plant Growing ⫮→ Harvest Rate

Modified from Mirchi, A., Madani, K., Watkins, D., Ahmad, S., 2012. Synthesis of system dynamics tools for holistic conceptualization of water resources problems. Water Resour. Manag. 26, 2421–2442.

Forrester (Forrester, 1968) as an approach to study problems of control and feedback in industrial systems. SDM is, therefore, ideally suited for studying complex systems governed by complexity, delay, and feedback, such as the WEF nexus. One of the earliest and perhaps well-known applications of SDM was in the classic *Limits to Growth* study of Meadows et al. (1972), which considered prospects of human growth and industrial development in the context of living on a planet with finite resources being degraded by pollution. Although at the time the Meadows et al. (1972) study was heavily criticized for not being realistic, more recent work has demonstrated that the *trends* predicted in the model simulations were broadly correct for many parameters (Turner, 2008). SDM has been applied to a vast diversity of environmental issues (e.g., Kojiri et al., 2008; Davies and Simonovic, 2011; Rehan et al., 2011; Sušnik, 2015, 2018; Sušnik et al., 2012, 2013a, b; Ghashghaei et al., 2014; Sahin et al., 2014; Mereu et al., 2016; Hayward and Roach, 2018; Bakhshianlamouki et al., 2020; Purwanto et al., 2021) and is useful for nonexpert communication (Tidwell et al. 2014).

SDMs typically comprise three main model elements, i.e., stocks, flows, and converters (Fig. 5.4). Stocks store material (e.g., water in a reservoir) and have units that are non-time dependent (i.e., they integrate over time; e.g., m^3, number of people). Flows move material into and out of stocks (e.g., river discharge, evaporation; $m^3\ s^{-1}$). Finally, converters alter the rates of flows (e.g., runoff coefficients or evaporation rate). Changes in stock levels are calculated through finite difference equations. Long-term trends in stock levels and

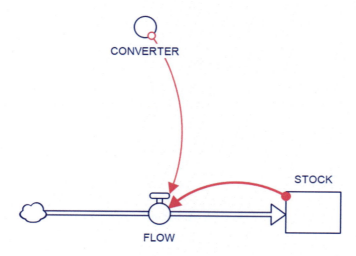

FIGURE 5.4

The three major SD modeling elements: stocks (square box), flows (large arrow with the "valve"), and converters (small circle). Connections (pink lines) transmit information between modeling elements. SD, system dynamics.

derived indicators (e.g., www.wefnexusindex.org) can be tracked according to changes in system variables. Model elements are linked to form feedback loops that can include delay and nonlinear functions. Mathematical, logical, statistical, or control expressions define element interaction and can be probabilistic, utilizing Monte-Carlo sampling.

One main advantage of SDM over other modeling approaches is the ability to build models from "the ground up." This means models can cross disciplines, allowing truly "systems" or "nexus" thinking to analyze the connections between WEF nexus sectors, and how these may respond to external driving forces such as population change or policy implementation. It can also be applied at almost any spatial (e.g., household, regional, national, global) or temporal (e.g., daily, monthly, yearly) scale, offering the flexibility required to study WEF nexus issues, which are operational at different scales (see Chapter 5). Another advantage of the bottom-up modeling approach is the ability to involve stakeholders in the modeling process (cf. Tidwell et al. 2014; Sušnik et al., 2018), which leads to models better representing the system under consideration as well as to better understanding of the model outputs by stakeholders, meaning that results and recommendations are more likely to be accepted and taken up. However, there are aspects that SDM is less able to deal with, including interactions between people and the environment, spatially distributed phenomena, and fine-detailed analysis of individual systems. In these regards, other approaches may be better suited.

2.3 Agent-based modeling

Agent-based modeling (ABM) has its roots in complexity science and is not dissimilar to cellular automata. ABM allows to capture or generate emergent phenomena that result from the interactions between individual entities. Two of the main characteristics of this tool are its flexibility and its ability to produce a natural description of the system, i.e., the most probable scenarios that can happen in reality (Bonabeau, 2002). In ABM, groups of "things" with similar characteristics (e.g., people, animals, classes of people such as farmers) are represented as agents. Each agent has its own set of decision-making rules. These might define, for example, how a farmer responds to changes in agricultural policy or how they change crops depending on rainfall patterns. Or the rules could define how a city spreads out as it grows, perhaps being guided by proximity to major infrastructure. Complexity and emergent behavior comes about through the interactions of agents with each other and with their background "environment." As such, ABM has been defined as "… *a computerized simulation of a number of decision-makers (agents) and institutions, which interact through prescribed rules*" (Farmer and Foley, 2009). Due to this ability of agents to respond to changes in the environment and vice versa, ABMs are adept at simulating human—environment interactions. Similar to SDM, bottom-up model development is favored to capture the detailed decision rules to be employed by each agent and to properly characterize the interaction with the environment. In some ABMs, the agents can learn, adapting to new circumstances or environmental changes, a feature again very useful in modeling human–environmental systems.

There are several ABM methods (see An, 2012 for a thorough review). *Microeconomic models* are focused on resource-related studies where the agents aim to "maximize" profit or revenue while not violating constraints (e.g., in resource availability). One major assumption is that agents always make rational decisions, which is not always the case. Applications include agents using land for different purposes (e.g., Reeves and Zellner, 2010). One major point to consider is the choice of variable and the form of those variables to enter in the utility functions of microeconomic ABMs. *Space theory models* concern decisions made when space, certain characteristics in space, or distance to other objects is of primary concern. For example, some decisions may be made based on ground slope or aspect relative to the sun. Many ABMs predicting city expansion use rules linking expansion-to-distance-to-infrastructure such as road and rail networks (Haase et al., 2010; Hosseinali et al., 2013; Firdausiyah et al., 2019). Distance to green spaces or to coastlines may also be considered as critical decision-making factors. However, there can be arbitrariness in deciding what environmental/socioeconomic elements and what relationships between the agent's decision and the chosen elements should enter the model. *Psychosocial models* are based on beliefs, concepts, memory, and experiences of a system. They tend to aim to represent the net effect of peoples' thought process and actions within a system. One subset is fuzzy cognitive mapping, using nodes and edges to represent relationships between elements in a system (e.g., Martinez et al., 2018). Another is actor-centred theory that postulates that actors influence and/or are influenced by changes in social structures. Although a rich area of research, more understanding is needed of the role that social networks play in human decision making. Closely related are *institution-based models*, which aim to assess the interactions of institutes with each other and in response to changes in their environment. *Experience/preference-based models* are based on real-world experiences and the decisions brought about from those experiences. They are, therefore, easier to communicate and understand as they represent more closely real-world choices. However, they can incorporate more uncertainty due to the diversity of choice options, which is where blending with fuzzy logic methodologies can come in useful to estimate the degree of likelihood of a particular course of action based on a ranking of the "desirability" of different options. Decision rules in this type of model are often updated. *Participatory models* are built with the express involvement of stakeholders, who help define how the models are to be built (cf. Sušnik et al., 2018). Through such involvement, stakeholders are more likely to trust model outputs, and the model may better represent the system under study due to the expert knowledge from the stakeholders. *Empirical rule models* derive their ABM rules through analysis of (statistical) trends and relationships in data, measurements, and observations. Occasionally, methods such as neural networks as used to learn the rules form complex data sets; however, this has the disadvantage of the user not knowing how the rules came about. While useful for deriving rules from large and complex data, this method suffers from the downside that one cannot understand why the rule is made—it is more a mechanistic procedure to be implemented. *Evolutionary programming*—based ABMs are a type of empirical model as described earlier, but utilizing concepts borrowed from the theory of natural selection. Agents contain various attributes regarding decision-making, and those agents with the attributes most likely to succeed and adapt will "survive." Just as in the natural world, agents and their attributes can copy, cross-breed, or mutate rules, leading to better chance of survival. The final major type of ABM are *assumption-based models*. These are implemented where hypothetical rules are used in the absence of sufficient data, knowledge, or information about a process to utilize one of the aforementioned approaches. They can be useful in modeling social systems, for example, when making assumptions about how many hours working adults are out of the house for in a workday (Perez and Dragicevic, 2009). Of course the main downside is that the rules may not be correct, and because of the lack of information, there is no possibility to test if the rule is correct or not. Another issue is that while the model may produce good results, it may be for the wrong reasons. This suite of models must be used with caution.

2.4 (Multiregion) input—output modeling

Multiregion input—output (MR)IO modeling is a top-down approach to environmental accounting. Such IO tables and analysis help demonstrate how much product a given economic sector produces (the output) and how much other product is needed (the input) to realize this output. As databases have become more comprehensive (e.g., the widely used EXIOBASE database; www.exiobase.eu/), it is possible to consider primary resource use and emissions within any given sector, allowing the wider intersectoral linkages within an economy, such as energy needed in the production of a given product, to be identified and quantified (Tukker and Vivanco, 2018). (MR)IO therefore combines all the information about economic relations, pressures on different nexus resources, and how consumption relates to these pressures in a consistent framework. When "flows" of good and resources are between regions and nations, it becomes a multiregional study. Such studies help elucidate the wider "footprint" of resource use in the production and consumption of products, as well as being able to assess the resource use of country or sector within a country, including where the input material originates from, and where the produced material is consumed, along with the wider environmental impact (Tukker and Dietzenbacher, 2013). A well-known example of a footprint is the so-called water footprint (e.g., Hoekstra and Mekonnen, 2012; and www.waterfootprint.org), which allows assessment of which countries/regions virtually "import" or "export" water through the trade of goods and services. Because the resource demands and impacts from production and consumption of products within and between countries is analyzed, nexus wide connections can start to be assessed. For example, how much water or energy fuels are embedded in the production or consumption of a specific product and are given countries net water or energy importers or exporters? These approaches are internally consistent, allowing for direct comparability.

As an example, Meng et al. (2019a) analyzed the urban water—carbon nexus in Beijing, showing that the electricity sector had the greatest absolute direct water consumption, followed by construction and metal smelting. However, in terms of the intensity of water use (defined as volume of water needed to generate a unit of economic return, m^3 US^{-1}), metal smelting was by far the most intense water user. In terms of carbon emissions, the electricity and transport sectors showed the greatest direct carbon emissions, while metal mining was the most carbon intensive sector. Embodied water and carbon consumption were also analyzed, with food and tobacco representing the greatest embodied water consumption, and metal mining and construction representing the greatest carbon emissions. Similarly, Wang et al. (2018) perform an MRIO analysis in China, showing that Beijing and Shanghai are resource "importers." Generally, embodied water was transferred from western to eastern and from northern to southern regions in China.

(MR)IO models are sensitive to sectoral price assumptions and sectoral aggregation (Meng et al., 2019b). Results are also sensitive to the weighting factors assigned for different resource use and impact, with many factoring approaches available (Tukker and Vivanco, 2018). Another downside is that the tables, though sophisticated, are static and must be regularly updated. They are not able therefore to deal with dynamically changing situations. The damage and impact to ecosystems and their services is not usually explicitly considered. Also, as the system boundary is expanded, the assessment becomes ever-more complex. It is to be recognized that (MR)IO analyzes connections within economic systems and is not a nexus analysis tool per se. Despite this, it can be useful to gain insight into certain nexus connections and relationships.

2.5 Life cycle assessment

Life cycle assessment (LCA) is related to MRIO, but is a bottom-up approach, allowing a finer resolution of the inputs and outputs of specific products through various stages of the life cycle, including from cradle-to-grave (i.e., all process related to the production, use, and waste management of a product; van der Voet and Guinee, 2018). LCA is a method that computes and evaluates inputs, outputs, and environmental impacts from design to disposition of a product or technology (Guinee, 2002) using detailed databases such as SIMA-Pro. LCA is composed of four main stages: (1) define the goal and scope; (2) inventory analysis; (3) impact assessment; and (4) interpretation (ISO, 2006). As with MRIO analysis, the definition of the system boundary is critical to ensure accurate results and the tractability of the analysis. LCA assesses the demands of materials throughout an entire product life span, or through parts of it. For example, in principle, an LCA could be undertaken for the whole chain depicted in Fig. 5.5, or just for individual elements in the chain. The majority of LCA studies are conducted to assess the environmental impact associated with a certain product or process.

As mentioned earlier, one critical aspect to consider is the definition of the system boundary as illustrated in Fig. 5.6. For example, on the manufacturing side, one LCA could consider the resources required to produce the cotton involved in making a shirt. This may include the land, water, and energy resources associated with the cotton harvesting, and the subsequent water, energy, and human resources involved in the production of the shirt. But in principle, a study could go another step "back" in the chain and attempt to assess the metals and energy consumption involved in the production of the machinery used to harvest the cotton. This adds another layer of complexity. One can imagine going ever-further "deeper" into the production system until the assessment is too complex to carry out, even in principle. Therefore, the boundary is critical to define, with everything outside of the boundary taken as a given exogenous input.

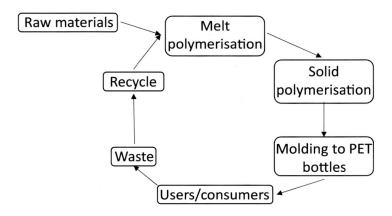

FIGURE 5.5

Flow diagram for PET bottle production. *Adapted from Marathe, K.V., Chavan, K.R., Nakhate, P., 2019. Life-cycle assessment (LCA) of PET bottles. In: Thomas, S., Kanny, K., Thomas, M.G., Rane, A., Abitha, V.K. (Eds.). Recycling of Polyethylene Terephthalate Bottles. Elsevier.*

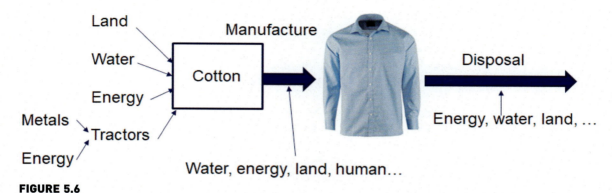

FIGURE 5.6
Where to draw the boundary of LCA studies to ensure tractability of assessments? *LCA*, life cycle assessment.

Because LCA attempts to account for all the resources consumed during a product life cycle, also accounting for environmental impacts (e.g., global warming potentials arising from the use of resources associated with a product during its life cycle), it can account for whole-nexus implications. However, like MRIO, the method is not dynamic, so changes in demand over time and space cannot be dynamically accounted for (van der Voet and Guinee, 2018). Rather, a current situation "snapshot" of a steady state is afforded. In addition, it is up to practitioners on which unit-process data to use, and what allocation choice to make in LCAs. This can lead to considerable uncertainty and variability in the results of LCA analysis for the same product and processes. Another point to be aware of is that LCA tends to use most useful for microprocesses, and it is typically not suitable for upscaling to larger systems (van der Voet and Guinee, 2018). Current research is attempting to extend LCA methodologies to include social and economic impacts, to extend the spatial applicability of LCA studies, and to allow for better dynamic interlinkage analysis, rather than only considering static snapshots of small systems.

2.6 Integrated assessment models

Integrated assessment models (IAMs), largely stemming from the climate and energy sciences (Hamilton et al., 2015), are used to attempt to assess multisectoral impacts of various pressures using scenarios. IAMs have undergone recent rapid development in terms of approaches, sophistication, resolution, and the sectors assessed (e.g., Hamilton et al., 2015; Krey et al., 2019). Accounting for feedback between processes and the ideas of integration and cross-cutting assessment are becoming more prevalent (Huppmann et al., 2019; Krey et al., 2019). Typically, IAMs either consist of the integration of many models to assess multisectoral impacts ("soft linking," the assemblage approach; Voinov and Shugart, 2013) or from developing models from the ground up to integrate different aspects (the integral approach). Therefore, IAMs and their results, data, approaches, and assumptions differ depending on the origins of the various models that are combined to form the IAMs. Some may have energy-based origins, while others are economic or climatic in origin for example. In terms of what is "integrated," Hamilton et al. (2015) identify 10 dimensions of integration, divided into three broad categories: (1) key drivers of integration (stakeholders, issues of concern, governance setting); (2) methodological aspects for integration (sources and types of uncertainty, methods, models, tools, and disciplines); and (3) system aspects to be integrated (spatial scale, temporal scale, natural setting, human setting). As a result of the fundamental underlying differences between IAMs and due to differences in, for example, the detailed implementation of different energy-generating technologies (including capital costs and operation and maintenance cost assumptions, and the relative carbon reduction impacts of the technologies), while IAMs tend to agree on broad high-level issues and trends, there tends to be disagreement on finer-scale details. This is explored in detail in Krey et al. (2019). As an example, while the electricity sector is generally projected to decarbonize under climate policy, the speed of this transition and especially the nature of the resulting technology mix in power generation can be very different across IAMs. These differences in data, assumptions, technical (model) implementation, and integration methods must be fully acknowledged and considered, and attempts could be made to add coherence between IAM results. Another issue is that some IAMs are so complex that it can be unclear as to how and why certain results are obtained, leading to a lack of transparency and trust in results. Indeed, IAMs have come under criticism for being "subjective" and having created their own "reality," which has been accused of being misleading (Ellenbeck and Lilliestam, 2019) and none as yet cover all WEF sectors comprehensively or coherently. Other potential issues with IAMs are that they may lack full representation of sectoral interconnections and that they tend to address more abstract high-level problems rather than shorter-term more applied issues (Bazilian et al., 2011). Some prominent IAMs include GCAM (http://www.globalchange.umd.edu/gcam/), IMAGE (models.pbl.nl/image/index.php/Welcome_to_IMAGE_3.0_Documentation), and WITCH (https://www.witchmodel.org/).

As an example of IAM application, Bijl et al. (2017) use the IMAGE model to assess the long-term water demand in the electricity, industrial, and household sectors. They show that water withdrawals and consumption are both expected to increase globally; however, highly aggressive measures to improve water use efficiency can lead to water use reductions. Such aggressive measures are not expected to be reasonable globally, however. Similarly, Admiraal et al. (2016) also use the IMAGE modeling framework to assess how the costs and benefits of climate mitigation strategies may change depending on the timing of their implementation. The study suggests that gradual change is most effective in terms of costs and net benefits, rather than delayed or early action; however, results are affected strongly by assumptions in the financial discount rates applied.

3. Indices for WEF nexus performance assessment (analysis)

The definition of outputs and indicators of success is of paramount importance in the development of models and tools associated with sustainability. Model outputs must be relevant in that they must provide requisite evidence to researchers, NGOs, policy- and decision-makers, and other stakeholders. The timeliness of the outcomes is also essential because the modeling results must provide information and knowledge to address critical current issues. This information can be generated by means of data, indicators, indices, and qualitative and quantitative studies (such as the models described earlier in this chapter), as demonstrated in Fig. 5.7.

To understand the level of attainment of specific sustainability goals, various indicators have been developed and monitored. These are necessary to benchmark a province, state, nation or region, or the state of a system. They are also invaluable for ascertaining progress and trends and identifying focus areas for policy or development interventions. These indicators are typically recorded on a basin, subnational, or national level, in accordance with an internationally

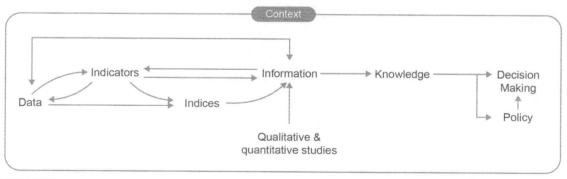

FIGURE 5.7

From data to decision-making. *Modified from Segnestam, L., 2002. Indicators of Environment and Sustainable Development: Theories and Practical Experience. The World Bank Environment Department (Environmental Economics Series. Paper No. 89). Waas T., Hugé J., Block T., Wright T., Benitez-Capistros F., Verbruggen A., 2014. Sustainability assessment and indicators: tools in a decision-making strategy for sustainable development. Sustainability 6, 5512–5534. in Simpson G.B., Jewitt G.P.W., Becker W., Badenhorst J., Neves AR., 2020. The Water-Energy-Food Nexus Index: A Tool for Integrated Resource Management and Sustainable Development. OSF Preprints.*

agreed-upon methodology. The indicator values, together with the underlying data, are subsequently audited by international bodies such as the World Bank, United Nations, International Energy Agency (IEA), or Food and Agriculture Organization (FAO) of the United Nations. In terms of the WEF nexus, indicators relevant to access to, and availability of, water, energy, and food are of particular interest.

Following from the development of individual indicators, composite indicators were developed to enable the understanding of complex concepts such as competitiveness, industrialization, and sustainability. This was necessary because of the difficulty in assessing, and drawing conclusions from, a myriad of indicators. Fig. 5.7 presents the complementary role that indices can fulfill in generating information and knowledge for policy- and decision-makers. A composite indicator is formed "when individual indicators are compiled into a single index on the basis of an underlying model" (OECD, 2008). Some actors, for example, advocacy groups, view composite indicators as a valuable tool to further their causes. Others, such as cautious professional statisticians, are wary of composite indicators due to the potentially subjective nature of the selection of the constituent indicators, the method of aggregation, and the weighting of the indicators. Because composite indicators are not universally accepted, they must be developed transparently and used responsibly.

3.1 Human development index

In 1990, the Human Development Index (HDI) was developed by Pakistani economist Mahbub ul Haq to provide a more comprehensive representation of wellbeing than the GDP. He included health and education indicators with the natural logarithm of the gross national income (GNI) per capita. The HDI was based on the premise that human development should focus on the three essential elements of human life, namely, longevity, knowledge, and decent living standards (UNDP, 1990). Although the method of calculating the HDI has changed with time, it has served as a valuable tool for the United Nations Development Programme (UNDP) and other organizations in evaluating developmental progress in many countries and regions under their jurisdiction.

3.2 Environmental sustainability index

Some composite indicators, in contrast to the HDI, are relatively complex. The Environmental Sustainability Index (ESI) integrates 76 data sets into 21 indicators, which are subsequently condensed into a single index (Esty et al., 2005). The ESI serves as a policy tool for identifying issues that require focused attention within national environmental protection programs and across societies more generally (Esty et al., 2005).

3.3 Sustainability development goals

At the beginning of 2016, the Sustainable Development Goals (SDGs) were launched. Associated with these goals are 230 individual indicators to monitor the 17 SDGs and 169 targets of the SDGs. Included in these goals are SDGs 2 (Zero Hunger), 6 (Clean Water and Sanitation), 7 (Affordable and Clean Energy), 12 (Sustainable Consumption and Production), and 13 (Climate Action), which are relevant to resource security and distributional justice associated with these resources. These goals are termed SDGs, and not simply development goals, because of the pervasive negative impact of humanity on the planet, and some such as access to electricity are not ends in themselves. How electricity is generated is, ultimately, of comparable importance to its availability. SDG 7 is, therefore, to "Ensure access to affordable, reliable, *sustainable* and *modern* energy for all." Similarly, SDG 13 requires that humanity must "Take urgent action to combat climate change and its impacts," while SDG 12 stresses the sustainable production and consumption of the materials and services we consume.

When these SDGs of the *2030 Agenda for Sustainable Development* were adopted, the United Nations stated that:

> "Indicators will be the backbone of monitoring progress towards the SDGs at the local, national, regional, and global levels. A sound indicator framework will turn the SDGs and their targets into a management tool to help countries develop implementation strategies and allocate resources accordingly, as well as a report card to measure progress towards sustainable development and help ensure the accountability of all stakeholders for achieving the SDGs".
>
> **(UN, 2015)**

Because of the large number of indicators associated with the 17 SDGs, an SDG Index was developed (Sachs et al., 2016, 2018; Schmidt-Traub et al., 2017). The SDG Index reports on 156 countries' progress toward all 17 goals and indicates areas where more rapid progress is required. All countries are ranked according to their percentage of achievement on the same group of indicators, and a dashboard has been generated to facilitate comparison between and within countries. Several indicators within the SDG Index are existing composite indicators, such as follows:

- Ocean Health Index
- Sustainable Nitrogen Management Index
- Universal Health Coverage Tracer Index
- Logistics performance index
- Climate Change Vulnerability Monitor

- Red List Index of species survival
- Corruption Perception Index
- Financial Secrecy Score
- Global Slavery Index
- PISA (Programme for International Student Assessment) score

The SDG Index and the associated dashboard apply equal weighting to each indicator and for each goal since all SDGs are considered to have equal importance in the 2030 Agenda (Sachs et al., 2019). Experts attempted to determine different weightings for some indicators at an earlier developmental stage of the SDG Index. However, a consensus on assigning different weights to the indicators could not be reached. The SDG Index values indicate that "no country is completely on track to achieve all SDGs" (Sachs et al., 2018). It also demonstrates that much work remains if equitable and sustainable global access to economic-enabling resources is to be realised.

3.4 WEF nexus index

Following the attention that the WEF nexus has garnered since 2011, various attempts have been made to define, conceptualize, model, and operationalize it, especially for policy- and decision-making. The challenge in obtaining a unified assessment of the WEF nexus is that the three resource sectors are measured in different units, e.g., percentage access, cubic meters, precipitation depth, metric tons of CO_2, kWh, kg per hectare, and international dollars per capita. To provide a coherent quantitative measure of the WEF nexus, a composite indicator that utilizes this framework as its guiding context was developed. The WEF Nexus Index was developed following an assessment of 87 globally available and relevant indicators. Utilizing the methodology espoused by the JRC's *Competence Centre on Composite Indicators and Scoreboards*, 21 indicators were selected to constitute this multicentric index. The WEF Nexus Index, together with its visualization website (www.wefnexusindex.org), provides a lens for assessing integrated resource management and security.

Does the SDG Index, which incorporates (among others) SDGs 2, 6, and 7, render the WEF Nexus Index redundant? El Costa (2015) suggested that since the SDGs seek to incorporate multiple development goals, identifying targets *at the nexus* of various sectors will be instrumental in yielding a more straightforward SDG framework. There is, therefore, a compelling argument in favor of developing an indicator framework for a *subsystem* within the SDGs, such as the WEF nexus. Boas et al. (2016) agree, arguing that "novel ways of cross-sectoral institutionalization" are required if the *2030 Agenda for Sustainable Development* is to be attained.

4. Conclusions

The WEF nexus is an extraordinarily complex system that operates at scales from local to global, and where the three resource sectors interact not just with themselves, but also with exogeneous drivers such as climate change, socioeconomic developments, and policy directions. Coherently modeling the nexus therefore poses a challenge, as numerous sectors, units, underlying philosophies, and data sets should be combined into an integrated framework. As such, there is no one-size-fits-all model capable of modeling and assessing the entire WEF nexus. This is also in part due to the vast diversity in nexus challenges, spatial and temporal scales, and foci of different (research) investigations. However, a number of modeling approaches are available, some of the more prominent of which are introduced in this chapter. Ultimately, it is up to the practitioner to select the tool best suited to the nature of the study being undertaken, and the examples given here offer a glimpse into some of the most used possibilities and their limitations. This chapter also introduces a number of composite indicators, which are developed to report on the performance of the WEF nexus as a whole, but which can also be interrogated to give sectoral (or "pillar")-level information. This information can in turn be used to track progress toward the SDGs or toward nationally or locally determined WEF-related policy objectives.

References

Admiraal, A.K., Hof, A.F., den Elzen, M.G.J., van Vuuren, D.P., 2016. Costs and benefits of differences in the timing of greenhouse gas emission reductions. Mitig. Adapt. Strategies Glob. Change 21, 1165–1179.

An, L., 2012. Modelling human decisions in coupled human and natural systems: review of agent-based models. Ecol. Model. 229, 25–36.

Bakhshianlamouki, E., Masia, S., Karimi, P., van der Zaag, P., Sušnik, J., 2020. A system dynamics model to quantify the impacts of restoration measures on the water-energy-food nexus in the Urmia Lake Basin, Iran. Sci. Total Environ. 708, 134874.

Bazilian, M., Rogner, H., Howells, M., Hermann, S., Arent, D., Gielen, D., Steduto, P., Mueller, A., Komor, P., Tol, S.J., Yumkella, K.K., 2011. Considering the energy, water, and food nexus: towards an integrated modelling approach. Energy Pol. 39, 7896–7906.

Bijl, D.L., Bogaart, P.W., Dekker, S.C., Stehfest, E., de Vries, B.J.M., van Vuuren, D.P., 2017. A physically-based model of long-term food demand. Global Environmental Change, 45, 47–62. https://doi.org/10.1016/j.gloenvcha.2017.04.003.

Binder, T., Belyazid, S., Haraldsson, H.V., et al., 2004. Developing system dynamics models from causal loop diagrams. In: Proceedings of the 22nd International Conference of the System Dynamics Society. Oxford, UK.

Boas, I., Biermann, F., Kanie, N., 2016. Cross-sectoral strategies in global sustainability governance: towards a nexus approach. Int. Environ. Agreements Polit. Law Econ. 16, 449–464.

Bonabeau, E., 2002. Agent-based modelling: methods and techniques for simulating human systems. Proc. Natl. Acad. Sci. Unit. States Am. 99, 7280–7287.

Collins, J., 2009. How the Mighty Fall - and Why Some Comapnies Never Given in. Harper-Collins, New York.

Davies, E.G.R., Simonovic, S., 2011. Global water resources modelling with an integrated model of the social-economic-environmental system. Adv. Water Resour. 34, 684–700.

El Costa D., 2015. Conceptual Frameworks for Understanding the Water, Energy and Food Security Nexus. Working Paper. Economic and Social Commission for Western Asia (ESCWA). Report number: E/ESCWA/SDPD/2015/WP.

Ellenbeck, S., Lilliestam, J., 2019. How modelers construct energy costs: discursive elements in energy system and integrated assessment models. Energy Res. Social Sci. 47, 69–77.

Esty, D.C., Levy, M.A., Srebotnjak, T., de Sherbinin, A., 2005. Environmental Sustainability Index: Benchmarking National Environmental Stewardship. Yale Center for Environmental Law & Policy, New Haven, CT.

Farmer, J.D., Foley, D.K., 2009. The economy needs agent-based modelling. Nature 460, 685–686.

Firdausiyah, N., Taniguchi, E., Qureshi, A.G., 2019. Modeling city logistics using adaptive dynamics programming based multi-agent simulation. Transport. Res. E Logist. Transport. Rev. 125, 74–96.

Ford, A., 2010. Modeling the Environment, second ed. Island Press, Washington, D.C.

Forrester, J., 1968. Urban Dynamics. Pegasus Communications, Waltham, MA.

Ghashghaie, M., Marofi, S., Marofi, H., 2014. Using System Dynamics Method to Determine the Effect of Water Demand Priorities on Downstream Flow. Water Resources Management.

Guinee, J., 2002. Handbook on Life Cycle Assessment. Operational Guide to the ISO Standards. Springer.

Haase, D., Lautenbach, S., Seppelt, R., 2010. Modelling and simulating residential mobility in a shrinking city using an agent-based approach. Environ. Model. Software 25, 1225–1240.

Hamilton, S.H., ElSawah, S., Guillaume, J.H.A., Jakeman, A.J., Pierce, S.A., 2015. Integrated assessment and modelling: overview and synthesis of salient dimensions. Environ. Model. Software 64, 215–229.

Hayward, J., Roach, P.A., 2018. Newton's Laws as an Interpretive Framework in System Dynamics. System Dynamics Reviews.

Hoeksta, A.Y., Mekonnen, M.M., 2012. The water footprint of humanity. Proc. Natl. Acad. Sci. Unit. States Am. 109, 3232–3237.

Hosseinali, F., Alesheikh, A.A., Nourian, F., 2013. Agent-based modelling of urban land-use development, case study: simulating future scenarios of Qazvin city. Cities 31, 105–113.

Huppmann, D., Gidden, M., Fricko, O., Kolp, P., Orthofer, C., Pimmer, M., Kushin, N., Vinca, A., Mustrucci, A., Riahi, K., Krey, V., 2019. The MESSAGE$_{ix}$ Integrated Assessment Model and the *ix modelling platform* (ixmp): an open framework for integrated and cross-cutting analysis of energy, climate and the environment, and sustainable development. Environ. Model. Software 112, 143–156.

ISO International Standard 14044, 2006. Environmental Management – Life Cycle Assessment – Requirements and Guidelines. International Organisation for Standardisation, Geneva.

Kojiri, T., Hori, T., Nakatsuka, J., Chong, T.-S., 2008. World continental modelling for water resources using system dynamics. Phys. Chem. Earth 33, 304–311.

Krey, V., Guo, F., Kolp, P., Zhou, W., Schaeffer, R., Awasthy, A., Bertram, C., De Boer, H.S., Fragkos, P., Fujimori, S., He, C., Iyer, G., Keramidas, K., Koberle, A., Oshiro, K., Reis, L.A., Shoai-Tehrani, B., Vishwanathan, S., Capros, P., Drouet, L., Edmonds, J.E., Garg, A., Gernaat, D., Jiang, K., Kannavou, M., Kitous, A., Kriegler, E., Luderer, G., Mathur, R., Muratori, M., Sano, F., van Vuuren, D., 2019. Looking under the hood: a comparison of techno-economic assumptions across national and global integrated assessment models. Energy 172, 1254–1267.

Marathe, K.V., Chavan, K.R., Nakhate, P., 2019. Life-cycle assessment (LCA) of PET bottles. In: Thomas, S., Kanny, K., Thomas, M.G., Rane, A., Abitha, V.K. (Eds.), Recycling of Polyethylene Terephthalate Bottles. Elsevier.

Martinez, P., Blanco, M., Castro-Coampos, B., 2018. The water-energy-food nexus: a fuzzy-cognitive mapping approach to support nexus-compliant policies in Andalusia (Spain). Water 10, 664.

Meadows, D.H., Meadows, D.L., Randers, J., Behrens, W.W., 1972. The Limits to Growth. Universal Books.

Meadows, D.H., Randers, J., Meadows, D.L., 2004. The Limits to Growth: The 30-Year Update. Chelsea Green Publishing, London.

Meng, F., Liu, G., Chang, Y., Su, M., Hu, Y., Yang, Z., 2019a. Quantification of urban water-carbon nexus using disaggregated input-output model: a case study in Beijing (China). Energy 171, 403–418.

Meng, F., Liu, G., Liang, S., Su, M., Yang, Z., 2019b. Critical review of energy-water-carbon nexus in cities. Energy 171, 1017–1032.

Mereu, S., Sušnik, J., Trabucco, A., Daccache, A., Vamvakeridou-Lyroudia, L.S., Renoldi, S., Virdis, A., Savić, D.A., Assimacopoulos, D., 2016. Operational resilience of reservoirs to climate change, agricultural demand, and tourism: a case study from Sardinia. Sci. Total Environ. 543, 1028–1038.

Mirchi, A., Madani, K., Watkins, D., Ahmad, S., 2012. Synthesis of system dynamics tools for holistic conceptualization of water resources problems. Water Resour. Manag. 26, 2421–2442.

OECD, 2008. Handbook on Constructing Composite Indicators: Methodology and User Guide. Organisation for Economic Co-operation and Development, Paris.

Perez, L., Dragicevic, S., 2009. An agent-based approach for modelling dynamics of contagious disease spread. Int. J. Health Geogr. 8, 50.

Purwanto, A., Sušnik, J., Suryadi, F.X., de Fraiture, C., 2019. The use of a group model building approach to develop causal loop diagrams of the WEF security nexus in a local context: a case study in Karawang Regency, Indonesia. J. Clean. Prod. 240, 118170.

Purwanto, A., Sušnik, J., Suryadi, F.X., de Fraiture, C., 2021. Quantitative simulation of the water-energy-food (WEF) security nexus in a local planning context in Indonesia. Sustain. Prod. Consum. 25, 198–216.

Reeves, H.W., Zellner, M.L., 2010. Linking MODFLOW with an agent-based land-use model to support decision making. Ground Water 48, 649–660.

Rehan, R., Knight, M.A., Haas, C.T., Unger, A.J.A., 2011. Application of system dynamics for developing financially self-sustaining management policies for water and wastewater systems. Water Res. 45, 4737–4750.

Sachs, J., Schmidt-Traub, G., Kroll, C., Durand-Delacre, D., Teksoz, K., 2016. SDG Index & Dashboards—Global Report. Bertelsmann Stiftung and Sustainable Development Solutions Network (SDSN), New York.

Sachs, J., Schmidt-Traub, G., Kroll, C., Lafortune, G., Fuller, G., 2018. SDG Index and Dashboards Report. Global Responsibilities - Implementing the Goals. Bertelsmann Stiftung and Sustainable Development Solutions Network, New York.

Sachs, J., Schmidt-Traub, G., Kroll, C., Lafortune, G., Fuller, G., 2019. Sustainable Development Report 2019. New York.

Sahin, O., Siems, R.S., Stewart, R.A., Porter, M.G., 2014. Paradigm shift to enhanced water supply planning through augmented grids, scarcity pricing and adaptive factory water: a system dynamics approach. Environ. Model. Software 75, 348–361. https://doi.org/10.1016/j.envsoft.2014.05.018.

Schmidt-Traub, G., Kroll, C., Teksoz, K., Durand-Delacre, D., Sachs, J.D., 2017. National baselines for the sustainable development goals assessed in the SDG index and dashboards. Nat. Geosci. 10, 547–555.

Segnestam, L., 2002. Indicators of Environment and Sustainable Development: Theories and Practical Experience. The World Bank Environment Department (Environmental Economics Series. Paper No. 89).

Simpson, G.B., Jewitt, G.P.W., Becker, W., Badenhorst, J., Neves, A.R., 2020. The Water-Energy-Food Nexus Index: A Tool for Integrated Resource Management and Sustainable Development. OSF Preprints.

Sterman, J.D., 2000. Business Dynamics: Systems Thinking and Modeling for a Complex World, sixth ed. McGraw Hill India, New Delhi.

Sušnik, J., 2015. Economic metrics to estimate current and future resource use, with a focus on water withdrawals. Sustain. Prod. Consum. 2, 109–127.

Sušnik, J., 2018. Data-driven quantification of the global water-energy-food system. Resour. Recycl. Conserv. 133, 179–190.

Sušnik, J., Vamvakeridou-Lyroudia, L.S., Savić, D.A., Kapelan, Z., 2012. Integrated system dynamics modelling for water scarcity assessment: case study of the Kairouan region. Sci. Total Environ. 440, 290–306.

Sušnik, J., Vamvakeridou-Lyroudia, L.S., Savić, D.A., Kapelan, Z., 2013a. Integrated modelling of the water-agricultural system in the Rosetta region, Nile delta, Egypt, using system dynamics. J. Water Clim. Change 4 (3), 209–231.

Sušnik, J., Molina, J.-L., Vamvakeridou-Lyroudia, L.S., Savić, D.A., Kapelan, Z., 2013b. Comparative analysis of system dynamics and object-oriented Bayesian networks modelling for water systems management. Water Resour. Manag. 27 (3), 819–841.

Sušnik, J., Chew, C., Domingo, X., Mereu, S., Trabucco, A., Evans, B., Vamvakeridou-Lyroudia, L.S., Savić, D.A., Laspidou, C., Brouwer, F., 2018. Multi-stakeholder development of a serious game to explore the water-energy-food-land-climate nexus: the SIM4NEXUS approach. Water 10, 139 (S.I. Understanding Game-based Approaches for Improving Sustainable Water Governance: The Potential of Serious Games to Solve Water Problems).

Tidwell, V.C., Moreland, B.D., Zemlick, K.M., Roberts, B.L., Passell, H.D., Jensen, D., Forsgren, C., Sehlke, G., Cook, M.A., King, C.W., Larsen, S., 2014. Mapping water availability, projected use and cost in the western United States. Environ. Res. Lett. 9, 064009.

Tukker, A., Dietzenbacher, E., 2013. Global multi-regional input output frameworks: an introduction and outlook. Econ. Syst. Res. 25, 1–19.

Tukker, A., Vivanco, D.F., 2018. Input-Output analysis and resource nexus assessment. In: Bleischwitz, R., Hoff, H., Spataru, C., van der Voet, E., VanDeveer, S.D. (Eds.), Routledge Handbook of the Resource Nexus. Earthscan (Routledge), Oxon.

Turner, G.M., 2008. A comparison of *the Limits to Growth* with 30 years of reality. Global Environ. Change 18, 397–411.

UN, 2015. Indicators and a Monitoring Framework for the Sustainable Development Goals. United Nations (Accessed June 2021). https://sustainabledevelopment.un.org/index.php?page=view&type=400&nr=2013&menu=35.

UNDP, 1990. Human Development Report. United Nation Development Programme, New York.

Vamvakeridou-Lyroudia, L.S., Sušnik, J., Masia, S., Indriksone, D., Bremere, I., Polman, N., Levin-Koopman, J., Linderhof, V., Blicharska, M., Teutschbein, C., Mereu, S., Trabucco, A., Castro, B., Martinez, P., Blanco, M., Avgerinopoulos, G., Laspidou, C., Mellios, N.,

Ioannou, A., Kofinas, D., Papadopoulou, C.-A., Papadopoulou, M., Conradt, T., Bodirsky, B., Pokorný, J., Hesslerová, P., Kravčík, M., Griffey, M., Ward, B., Evans, B., Hole, N., Khoury, M., Petersen, C., Fournier, M., Janse, J., Doelman, J., 2019. D3.4: Final Report on the Complexity Science and Integration Methodologies. SIM4NEXUS Deliverable 3.4 (Accessed June 2021). https://www.sim4nexus.eu/page.php?wert=Deliverables.

van der Voet, E., Guinee, J.B., 2018. Life cycle assessment for resource nexus analysis. In: Bleischwitz, R., Hoff, H., Spataru, C., van der Voet, E., VanDeveer, S.D. (Eds.), Routledge Handbook of the Resource Nexus. Earthscan (Routledge), Oxon.

Voinov, A., Shugart, H.H., 2013. 'Integronsters', integral and integrated modeling. Environ. Model. Software 39, 149–158.

Waas, T., Hugé, J., Block, T., Wright, T., Benitez-Capistros, F., Verbruggen, A., 2014. Sustainability assessment and indicators: tools in a decision-making strategy for sustainable development. Sustainability 6, 5512–5534.

Wang, S., Liu, Y., Chen, B., 2018. Multiregional input-output and ecological network analysis for regional energy-water nexus within China. Appl. Energy 227, 353–364.

Wolstenholme, E.F., 1999. Qualitative vs quantitative modelling: the evolving balance. J. Oper. Res. Soc. 50, 422–428.

CHAPTER 6

Transboundary WEF nexus analysis: a case study of the Songwe River Basin

Sara Masia[1,2], Janez Sušnik[1], Graham Jewitt[1,3,4], Zolo Kiala[5,6] and Tafadzwanashe Mabhaudhi[4,6,7]

[1]IHE Delft Institute for Water Education, Delft, The Netherlands; [2]CMCC Foundation — Euro-Mediterranean Centre on Climate Change, IAFES Division, Sassari, Italy; [3]Civil Engineering and Geosciences, Delft University of Technology, Delft, The Netherlands; [4]Centre for Water Resources Research, School of Agricultural, Earth and Environmental Sciences, University of KwaZulu-Natal (UKZN), Pietermaritzburg, South Africa; [5]Origins Center, School: Geography, Archaeology and Environmental Studies, University of the Witwatersrand, Johannesburg, South Africa; [6]Centre for Transformative Agricultural and Food Systems (CTAFS), School of Agricultural, Earth and Environmental Sciences, University of KwaZulu-Natal, Pietermaritzburg, South Africa; [7]International Water Management Institute, West Africa Regional Office, Accra, Ghana

1. Introduction

One of the main challenges of the 21st century is to cope with the rising pressures on resource demand due to the world's rapid population growth and socioeconomic development. By 2050, global water and energy demand are expected to increase by 55% and 80%, respectively (OECD, 2012), while to meet food demand, agricultural production needs to increase by almost 50% more than in 2012 (FAO, 2017). These trends threaten water, energy, and food (WEF) security putting at risk their access and availability. Over the past decade, the call to move from a "silo-thinking approach" to an "integrated approach" to understand and analyze these sectors and better address resource management and decision-making has been growing worldwide. The WEF nexus is recognized as an effective approach to highlight interlinkages, enhance synergies, and minimize trade-offs among the components in a system. The WEF nexus approach is emerging as an important pillar of the global 2030 Agenda for Sustainable Development in that progress toward the majority of the Sustainable Development Goals (SDGs) is closely related to the water, energy, and food sustainable management (FAO, 2018). Recently, in view of the need to accelerate progress toward meeting the SDGs, the number of stakeholders such as nongovernmental organizations, governmental ministries, private and public sectors, and academic institutions expressing their support for a WEF nexus approach is increasing.

Water, energy, and food are at the core of developing countries' development goals and strategies, and interest in the WEF nexus approach is rapidly growing (SADC, 2016; GWP-SA, 2019). For example, the Southern African Development Community (SADC) has adopted the WEF nexus approach as a framework to achieve national goals aligned with the National Development Plan and the SDGs. In SADC, this approach has evolved as a focus for integrated resources development and is strongly aligned with activities under SADC Regional Strategic Action Plan (RSAP IV) for Water Resources Management and the SADC Industrialization Strategy and Roadmap. The SADC regions largely rely on goods derived from natural resources, which are essential for eradicating poverty. In these countries, food security often depends on ecosystem goods and services; thus the integrated management of these resources is at the basis of sustainable development (SADC and GWP, 2019). However, a lack of empirical evidence and a need for appropriate methods, and qualitative and quantitative WEF nexus assessment tools have been highlighted.

This chapter presents an overview of a WEF nexus analysis approach to support sustainable socioeconomic development in the Songwe River Basin (SRB) located on the border of two SADC countries, i.e., Malawi and Tanzania and a detailed description of the first component of this approach. This research is currently ongoing within the WEF Nexus Toolkit (WEF-Tools) project (https://wef-tools.un-ihe.org, 2020−23). The work aims at assessing the SRB Development Programme's (SRBDP's) expected outcomes by applying an approach that follows from conceptual mapping of the SRB nexus system to the development of quantitative tools such as system dynamics models (SDMs), and identification of indicators for the assessment of different scenarios and management strategies, which can contribute to information for decision-makers to assess feasible development pathways.

The expected outcomes of the SRBDP assessment will be a structured knowledge base, simulation tool, dashboard, and a composite nexus index codeveloped, tested, validated, and refined through interactive collaboration with stakeholders and local experts. Ultimately, this toolkit is intended to support the development of short-, medium-, and long-term strategies for sustainable integrated resource management and policy development in this and similar basin development initiatives. Outcomes will provide a means for government ministries, NGOs, and development agencies to assess progress toward relevant SDGs, particularly SDGs 2, 6, and 7.

2. Case study description

The SRB is located in southwest Tanzania and northern Malawi. The Songwe River creates an international border between the two countries and is 200 km in length (Fig. 6.1). The basin area is 4243 km^2, and the population is over 341,000 of which about 52,000 are reported to suffer from flooding and land losses. The basin is composed of six districts: Ileje, Mbozi, Mbeya,

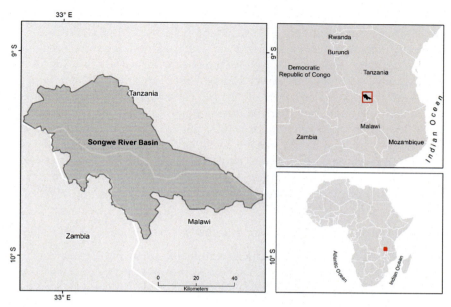

FIGURE 6.1
Location of the Songwe River Basin.

and Kyela (Tanzania side) and Karonga and Chitipa (on the Malawi side) (Munthali et al., 2011; SRBDP, 2018). The SRB is characterized by fertile alluvial soil and abundant water resources on which people rely for their living. Rural people represent about 80% of the total basin's population. Average annual income is about 386 USD per capita. In the basin, about 50% and 75% of the people lack access to safe water supply and electricity, respectively. Both Malawi and Tanzania are currently experiencing electricity shortages. The increasing population growth is having a negative impact on the environment and resource availability (CRIDF). Reducing poverty, improving human health and livelihoods, ensuring water, food, and energy security, mitigating floods, and enhancing sustainable river basin management are the main challenges that the two countries are currently facing in the SRB.

2.1 The Songwe River Basin Development Programme

The governments of Malawi and Tanzania have decided to collaborate to develop the SRBDP, which includes 26 multisectoral projects (CRIDF). Socioeconomic development, poverty, electricity, and clean water access, and riverbank instability are among the main challenges that the SRBDP aims to address in the near future (SIWI, 2019). The core of the Programme is the construction of a multipurpose reservoir located in the Lower Songwe. The reservoir will have a capacity of 330 Mm^3 and a hydropower plant capacity of

180.2 MW and is planned to be managed as a public–private partnership that can feed the Southern Africa Power Plant (CRIDF, SRBDP, 2018). The Programme includes the development of two irrigation schemes with a total area of 6200 ha made up of cover 3050 ha in the Lower Songwe River Malawi (LSRM) and 3150 ha in the Lower Songwe River Tanzania (LSRT) to which the reservoir will supply water (SRBDP, 2018). The irrigated land will be beneficial to over 5500 farming families. Two urban water supply projects will serve a total of 450,000 people. Livelihoods will be enhanced by increasing access to water and irrigated land, but also to electricity thanks to the planned Rural Electrification Project, which will benefit around 120,000 people in 22,200 households and fisheries. Additional Tourism Development Projects are planned to boost socioeconomic development in the basin (SIWI, 2019; CRIDF).

Some of the SRBDP outcomes are as follows:

- **"Increased hydropower production** to facilitate the development of small and medium industries (SMI) and improve energy source in the basin, and electricity grids in Malawi and Tanzania (increased electricity access for 60% of the SRB population)
- **Increased food production** through irrigated agriculture (a benefit for 5500 farm families)
- **Increased access to water supply and sanitation** in the basin (more than 260,000 people by 2025)
- **Water conservation/storage** to improve water access during droughts
- **Socioeconomic improvement** of the SRB inhabitants (up to 5244 full-time jobs per year in agriculture. 5560 and 3000 people per year for the infrastructure construction and operation and maintenance for 50 years for agriculture and HHP, respectively)
- **Mitigation of floods** (more than 52,000 people will be relieved)
- **Small-scale fisheries activities** to enhance protein intake and provide an alternative source of income to the inhabitants
- **Sustainable management of the SRB**
- **Improved management information system** through water resources monitoring, development, and management
- **Improved cooperation in transboundary WRM** through a formal framework
- **Enhanced cooperation between Malawi and Tanzania"**

(SRBDP, 2018; SIWI, 2019).

The application of the aforementioned WEF nexus analysis approach will help to address the main expected outcomes identified in the SRBDP both qualitatively and quantitatively.

2.2 WEF nexus analysis approach for the Songwe River Basin

The approach proposed to assess the outcomes of SRBDP consists of four main steps:

(1) Case study nexus system conceptualization
(2) Data mapping and collection
(3) System dynamics modeling
(4) Composite nexus index development

The *first step* aims at developing a conceptual nexus map where the main nexus issues, sectors, subsectors, and interlinkages between WEF components are highlighted. The conceptual model represents a qualitative assessment of the case study. It usually starts relatively simply, gradually building up in complexity according to the information that can be collected and the needs of stakeholders. The conceptual model represents WEF interactions at a high level and should be developed and validated by stakeholders and local experts (see details about the conceptual framework in Chapter 6, Susnik et al., 2018; Vamvakeridou-Lyroudia et al., 2019). The *second step* is focused on data identification and collection. Data can be collected in different units and formats from various sources (Eddy covariance stations, Earth observations, thematic models, statistics, etc.). They can be at different temporal and spatial scales. The collected data need to be used in a quantitative model (see details about data and scale in Chapters 3, 4, and 5). These data are then used in the *third step*, which is to develop the quantitative model as an SDM (Ford, 2010). SDM is a widely known modeling approach used to understand and quantify complex systems. The SDM can be developed with local experts to ensure that it is as representative of the case study as possible. Once the SDM structure is ready and the model runs, obtained results should be discussed with local experts and stakeholders for maximum impact (see details about the SDM in Chapter 5, Sušnik et al., 2020). Once results are validated, they can be used in step 4 to develop a composite nexus index, which, again, should be discussed and validated by local experts and stakeholders (see details about the composite indicators in Chapter 6; https://wefnexusindex.org).

The approach described in this chapter considers the role of local experts and stakeholders as crucial for the achievement of the final result. Indeed, stakeholders and local experts are essential to guide and validate the work developed in each step of the approach. The approach applied to analyze the WEF nexus system in the SRB is intended to directly address the expected outcomes identified in the SRBDP. The final results in steps 3 and 4 depend on data availability. A similar approach to that described here was adopted in the SIM4NEXUS project (https://www.sim4nexus.eu/) where it has been applied successfully from regional to global scale in 12 case studies (https://www.sim4nexus.eu/, Susnik et al., 2018).

In this chapter, we focus on step 1, i.e., case study nexus system conceptualization.

2.3 Conceptualizing the WEF nexus in the Songwe River Basin

The qualitative assessment of the SRB was undertaken by developing conceptual maps, which consist of two main parts: a "high level conceptual model" where only the main sectors and the major links among them are highlighted, and an "extended conceptual model," which describes in detail each nexus sector and the main links among its subsectors and all the other sectors of the system.

The analysis of the SRB has been carried at basin scale. The geographical boundary of the basin itself has been set as the boundary for the WEF nexus system assessment (Fig. 6.1).

2.3.1 High-level conceptual model

In-depth desktop analysis of the SRB was carried out and used as a base to build the high-level nexus system conceptual model. The analysis identified six main nexus components/sectors and how they interact with each other. The six sectors, i.e., water, land, food, energy, climate, socioeconomic system, human health, ecosystem health, and the main interlinkages between them are illustrated in the high-level conceptual model (Fig. 6.2). The human and ecosystem health sectors are part of the socioeconomic and land sectors, respectively, but given their crucial relevance in this case study, they have been explicitly incorporated in the map (dotted box in Fig. 6.2). The qualitative map highlights the strong link between the socioeconomic system and the land sector. Indeed, it is evident in the basin that there is considerable land use change, in particular from wetlands to cropland to accommodate growing food demand. An unavoidable consequence of increasing agricultural production is water pollution due to the use of fertilizers and pesticides needed to enhance food production and prevent crop diseases. This issue is projected to worsen due to projected increasing food demand.

One of the main challenges of the SRBDP is to increase clean water access and supply. In this regard, a new reservoir is planned, which will increase water storage, access, and supply, in particular during droughts. The reservoir is intended to supply water to irrigate fields and allow for diversification of crop types and increased yields. The construction of a reservoir will be essential for flood control, potentially helping to protect more than 52,000 people who live in the

2. Case study description

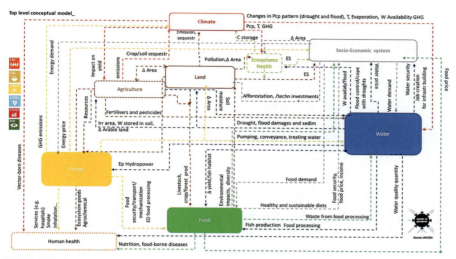

FIGURE 6.2
High-level nexus conceptual model for the Songwe River Basin. The main tentative SDGs are shown (left). *SDG*, Sustainable Development Goal.

flood plain and that currently suffer from flooding and land loss. In addition, the reservoir will increase energy production and, with the improvements in water and food availability and access, is intended to improve livelihoods, boost the economy, reduce poverty, and improve the quality of life and human health. The reservoir is therefore expected to have significant impact on the socioeconomic system in terms of job employment in the different sectors and income generation. The main SDGs that are addressed in the SRB qualitative analysis are 1, 2, 6, 7, 8, and 13.

2.3.2 Extended conceptual model
Following the development of a high-level conceptual model, a more detailed analysis of each of the components in the high-level conceptualization was undertaken.

2.3.2.1 Water
In the water sector, three main subsectors, i.e., water availability, water use, and water quality, have been identified in the SRB. The links between them and the other nexus sectors are shown in Fig. 6.3. In the basin, not everyone has access to clean water, and ways of enhancing supply are being sought. In the water availability subsector, access to basic water requirements as well as agricultural supply is limited. As a consequence, basic WHO health, e.g., prevention measures recommended during the COVID-19 pandemic may not be met. Blue and green water have been explicitly represented to emphasize the importance

CHAPTER 6: Transboundary WEF nexus analysis: a case study of the Songwe River Basin

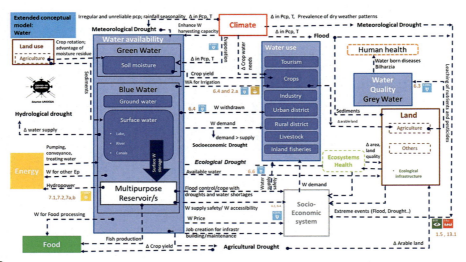

FIGURE 6.3
Extended conceptual model for the water sector in the Songwe River Basin. The main tentative SDGs and SDG targets (in orange) are shown. *SDG*, Sustainable Development Goal.

of assessing soil moisture in the agricultural production. Groundwater and surface water are represented in blue water (Fig. 6.3), highlighting the potential role of reservoir construction for increasing surface water storage and water availability. The natural variability of rainfall and its likely changes in the future result in imbalances between water supply and demand and have a knock-on effect on the other sectors, in particular on the land sector on which the food production depends. Indeed, the SRB currently relies mostly on rain-fed agriculture. Therefore, prolonged drought conditions lead to increase pressure on food security. These and the impacts of climate change on the SRB are intended to be mitigated through the implementation of multisectoral solutions (SRBP, 2019). The potential impact of meteorological, hydrological, agricultural, ecological, and socioeconomic drought is shown in Fig. 6.3. The qualitative analysis of the SRB system confirmed potential positive impact of the reservoir on water, energy, and food availability and access, irrigated land, food diversification, job creation, income generation, and ecosystem health (Fig. 6.3). The downside of the increasing land use and crop yield, as expected from the SRBDP, for growing food (cropland and wetlands in particular), is a likely negative impact on water quality (gray water, Fig. 6.3) due to the amount of nutrient and pesticides loads in the river. The role of small-scale fisheries in wetlands has been highlighted in the basin. Increasing these activities is one of the main Programme outcomes. Indeed, wetlands contribute to food security, income, and job creation in rural communities that live in these areas.

The research included the identification of relevant SDGs and the related targets that may be possible to address and compute in a future analysis of the basin (Fig. 6.3). From the number of SDGs, objectives, and indicators tentatively identified in the water sector is already possible to understand the important role of the nexus assessment in contributing to achieving the SDGs in the SRB. The main SDGs identified in the water sector are 1, 2, 6, 7, and 13. The tentative SDG targets are shown in Fig. 6.3 (in orange) (https://sdgs.un.org/goals).

2.3.2.2 Land and food

The land and the food sector are closely linked. The food sector is one of the most influential due to the high demand for water and energy. Food production influences all nexus sectors in the SRB, but in particular the land sector where both irrigated and rain-fed crops are cultivated. Most crops are currently cultivated in rain-fed, and the main crop is rice. Due to the increasing demand for rice, farmers have started to cultivate in wetlands (Kalisa et al., 2013). The cultivation of these lands is not controlled, and it is rapidly increasing due to the high pressure from food demand/crop production. These changes are having a considerable impact on socio-economic activities and the livelihood of local people (Gwambene, 2017). Income is generated from the expansion of agricultural land (for rice production in particular). However, the rapid land use change is undermining other sources of livelihood and is having a significant impact on biodiversity (Fig. 6.4) (Kalisa et al., 2013). For example, to increase rice production, permanent wet areas have been converted to arable land to ensure this cultivation. This turns out to be one of the main causes of the loss of four different fish species and the reduction in macrophytes (Kalisa et al., 2013).

Hunting, which is a means to ensure food security in the basin, is also greatly affected. The increasing population is having a considerable impact on the use of wetland resources and on the sustainability of the wetland ecosystem. In part of the study area, it has been reported that, due to resource overexploitation, natural vegetation was removed and permanent wet areas have disappeared. In the lower plain of the basin, it is noticed that over 95% of the land has been converted to cultivated area (Kalisa et al., 2013) to meet the increasing food demand. The intensive agriculture on available cropland is leading to soil fertility decline that, together with the highly variable climate and associated water supply, and an increasing need for arable land, is one of the main reasons for farmers' migration to wetlands. Indeed, people who are living in the wetlands have different sources of livelihoods including fishing, crop production and livestock keeping, and handcraft production (Kalisa et al., 2013). Wetlands are a source of income in particular for fisheries (Fig. 6.4). They contribute to improving the socioeconomic development in the basin by

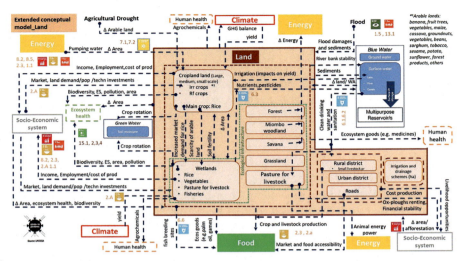

FIGURE 6.4

Extended conceptual model for the land sector in the Songwe River Basin. The main tentative SDGs and SDG targets (in orange) are shown. *SDG*, Sustainable Development Goal.

generating income and employment, thus enhancing the quality of life of the inhabitants of these areas. The future impact of climate change on crop yield is expected to exacerbate existing climate variability and threaten food security given the significant rain-fed production in the basin (Gwambene, 2017). From this perspective, the construction of the multipurpose reservoir, a key component of the SRBDP, is expected to significantly reduce this risk.

Relevant issues directly linked to food security are related to energy and water access and availability, and to ecosystem health, which are strongly linked to the sustained provision of ecosystem goods and services in the basin. The two irrigation schemes included in the SRBDP are expected to increase food production, as well as allow food diversification, resulting in a positive impact on human health and socioeconomic activities. However, it is important to consider that the increase of agricultural products may lead to the use of more chemicals (Gwambene, 2017) that can adversely influence human and ecosystem health (Fig. 6.4). The sustainable management of the SRB has been emphasized by the Programme and needs to be guaranteed. Flooding is the main cause of damages to the population living in the plain and to fertile land (Munthali et al., 2011). For that reason, flood mitigation is an important goal of the Programme. In this regard, the construction of the dam may be relevant to reduce the flood risk in the study area. In addition, this analysis highlighted the population whose access to food is threatened by a lack of services, such access to markets (Gwambene, 2017). The SRBDP also includes the improvement of roads, which could enhance food accessibility and production (Fig. 6.4).

2. Case study description   101

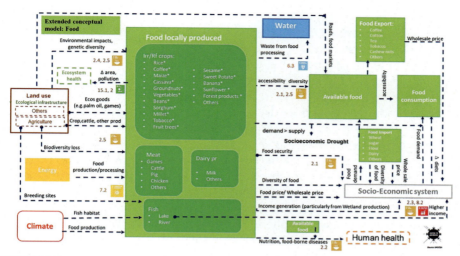

FIGURE 6.5
Extended conceptual model for the food sector in the Songwe River Basin. The main tentative SDGs and SDG targets (in orange) are shown. *SDG*, Sustainable Development Goal.

Fig. 6.5 shows the food balance in terms of food locally produced, imported, exported, and consumed food. The importance of increasing and ensuring food availability, accessibility, diversity, and price is considered. The increasing demand for food driven by the increasing population growth in the basin is having a considerable impact on the food balance. To meet the increasing local demand, more intensive crop production is needed. The rapid land use change in the basin is undermining biodiversity and threatening ecosystem health (Figs. 6.4 and 6.5). The local food production and food export are sources of income for the people who live in the basin (PO-RALG, 2019). The diversification of food produced and available/accessible for consumption is crucial for coping with nutrition issues and food-borne diseases (Fig. 6.5). Due to COVID-19, the already relevant difficulties related to food security are expected to potentially increase due to a disruption of food production and distribution (Figs. 6.4 and 6.5). The main tentative SDGs identified in the land sector are 1, 2, 6, 7, 8, 13, and 15 (https://sdgs.un.org/goals).

2.3.2.3 Energy
The analysis of the energy sector in the SRB highlighted the importance of available energy resources, available secondary energy, and energy use (NBS, 2016). Both available energy resources and available secondary energy have been divided into renewable and nonrenewable, and the use of both is also indicated. This is crucial to ensure sustainable development in the case study. This issue is also included in the SRBDP where the construction of dams is

directly linked to hydropower production, and thus to the possibility of increasing availability, access, and use of renewable energy. Energy is crucial in food production and processing. The use of fossil fuels has an impact on climate in terms of greenhouse gas emissions (Fig. 6.6). The available energy is mainly used in rural and urban districts, mining, and the agricultural sector, but not everybody has electricity access. The imported and exported available energy resources, as well as the cost of energy from various sources, which has an impact on the socioeconomic system, have been also accounted for. The availability of energy is expected to contribute to increasing human empowerment and services (like hospitals and so on) with an impact on socioeconomic development and human health. The need to build infrastructure to provide and increase access to affordable, reliable, and sustainable energy is indicated in the analysis. The link between firewood and land is relevant in this case study because, despite firewood being a source of income, its collection is also a main cause of deforestation. Health problems are caused by kerosene, coal, and oil smoke inhalation, and this is an important aspect in terms of impact on human health.

The impact of mining on water quality is addressed in terms of water pollution. Personnel and facilities shortages due to COVID-19 can lead to disrupted access to electricity, further weakening health system response and capacity (Fig. 6.6). The people that most suffer from energy access are located in rural districts; thus the rural electrification project included in the SRBDP will be beneficial to improve the socioeconomic system in these areas. SDGs 2, 6, and 7 have been shown in Fig. 6.6 (https://sdgs.un.org/goals).

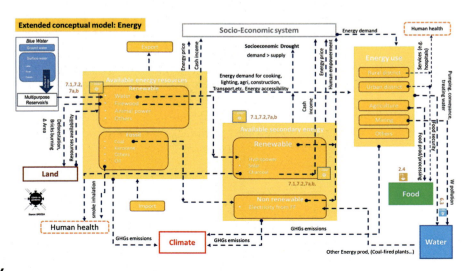

FIGURE 6.6

Extended conceptual model for the energy sector in the Songwe River Basin. The main tentative SDGs and SDG targets (in orange) are shown. *SDGs*, Sustainable Development Goals.

2.3.2.4 Climate

Climate variables such as precipitation, wind, and temperature were assessed to identify the impact of increasing greenhouse gas emissions, particularly in the land and energy sectors (Fig. 6.7). Climate change is expected to amplify the natural variability of the climate and further compromise the availability and timing of water in the basin, and thus the available water for domestic use, crop productions, and hydropower production. Changes in the frequency and intensity of extreme events are expected with a significant impact on the socioeconomic system (Gwambene, 2017). The role of the reservoir in reducing the risk of flooding may be crucial to mitigate damage to people, villages, and food production. Investments in increasing water storage are key to ensuring water, food, and energy security in the basin. The conservation and restoration of ecological infrastructure is crucial for coping with current climate variability as well as climate change. SDGs 1 and 13 have been tentatively indicated as the main goals that can be addressed through the nexus analysis of the SRB (https://sdgs.un.org/goals).

2.3.2.5 Socioeconomic system

The main sectors that characterize the socioeconomic system of the study area include agriculture, fisheries, and tourism. Increasing population growth and the changes in food demand and diet are having an impact on the food sector and therefore in the land sector (Fig. 6.8). Investments in technologies can

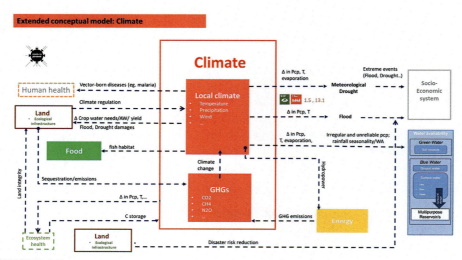

FIGURE 6.7

Extended conceptual model for the climate sector in the Songwe River Basin. The main tentative SDGs and SDG targets (in orange) are shown. *SDGs*, Sustainable Development Goals.

104 CHAPTER 6: Transboundary WEF nexus analysis: a case study of the Songwe River Basin

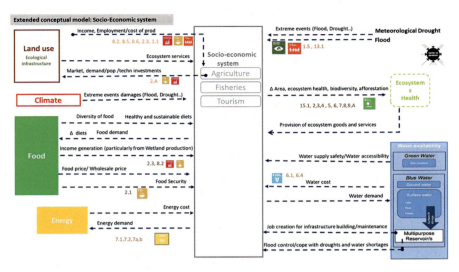

FIGURE 6.8
Extended conceptual model for the socioeconomic sector in the Songwe River Basin. The main tentative SDGs and SDG targets (in orange) are shown. *SDG*, Sustainable Development Goal.

help to ensure water, food, and energy security. Technology is also expected to increase sustainable resource management in the basin (e.g., monitoring systems). Higher income is expected from increasing food production, in particular from the increasing productions in wetlands, with a beneficial effect on the socioeconomic system, but a detrimental impact on ecosystem. Changes in food, water, and energy prices due to changes in production and consumption are also expected. Increasing water, energy, and food demand is expected to have a negative impact on ecosystem health, which in turn is fundamental to provide goods and services. Extreme events are expected to change in intensity and frequency and to impact the socioeconomic system causing damages to built-up areas and cultivated land. The SRBDP aims at reducing the impact of drought and floods in the basin. The links between these sectors and all the others identified in the basin are particularly evident from the number of SDGs shown in Fig. 6.8. The tentative SDGs indicated are 1, 8, 13, and 15 (https://sdgs.un.org/goals).

2.3.2.6 Ecosystem and human health

The ecosystem health (Fig. 6.9) in the SRB is threatened in particular by water availability and land use changes. Degradation of the landscape and climate change are threatening ecosystem health, and concerns of an ecological drought, which would impact negatively on the socioeconomic system, have been raised. Increasing agricultural activities have an impact in terms of pollution, loss of biodiversity, provision of ecosystem goods and services, land

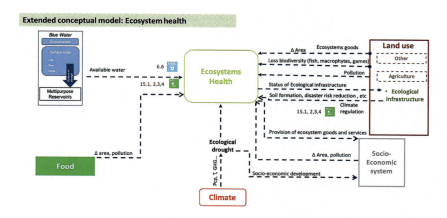

FIGURE 6.9
Extended conceptual model for the ecosystem sector in the Songwe River Basin. The main tentative SDGs and SDG targets (in orange) are shown. *SDG*, Sustainable Development Goal.

degradation, and soil fertility (Gwambene, 2017). The status of ecological infrastructure is crucial to ensure ecosystem health, so actions to preserve it are fundamental.

Human health is threatened by activities in all nexus sectors (Fig. 6.10). Actions to improve availability and accessibility to water, food, and energy are crucial to

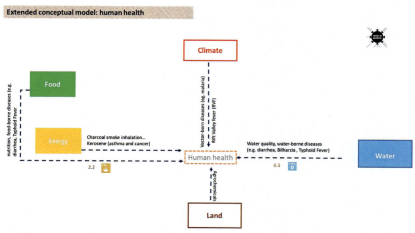

FIGURE 6.10
Extended conceptual model for the human health sector in the Songwe River Basin. The main tentative SDGs and SDG targets (in orange) are shown. *SDG*, Sustainable Development Goal.

ensure WEF security, as well as maintain human health. Improving the sustainable resources use in the nexus sectors will contribute to reduce the risk of diseases. Technology and knowledge have a key role in this sector.

The ecosystem and human health system have been analyzed separately because their importance in the basin is crucial. The SDGs, the objectives, and the indicators shown in Fig. 6.9 are already identified in the land sector (Fig. 6.4), while SDGs 2 and 6 come into focus in the human health system (Fig. 6.10) (https://sdgs.un.org/goals).

3. Conclusions

The approach outlined in this chapter aims at supporting sustainable socioeconomic development. The analysis carried out focuses on the assessment of the WEF nexus in the SRB located on the border between Malawi and Tanzania. Reducing poverty, improving human health and livelihoods, ensuring water, food, and energy security, mitigating natural climate variability and associated floods and droughts, and enhancing sustainable river basin management are the main challenges recognized by the SRBDP jointly developed by the governments of both countries. The construction of a multipurpose reservoir is a key objective of the SRBDP. The reservoir will supply water for 180 MW hydropower plant, 3000 ha of irrigation schemes in each country, and control floods in the lower part of the basin. The assessment of SRBDP's expected outcomes will be carried out by applying an approach that starts from conceptual mapping of the SRB nexus system to the development of quantitative tools such as SDMs, and identification of suitable indicators for the assessment of different scenarios and management strategies, subsequently providing decision-makers with feasible development pathways. This research is currently ongoing within the WEF Nexus Toolkit (WEF-Tools) project (https://wef-tools.un-ihe.org, 2020–23).

In this chapter, the qualitative nexus analysis of the SRB is applied to illustrate the main sectors and subsectors involved in the SRBDP and to identify the main interlinkages between them. The analysis showed how a potential decision made in a sector may have an influence on multiple sectors. The qualitative assessment can help to understand where there can be synergies and trade-offs, and thus to work on strategies that enhance the former and reduce/avoid the latter. The cooperation between Malawi and Tanzania is critical to successfully achieve the outcomes of the Programme expected in the basin by the two countries and to guarantee the sustainable development of the case study. In this regard, the first attempt to identify the potential SDGs, objectives, and indicators that may be addressed in the basin through the nexus analysis has been made (Figs. 6.2–6.10). This highlighted the importance of the application of the holistic approach on the SRBDP to enhance and boost the achievement of SDGs in the basin. The next step, i.e., a quantitative analysis depends on the data that are possible to collect. The application of the WEF nexus approach for analysis of the SRB will provide structured knowledge base, tools, dashboard, and a composite nexus index. The approach considers the role of local experts and stakeholders as essential for the achievement of the final results of each step which are codeveloped, tested, validated, and refined with interactive collaboration. The SRB assessment is intended to support decision-making and, therefore, the development of short-, medium-, and long-term strategies for sustainable integrated resource management in the basin itself and in others with similar characteristics. The outcomes will provide a means to assess progress toward relevant SDGs, in particular SDGs 2, 6, and 7.

References

FAO, 2018. Policy Brief #9 Water-Energy-Food Nexus for the Review of SDG 7. https://sustainabledevelopment.un.org/content/documents/17483PB_9_Draft.pdf.

FAO, 2017. The Future of Food and Agriculture - Trends and Challenges. Rome. http://www.fao.org/3/i6583e/i6583e.pdf.

Ford, A., 2010. Modeling the Environment, second ed. Island Press, Washington, D.C.

Gwambene, B., 2017. Potential corollaries of land degradation on rural livelihoods in upper Songwe transboundary river catchment, Tanzania. J. Agric. Ext. Rural Dev. 3 (1), 139–148.

GWP, 2019. Promoting the Water, Energy and Food Nexus Approach and Youth Empowerment for Sustainable Development. Pretoria, South Africa. https://www.gwp.org/globalassets/global/gwp-saf-images/sadc-giz-twm/9th-dialogue-technical-background–paper_final.pdf.

Kalisa, D., Majule, A., Lyimo, J.G., 2013. Role of wetlands resource utilisation on community livelihoods: the case of Songwe River Basin, Tanzania. Academic J. 8 (49), 6457–6467.

Munthali, K.G., Irvine, B.J., Murayama, Y., 2011. Reservoir sedimentation and flood control: using a geographical information system to estimate sediment yield of the Songwe river watershed in Malawi. Sustainability 3, 254–269.

NBS, 2016. Basic Demographic and Socio-Economic Profile. 2012 Population and Housing Census. Mbeya Region.

OECD, 2012. OEsCD Environmental Outlook to 2050: The Consequences of Inaction Key Facts and Figures. www.oecd.org/env/indicators-modelling-outlooks/49910023.pdf.

PO-RALG President's Office - Regional Administration and Local Government, 2019. Songwe Region Investment Guide. http://www.songwe.go.tz/storage/app/uploads/public/5e3/a61/249/5e3a612499563950020443.pdf.

SADC and GWP, 2019. Fostering Water, Energy and Food Security Nexus Dialogue and Multi-Sector Investment in the SADC Region. www.gwp.org/globalassets/global/gwp-saf-images/nexus/sadc-wef-nexus-project_v5-1.pdf.

SADC, 2016. SADC, Regional Strategic Action Plan on Integrated Water Resources Development and Management Phase IV, RSAP IV, Gaborone, Botswana. www.sadc.int/files/9914/6823/9107/SADC_Water_4th_Regional_Strategic_Action_Plan_English_version.pdf.

SIWI, 2019. Invitation to Tender: Consultancy Services to Conduct an Agribusiness Case Feasibility Study in the Songwe River Basin Development Programme, Both in the Designated Areas of Malawi and Tanzania. https://www.siwi.org/wp-content/uploads/2019/03/Invitation-to-tender-Songwe-Agri_Consultant-15.03.2019.pdf?.

SRBDP, 2019. Tanzania/Malawi: Strengthening Transboundary Cooperation and Integrated Natural Resources Management in the Songwe River Basin. African Development Bank Group. www.afdb.org/en/documents/document/multinational-strengthening-transboundary-cooperation-and-integrated-natural-resources-management-in-the-songwe-river-basin-project-summary-109895.

SRBDP, 2018. Presentation on the Status of the Songwe River Basin Development Programme to Districts Prior to the Project Preparation Mission by AfDB. http://www.ilejedc.go.tz/storage/app/uploads/public/5b2/fe2/754/5b2fe275431d2616024163.pdf.

Susnik, J., Chew, C., Domingo, X., Mereu, S., Trabucco, A., Evans, B., Vamvakeridou-Lyroudia, L.S., Savic, D.A., Laspidou, C., Brouwer, F., 2018. Multi-stakeholder development of a serious game to explore the water-energy-food-land-climate nexus: the SIM4NEXUS approach. In: Water (S.I on Understanding Game-based Approaches for Improving Sustainable Water Governance: The Potential of Serious Games to Solve Water Problems), 10, p. 139.

Sušnik, J., Masia, S., Indriksone, D., Bremere, I., Polman, N., Levin-Koopman, J., Linderhof, V., Blicharska, M., Teutschbein, C., Mereu, S., Trabucco, A., Blanco, M., Martinez, P., Avgerinopoulos, G., Henke, H., Laspidou, C., Mellios, N., Ioannou, A., Kofinas, D., Papadopoulou, C.-A., Papadopoulou, M., Conradt, T., Bodirsky, B., Pokorný, J., Hesslerová, P., Kravčík, M., Griffey, M., Ward, B., Vamvakeridou-Lyroudia, L.S., Evans, B., Hole, N., Khoury, M., Petersen, C., Fournier, M., Janse, J., Kram, T., Doelman, J., 2020. D3.6 Complexity Science Models Implemented for All the Case Studies. SIM4NEXUS Deliverable 3.6. https://www.sim4nexus.eu/userfiles/Deliverable_D3.6.pdf.

Vamvakeridou-Lyroudia, L.S., Sušnik, J., Masia, S., Indriksone, D., Bremere, I., Polman, N., Levin-Koopman, J., Linderhof, V., Blicharska, M., Teutschbein, C., Mereu, S., Trabucco, A., Castro, B., Martinez, P., Blanco, M., Avgerinopoulos, G., Laspidou, C., Mellios, N., Ioannou, A., Kofinas, D., Papadopoulou, C.-A., Papadopoulou, M., Conradt, T., Bodirsky, B., Pokorný, J., Hesslerová, P., Kravčík, M., Griffey, M., Ward, B., Evans, B., Hole, N., Khoury, M., Petersen, C., Fournier, M., Janse, J., Doelman, J., 2019. D3.4: Final Report on the Complexity Science and Integration Methodologies. SIM4NEXUS Deliverable 3.4. www.sim4nexus.eu/page.php?wert=Deliverables.

Further reading

FAO, 2012. World Agriculture towards 2030/2050: The 2012 Revision. http://www.fao.org/fileadmin/user_upload/esag/docs/AT2050_revision_summary.pdf.

IUCN ROWA, 2019. Nexus Comprehensive Methodological Framework: The MENA Region Initiative as a Model of Nexus Approach and Renewable Energy Technologies (MINARET). IUCN, Amman, Jordan. www.iucn.org/sites/dev/files/content/documents/water_publication_2019_1.pdf.

Simpson, G., Jewitt, G., Becker, W., Badenhorst, J., Neves, A., Rovira, P., Pascual, V., 2020. The Water-Energy-Food Nexus Index: A Tool for Integrated Resource Management and Sustainable Development.

WRC, 2020. Development of Water-Energy-Food Nexus Index and its Application to South Africa and the Southern African Development Community. Water Research Commission Report No. 2959/1/19. WRC, Pretoria.

Websites

CRIDF. http://cridf.net/project-pipeline/songwe-river-basin-development-programme/.
SIM4NEXUS. https://www.sim4nexus.eu/.
UNDESA-DSDG. https://sdgs.un.org/goals.
WEF Nexus Index: wefnexusindex.org.
WEF-Tools. https://wef-tools.un-ihe.org.

CHAPTER 7

Applying the WEF nexus at a local level: a focus on catchment level

S. Walker[1,3], I. Jacobs-Mata[2], B. Fakudze[2], M.O. Phahlane[1] and N. Masekwana[1]

[1]*Agricultural Research Council — Natural Resources & Engineering, Pretoria, Gauteng, South Africa;* [2]*International Water Management Institute, Pretoria, Gauteng, South Africa;* [3]*Department of Soil, Crop & Climate Sciences, University of the Free State, Bloemfontein, Free State, South Africa*

1. Introduction

Global threats such as population growth, climate change, and increasing urbanization present a huge challenge to water managers in the future. However, solutions need to be found at a local level to ensure the viability and stability of local communities (Amey, 2010) by implementing local water interventions and actions. Societal megatrends, coupled with environmental, technological, economic, and demographic changes, continue to exert pressure on already-depleted natural resources, thus threatening their sustainability and diminishing resilience of communities, as well as delaying the achievement of the 2030 global Sustainable Development Goals (SDGs) (https://sdgs.un.org/goals; Gelsdorf, 2010). Livelihoods can continually be improved if a focus shifts to service provision, thus ensuring the ongoing reliability, resilience, and sustainability of resources (Cervigni et al., 2015). An integrated water–energy–food–agriculture management approach addresses these challenges using a system-based, transdisciplinary, and transformative approach to resource management, development, and utilization. It allows for inclusive and equitable development and coordinated resource planning and management (Nhamo et al., 2018). As climate change is cross-sectoral and multidimensional, managing these sectors using a nexus approach is seen as an important tool offering cross-sectoral mitigation and adaptation opportunities to harmonize interventions and build resilience in communities (Conway et al., 2015; Nhamo et al., 2019a).

The "water–energy–food" (WEF) nexus is a current philosophy, framework, or way of thinking about such problems from a systems perspective to address the integration of arenas addressed in sustainable development. Several reviews of available nexus tools and frameworks have already been conducted (Chang et al., 2016; Endo et al., 2017; Kurian, 2017; Albrecht et al., 2018). The vast majority of WEF nexus models and tools are technical, and few are in a format that is easily understood by the proposed end users whom they were supposedly designed for. Therefore, attention needs to be given to interpreting the WEF nexus tool output from the technical language into more easily understandable terminology. Such answers will be found in developing facilitative processes used in this project, which accompany the communication and translation of specific WEF nexus tools to a range of decision-makers and stakeholders (Fakudze et al., 2021). As Daher et al. (2017) noted, decision-makers differ in scope and capacity-making decisions at the small association, local, regional, national, or international levels, with different interests and complexities of their critical questions. The challenge of modeling the WEF nexus is to provide a clear, simple, yet comprehensive way of unpacking the interdependencies and trade-offs. An accompanying facilitative method supports the decision-making process without removing decision-makers autonomy by enabling the decision-maker to ask different questions, considering the nonlinear trajectory that evidence takes in informing policy while grappling with power asymmetries and other sociopolitical dynamics that are not able to be modeled (Fakudze et al., 2021). In essence, WEF nexus models allow trade-offs to be presented to the decision-makers who prioritize them and make choices based on simplified results (Daher et al., 2017). This accompanying facilitative process allows the decision-maker to grapple with their local social complexities that are not easily simplified.

The WEF nexus approach has been applied at the national and regional levels across southern Africa (Nhamo et al., 2019a,b). Similar results are summarized for a range of countries on the recently developed WEF Nexus Index webpage (www.wefnexusindex.org; Simpson and Jewitt, 2019). For example, for South Africa, it shows that using the globally available indicators, the WEF Nexus Index value is 56.2. So at a national level, South Africa has a water pillar value of 55.3, with 59.1 for the energy pillar and 54.1 for the food pillar. However, these same indicators cannot be used at a catchment level, as an individual catchment in a country can be self-sufficient in water or energy or food but has inter-catchment transfers for each of these commodities. So although water resources in South Africa are key to livelihoods and development, water scarcity limits socioeconomic development in semiarid regions, while another factor, energy, is vital for economic development (Zhang et al., 2019). Therefore, linkages across these three sectors, namely water, energy and agri-food, should be considered when planning interventions or development at various levels, namely regional, national, provincial, catchment, and community levels. Although the WEF nexus has been documented at a larger scale, there are few studies at the catchment or community level in southern Africa (Mabhaudhi et al., 2016, 2018). The WEF nexus offers significant opportunities for coordinated approaches to increase resilience in the future, as Mpandeli et al. (2015) recommended using the WEF nexus approach to alleviate poverty, improve livelihoods, and increase economic development together with job creation at a country level. However, this needs to begin at a catchment, municipal, and community level to be effective. The application of the WEF nexus at these levels is addressed in this chapter for the selected catchments of the Crocodile River and Lower Komati catchment in Mpumalanga as part of the Inkomati-Usuthu catchment management area.

The South African National Development Plan (NDP) intends to increase agricultural land by expanding areas under sustainable land management and reliable water control systems (including rural infrastructure and market access) to increase food supply and reduce hunger. The WEF nexus approach can be applied to this important decision-making process to achieve these NDP targets. A WEF decision support tool would assist in assessing the optimal combination or balance of resource allocation to reach these goals. Such an approach can protect the vulnerable communities, landscapes, and biodiversity from degradation, as WEF nexus analytical tools analyze complex, interrelated resource systems while providing tools to manage resources in a cohesive manner (Nhamo et al., 2019b). This provides recommendations for innovative policies concerning linkages between the water, food, and energy sectors while ensuring livelihood improvements and sustainable use of resources for human well-being. Subsequently, well-outlined evidence-based policies have the potential to improve resilience to natural disasters and extreme events.

At present, South African water utilization and conservation policies are formulated by individual sectors, namely agriculture, domestic/municipal use, industry, recreational, and ecotourism. However, as there are many conflicting demands for the limited water resources, there needs to be a negotiated balance between the demands and benefits of each sector. This type of WEF nexus approach allows for comparative studies of quantitative relationships across sectors, enabling one to account for cross-sectoral synergies and trade-offs using specially developed tools and indices (Nhamo et al., 2019b). Various WEF nexus tools have been developed worldwide for various users, at different levels and for different purposes (Dai et al., 2018). However, one of the main stumbling blocks is the incompatibility, inaccessibility, and unavailability of data (McCarl et al., 2017a,b), together with limitations in data sharing and the cost thereof, with the inconsistency of time and spatial scales across the selected area (Cash et al., 2006; Bhaduri et al., 2015). An important consideration worth noting is that a rationalist scientific endeavor assumes that these models show an objective truth. However, the politicization of scientific evidence shows that the use of science can be politically driven and is not necessarily objective and that power is an important influence in determining which knowledge will be considered legitimate and which not, and that data is never neutral.

Therefore, the imperfect and nonlinear process of communication and translation of WEF nexus information must be acknowledged when communicating nexus scenarios to both intended and unintended audiences.

Livelihood is the ability to access the basic needs in life, including food, water, energy, and clothing (Krantz, 2001). Therefore, the WEF nexus approach integrates these aspects. The livelihood of all South Africans is dependent on water resource use not only for domestic purposes, agriculture, and/or mining industries but also to provide salient services such as ecosystem services (Conway et al., 2015, 2019). The widely used sustainable rural livelihoods framework approach (Carney, 2003) emphasizes how people use their assets (natural, physical, social, human, and financial) to maintain viable livelihoods with positive outcomes. A detailed analysis of factors that influence water, energy, and food security is conducted at a local or community level in a livelihood approach. Since livelihood approaches capture the processes and contextual factors that shape adaptive capacity, the WEF nexus analytical livelihoods framework (ALF) assists in integrating effects across three resources (Mabhaudhi et al., 2019a,b). Such a framework should be evaluated at a local level to incorporate findings into the management and evaluation systems at a municipal level.

Therefore, a need exists for developing stakeholder-centric WEF analytical tools for a particular audience of stakeholders with unique needs. This must be accompanied by developing appropriate facilitative processes that allow for communication, knowledge transfer, and uptake of such tools and approaches by users. This will enable them to achieve their intended purpose of either advancing our understanding, informing planning processes, policy development, and/or helping to facilitate decision-making at an operational level. In this chapter, the components of a WEF nexus are identified together with water stressors in each sector represented in the Crocodile River catchment. The WEF framework approach will be applied at a catchment level using representative quantifiable indices to characterize each sector, thus taking into account the nexus interlinkages between the WEF parameters and how it can be used to influence development decisions relating to maintaining rural livelihoods.

2. Methodology and data

A comprehensive review of the available literature on WEF nexus models, frameworks, and tools was conducted. A general Internet search was done on databases such as Web of Science, Scopus, Research Gate, and Google Scholar. The search was limited to papers published in English, but no limitation was put on the year of publication nor geographical distribution. Following the identification of articles about WEF nexus models, further criteria were used to narrow the search for those suitable for the current local catchment level application. These criteria include whether the framework/tool can represent each factor (food/agriculture, water, and energy) in rural and urban areas and at similar levels, time and spatial scales, and whether the necessary data are available. It must have parameters relevant to the stakeholders and describe the physical and social systems within the catchment. Finally, there must be pertinent output parameters that apply to the catchment decision-making process and a means of clearly visualizing the outputs for ease of communication to stakeholders. A comparison table of WEF nexus tools was compiled, showing aspects such as input data requirements, expected output for different sectors and managers, and format and usability of the availability of the output of the models' source code. In this way, the choice of suitable tools was reduced to a manageable number that could be used for specific applications in the Crocodile River catchment.

This project addresses the application of the WEF nexus framework to the Inkomati-Usuthu catchment to address the decision-making around water allocation between users and stakeholders by the Inkomati-Usuthu Catchment Management Agency (IUCMA). The project is funded by the South African Water Research Commission (C2019/2020-00326). The Inkomati-Usuthu catchment is situated in the eastern part of South Africa, mainly in the Mpumalanga province ranging from the highveld where the rivers start and stretch to lowveld and border with Mozambique and Eswatini (Fig. 7.1). The Inkomati-Usuthu Water Management Area comprises the following river basins: Sand River; Sabie River; Crocodile River (East); Komati and Lomati Rivers; and the Usuthu River (Fig. 7.1).

The main use of the water is for irrigated crops (31%), ecological reserve (23%), and forestry (21%) that all fall within the agriculture—food sector. The other sectors where the water demand is increasing include the domestic flows and industrial use that are both less than 5% (Fig. 7.2). Following the review of the available WEF nexus tools and their data requirements, the outputs provided will be assessed in relation to the aim of using the WEF framework to address water allocation in this diverse catchment. The selected framework needs to provide useful indicators that can be used by IUCMA in their decision-making, with the data available from a variety of sources. The descriptive parameters for each of the segments of both availability and accessibility of water, energy, and food—agriculture will be developed according to the specific activities within the catchment. This characterization of the particular activities in this catchment will be approached in a facilitative manner with the stakeholders during the focus group discussion meetings. Representatives from all three sectors will also be interviewed concerning their approach to decision-making and management of the water and energy requirements in their sector, together with the projected changes under future scenarios.

CHAPTER 7: Applying the WEF nexus at a local level: a focus on catchment level

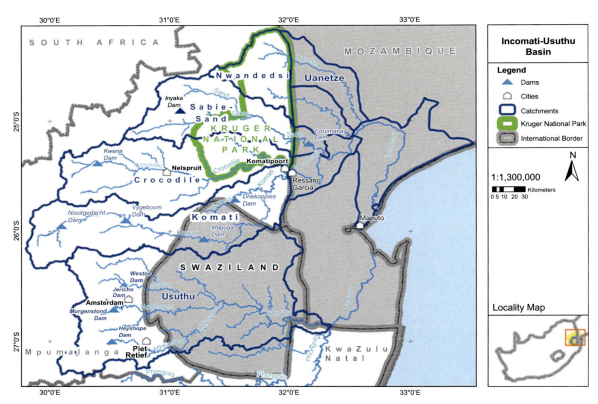

FIGURE 7.1

Map of Inkomati Catchment showing the subcatchments, location of cities, and wildlife area within Kruger National Park (IUCMA website https://www.iucma.co.za/).

Many WEF nexus parameters are closely linked to natural resources, so they play a vital role in the WEF nexus modeling activities. Natural resource data are probably the most readily available and retrievable from many reliable national archives. The scientific relationships between the climate, water, and agricultural parameters are well developed and well documented. For the food aspects, the amount of water needed for crop and livestock production and productivity efficiencies are known and available from numerous sources. However, only selected commodities that are a major part of the agri-industry in the selected catchment will be addressed.

Similarly, water used for energy production is documented particularly for the coal mining and thermoelectric steam—generated electricity in the highveld area of the catchment. As WEF nexus models consider a range of aspects of daily human life, it is important to have the correct information about the population in the selected catchment for this study. There are a wide range of

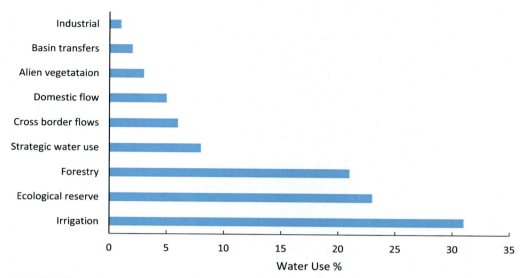

FIGURE 7.2
Distribution of water uses in the Inkomati-Usuthu Water Management Area, 2016 data (IUCMA, 2019).

population densities and socioeconomic situations across the catchment from rural to periurban to density urban areas, although there are few high-density areas. The availability of the necessary data, particularly with spatial distribution, will determine the settings and selection of the indices and inputs to the appropriate WEF nexus framework or model.

3. Progress with WEF nexus application at catchment level
3.1 WEF nexus available models

Over the past decade, various tools, approaches, and models have been developed to conceptualize, illustrate, and analyze the interdependencies within the WEF nexus. Due to this wide range of WEF nexus assessment tools and approaches already developed, researchers can evaluate what is available and select something fit for their purpose. There are several reviews of available nexus tools and/or frameworks (Chang et al., 2016; Endo et al., 2017, 2020; Kurian, 2017; Albrecht et al., 2018). Therefore, the use of such integrated tools and models, particularly for WEF nexus decision-making at the local level, is considered here, together with the processes of communicating these tools to decision-makers and elements needed to ensure successful uptake of said tools.

Dai et al. (2018) and Aboelnga et al. (2018) summarized the most widely published nexus tools and approaches by dividing them into three types of model, namely:

- Quantitative analysis models—quantify the resource flows but without modeling scenarios over temporal scales
- Simulation models—a single model for simulating scenarios over temporal scales
- Integrated models—a combined model with both quantitative and scenario functions

It also includes three different categories of application of the model results:

- Understanding and generating new knowledge: data promote a greater understanding of the nexus by demonstrating linkages and quantifying risks or opportunities
- Governing: with a purpose to guide an institutional or policy response
- Implementing: with the purpose to guide technical interventions and/or policy to improve efficiency or effectiveness or management of resources

Models with the targeted purpose of supporting governance or implementation processes are probably relevant to this Crocodile catchment. Several models in this category fall into a trap that their application area is either too vague or too broad for it to be useful to policy and decision-makers. Furthermore, many tools are not practical as they were developed to enhance our "understanding" (as application category) or to generate new knowledge with an emphasis on the quantitative analysis of the WEF nexus (Dai et al., 2018). Endo et al. (2017) agree, arguing that the nexus is more likely to be recognized at the research level (77 out of 137 organizations reviewed in their study) but is not fully acknowledged on the ground (out of 137 organizations, only 16 were from business and industry and only 20 were from civil society). So researchers often make the error of either trying to provide some sort of tool to policymakers that is not fit for their own purposes or claim that such tool originally used to improve the understanding can later be used to facilitate decision-making. Therefore, there is a need for the clear articulation of the purpose of a particular tool, its intended audience, and the development of appropriate facilitative processes that allow for the communication, knowledge transfer, and uptake of these tools and approaches by users. This will help the models achieve their intended purpose of helping to facilitate decision-making at the operational level and informing the planning processes and policy development at a catchment level.

For such a practical tool to be selected, it must meet the aims of this particular study. Criteria can be used, such as those defining the specific boundaries, including spatial and temporal scale needed, relevant stakeholders, description of the systems, and data availability. Finally, the output parameters and means of communication to relevant decision-makers are also important criteria (Daher et al., 2017). In addition, one needs to address uncertainties (e.g., in future scenarios) and possible new technological and resource development alternatives that have not previously been adopted in the region (McCarl et al., 2017a). As this particular project does not make provision for tools that are still in a developmental mode and cannot be directly applied, some of the more generic approaches have not been included despite being adopted by some WEF nexus researchers. Such approaches, including agent-based modeling, system dynamics modeling, life cycle assessment methods, and multiregion input—output models, have not been considered in this study.

Dai et al. (2018) and Aboelnga et al. (2018) made a comprehensive list of available models by dividing them according to the factors considered—some only cover two of the three WEF factors, namely water—energy or including environmental factors. Still, they do not meet all criteria in this study, so they are not included in Table 7.1 and will not be considered further. Some WEF tools include the additional climate and land factors influencing the WEF nexus. As that can be directly related to the food components, those are listed in Table 7.1 as they meet the selection criteria. It can be seen that many of the available methods are focusing on the governance and policy level compared with several generating new knowledge and understanding and relatively few that have been built for implementation. This emphasizes that many applications are still in the research and development phase and have not been used by practitioners for implementation or area decision-making. That is why a second step was necessary to shortlist some tools. A range of scales have been used for the various models—namely multiple scales, regional (including transboundary), national, and city scales. However, none are explicitly listed as being applied at catchment scale as in this project, which is surprising as the water is often the driving force as the most severely limiting factor. Then they should be considered within catchment boundaries. Perhaps, this is because most water management systems are based on the political and governance boundaries and not on the natural resource boundaries. The integrated models dominate the list, although a further useful division is provided in Table 7.2. In contrast to the lack of implementation methods, there are many quantitative methods, and only one simulation method is listed in Table 7.1.

3.2 Model selection and description

From this extensive list in Table 7.1, five tools or models were then selected that meet the criteria for this catchment level application or appear to be adapted to a catchment level application. This study is local, and decisions need information at a weekly and/or monthly or quarterly time scale. These selected WEF

Table 7.1 Summary of available methods to model the WEF nexus showing different sections for the combination of subcomponents.

Method	Geographical scale	Model type	Software	Purpose	Application category	Reference
Methods covering the water–energy–food nexus						
Water footprints—WEF nexus	Multiscales	Footprint method with indices	No	Ecological footprint, diet/food, virtual water trade, and water governance	Governing	Zhang et al. (2018), 2019, Chang et al. (2016), Yu et al. (2020), Hang et al. (2020), Li et al. (2019)
Biophysical and economical modeling of WEF nexus systems	Multiscales	Uses indices (physical and socioeconomic)	No	Use crop and economic models of land use, with water quality model	Knowledge generation	Giampietro et al. (2009), FAO (2014), Daher and Mohtar (2015), Daher et al. (2017)
Analytical livelihoods framework—ALF	Multiscales	Integrated model	No	Trade-off analysis, integrated use	Knowledge generation	Mabhaudhi et al. (2019b), Nhamo et al. (2020a,b)
ZeroNet DSS—decision supporting system	Regional level	Integrated model	Several free software	Decision support in resource management in the basin	Governing	Rich et al. (2005)
Nexus Assessment 1.0 FAO	Regional and national levels	Quantitative analysis model	Online rapid appraisal tool	Qualitative and quantitative assessment of nexus	Governing	FAO, 2014 and 2020
IAD-NAS—Institutional Analysis and Development Framework with value chain analysis	National level, periurban	Quantitative analysis model	No software	Impacts of institutions and policies on the sustainability of water, food, and energy	Governing	Villamayor-Tomas et al. (2015)
WEF Nexus Tool 2.0	National level	Simulation model	Online tool	Quantitative assessment and forecast of WEF nexus	Governing	Daher and Mohtar (2015)
DEA—Data Envelopment Analysis	Multiscales	Quantitative analysis model	No software	Evaluate regional input–output efficiency of resources holistically	Understanding	
WEFO—Water, Energy, and Food security nexus Optimization model	Multiscales	Integrated model	WEFO tool	Quantitatively assess the interconnections and trade-offs among resource systems as well as environmental effects	Governing	Zhang and Vesselinov (2017)

Basic Linked System (BLS) model	Multiscales	Integrated model		A world food system model developed by the International Institute for Applied Systems Analysis (IIASA)	Understanding	
Urban agriculture (rain harvesting)	City scale	Quantitative analysis model	No	Apply economic analyses and modeling	Knowledge generation	
WWEF Nexus Framework	National scale	Quantitative analysis model	No	Define and quantify interconnectivity between water–energy–food	Governing	

Methods covering the water–energy–food nexus/water–energy–land–climate nexus

CLEWs—climate, land, energy, and water	Multiscale	Integrated model	open-source tool OseMOSYS	Assess climate impacts on resources and supply help in policies evaluation	Implementing	Hermann et al. (2011)
TRBNA—Transboundary River Basin Nexus Approach	Transboundary	Integrated model	UNECE, NS	Assess the WEF + Ecosystem Nexus in transboundary river basins	Implementing	
Nexus Trade-off Assessment Tool—SWAN	Multiscales	Integrated model	World Wind visualization technology	Agricultural water visualization platform, focusing on irrigation of sugarcane in Eswatini	Understanding	
MSA—multisectoral systems analysis	City level	Quantitative analysis model	MATLAB tool	Understand resource flows as well as human effects on the urban metabolism	Understanding	
GCAM-USA—Global Change Assessment Model	Regional level	Integrated model	Open-source tool	Long-term analysis of water withdrawal and demand in the electricity sector of US states	Governing	

Continued

Table 7.1 Summary of available methods to model the WEF nexus showing different sections for the combination of subcomponents. *Continued*

Method	Geographical scale	Model type	Software	Purpose	Application category	Reference
PRIMA—Platform for Regional Integrated Modeling and Analysis	Regional and national levels	Integrated model	Velo	Simulate the interactions among climate, energy, water, and land at the decision-relevant spatial scale	Implementing	
MuSIASEM—Multiscale Integrated Assessment of Society and Ecosystem Metabolism	Regional and national levels	Integrated model	FAO, free online tool	Assess metabolic pattern of energy, food, and water-related to socioeconomic and ecological variables	Governing	Giampietro et al. (2009)
Foreseer	National and transboundary	Integrated model	University of Cambridge; online tool	Map flows of water, energy, land use, and GHGs	Understanding	
Modified SWAT—Soil and Water Assessment Tool	Transboundary	Integrated model	Open-source model	Water provisioning to each economic sector in a transboundary context	Understanding	El-Nasr et al. (2005), Arnold et al. (1998)

Adapted from Dai, J., Wu, S., Han, G., Weinberg, J., Xie, X., Wu, X., Song, X., Jia, B., Xue, W., Yang, Q., 2018. Water-energy nexus: a review of methods and tools for macro-assessment. Appl. Energy 210, 393–408 and Aboelnga, H.T., Khalifa, M., McNamara, I., Sycz, J., 2018. Water-Energy-Food Nexus Literature Review. A Review of Nexus Literature and Ongoing Nexus Initiatives for Policymakers. Bonn: Nexus Regional Dialogue Programme (NRD) and German Development Agency for International Cooperation (GIZ).

Table 7.2 Summary of quantitative WEF tools and models including scale and application method.

Tool	Scale		Inputs			Application method	Output			Users	Authors	Other article
	Time and spatial		Water	Food	Energy		Water	Food	Energy			

Integrative analytical model

| Analytical livelihoods framework | National | Annual, snapshot | Proportion of available freshwater resources per capita; proportion of crops produced per unit of water used | % moderate or severe food insecurity in population; proportion of sustainable agricultural production per unit area | Proportion of population with access to electricity; energy intensity, primary energy, and GDP | Process and multicriteria decision-making and analytic hierarchy to quantify interconnectivity resources with technical and economic parameters | Composite indices for each of the indicators | Composite indices for each of the indicators | Composite indices for each of the indicators | Government at national and regional levels | Nhamo et al. (2020a,b) | |

WEF resources

| Climate, land-use, energy, and water | Catchment, municipal; farm | Annual Decade | Water withdrawal at different scales municipal, catchment and farm Water demand Annual rainfall | Production costs fertilizer, machinima Food consumption Municipal GDP Population | Power generation and refining Energy UV radiation | Technical and economic parameters of power plants, farming; machinery, water supply chain, desalination terminals; irrigation technologies, fertilizer; production | Water pumped; different at scales Rainfall (mm) | Crop yield, GDP ratio, forestry | Energy for pumping catchment; farm and municipal; water use for energy production | Farmers Municipal/household Catchment management Nature conservation | Ferroukhi et al. (2015), Mpandeli et al. (2018) | Mannan et al. (2018), Howells et al., (2013), Engström et al. (2017), Sušnik et al. (2018) |

Continued

Table 7.2 Summary of quantitative WEF tools and models including scale and application method. *Continued*

Tool	Scale — Time and spatial	Inputs — Water	Inputs — Food	Inputs — Energy	Application method	Output — Water	Output — Food	Output — Energy	Users	Authors	Other article
Free access web-based tool											
WEF Nexus 2.0 Tool	Country, catchment, farm / Daily, Monthly, Annual	Identify different sources of water sources; groundwater withdrawal; distillation	Local food production levels versus import; technologies in agricultural crop production	Identify sources of energy for water, energy for agricultural production	Online tool allows the user to create different scenarios with varying food self-sufficiencies	Water use and water pumping requirements (m³)	Land requirements (ha); selected crop yield (ton)	Energy requirements (kJ); energy consumption through import (kJ); type of energy to suit a selected scenario	Import and export management; government officials Policy and regulation of export and import of agricultural products; Eskom	Daher and Mohtar (2015), Brouwer et al. (2018)	http://www.wefnexustool.org/user.php, Nhamo et al., 2019b, Sušnik et al. (2018)
Concept-based method											
WEF Nexus— Framework NexSym	Local to regional, national, or global / Daily, Monthly, Annual, Decadal	Surplus or excess sources of water and water resource hot spots	Agricultural resource allocation strategy choices	Energy allocation resources through the supply chain	Define and quantify interconnectivity resources, including integrative and holistic management strategies	Cost pumping and supply Purification	Food production yield per sector	Energy generation (KJ) Irrigation energy	Government at different levels (WEF); natural resource; food production; industries; Eskom	Martinez-Hernandez et al. (2017)	Mohtar and Daher (2016), Nhamo et al. (2020a,b)
Optimal management method											
MuSIASEM WEF nexus systems—crop and water quality and economic modeling of land use	Domestic, farm, industrial, municipal, catchment High risk Low risk / Day, Month, Year, Season	Water for drinking, domestic use; irrigation; industrial processes; rain	Land use; income (rent) and crops; water origin; labor costs	Energy flow in society	Biophysical, economical, and water quality and crop modeling, land use, and model Socioeconomic indicators and workforce evolution	L/day/person Cost pumping Energy used Turnover city $	Crop estimates yield Cost of production	Cost of pumping; energy used; turnover city different scales;	Household Local government City council Local government Water management agents SAParks Eskom Farmers	FAO (2014), Giampietro et al. (2009)	Daher and Mohtar (2015)

tools were developed and used over the past several years, and have also been applied at a variety of scales, both temporal and spatial scales (Table 7.2), ranging from local spatial scale to national, regional, and continental spatial scales accompanied by time scales from daily through to annual and decadal time scales for a variety of applications. The inputs and outputs needed for such models and tools will depend on both these scales of application. The outputs from such models can be used for day-to-day management decisions by various role players, medium-term monthly or quarterly planning activities, or tactical seasonal or long-term strategic planning for the catchment. Therefore, selecting a WEF tool or approach will depend upon the stakeholders' or clients' needs and requirements and the aims of the project. Some of the most useful WEF tools have been investigated. They will be described in detail in the following sections, summarizing the various aspects for comparison in Table 7.2.

3.2.1 Analytical livelihoods framework

ALF is a WEF nexus analytical tool that was developed by first defining indicators for each component—water, energy, and food—and then calculating the indicators into a composite score (Nhamo et al., 2020a,b). The water indicators describe water availability as the proportion of available freshwater resources per capita (m^3/capita) compared with the water productivity as the value of crops produced per unit of water used ($\$/m^3$). For the energy part, the indices are calculated from the proportion of the population with access to electricity compared with the productivity calculated from the energy intensity in terms of primary energy produced and GDP (MJ/GDP). For the food section of the calculation, one uses an indicator of self-sufficiency from the prevalence of moderate or severe food insecurity in the population (%) compared with the cereal productivity as sustainable agricultural production per unit land area (kg/ha). However, the indicator of cereal productivity is not applicable for this catchment. These indicators are then presented in the form of a spider graph to reveal the relative strengths and weaknesses, which can guide priority areas where interventions are needed to bring balance and cohesion across the sectors. As water and food together with an energy supply are vital to human livelihoods, the integration of these "access" and "availability" indicators is relevant to the area's livelihood status. This analytical framework enables one to evaluate synergies and trade-offs in resource planning and utilization in a way that other WEF tools had not achieved. However, this needs to be further developed to allow for input on governance issues, particularly cross-sectoral policy coordination (Rasul and Neupane, 2021), as missing links to policymaking, decision-making, lack of synergies, and trade-offs have already been identified. So, the ALF model has established quantitative relationships across the WEF sectors, thus simplifying the intricate interlinkages among resources by using relative indicators. It was previously applied at a national level for South

Africa on an annual basis as a case study (Nhamo et al., 2020a,b). Still, it needs to be adapted for other scales and time frames by developing suitable indicators at a smaller spatial local scale, such as a catchment level.

3.2.2 Water, energy, and food nexus tool 2.0

The WEF Nexus Tool 2.0 is a common platform that brings together scientific know-how and policy input to identify current and anticipated bottlenecks in resource allocation trends while highlighting possible trade-offs and opportunities to overcome resource stress challenges (Daher and Mohtar, 2015). The tool is scenario-based and attempts to explicitly quantify the interconnections between different resources while specifically capturing the effects of population growth, changing economies and policies, climate change, and other stresses. The WEF Nexus Tool 2.0 enables users to visualize and compare the resource requirements of their scenarios and calculate the "sustainability index" of each scenario (Ness et al., 2007). The model provides the user with the ability to create scenarios for a given country by defining the inputs as follows:

(a) Food portfolio: identifying local food production levels versus imports and technologies in agricultural production
(b) Water portfolio: identifying different sources of water and amounts needed
(c) Energy portfolio: identifying energy sources used for water and energy for agricultural production

Even though the WEF framework is generic and has previously been used at a national level, scenarios created by the tool are site-specific and defined by the local characteristics of the area of study and so can be applied at the catchment level. These may include local yields of food crops/products, water and energy availability and requirements, available technologies, and land requirements. The characteristics are defined by the user and allow for the creation of country-specific profiles with data selection being either daily, monthly, or annually. This tool identifies different sources of water needs, consumption, and withdrawal requirements from each source. In terms of food, it considers local food production (e.g., sugarcane) levels versus imported food and other agricultural crop production. The tool identifies energy sources used for water and energy used in agricultural production versus energy for other sectors such as industry. The tool interprets data and creates different scenarios with varying WEF self-sufficiencies. This assists in decision-making on land requirements (ha); selected crop yield (ton), energy requirements (kJ); energy consumption (kJ); type of energy for the selected scenario, as well as import and export management by government officials and other decision-making stakeholders such as the municipalities. The output results address policy and regulation of export/import of agricultural products and the energy regulators such as Eskom in the South African case. However, this may not ensure a proper trade-off between sectors as priorities and implementation strategies are different for government spheres.

3.2.3 Climate, land-use, energy, and water systems approach

The climate, land (food), energy, and water systems approach (CLEWs) (Hermann et al., 2011; Howells et al., 2011, 2013) focuses on assessing interlinkages between resource systems to understand how these are related to each other, where pressure points exist, and how to minimize trade-offs while enhancing and increasing the power of synergies (Ferroukhi et al., 2015; IRENA, 2015). This type of integrated assessment usually involves a strong quantification process, which can be performed in different scales of complexity, either by

(a) the use of accounting frameworks; or
(b) via the development of sectoral models (for water, energy, and land use) and subsequent soft-linking of tools in an iterative process; or
(c) using a single modeling tool that accounts for the representation of several cross- and intersystems interactions.

The models are then used to investigate questions related to the relevant nexus interactions for a specific location. The framework applies to different geographical scales, from global to regional (including transboundary), national, and urban levels (Strasser et al., 2016).

3.2.4 Multiscale Integrated Assessment of Society and Ecosystem Metabolism

Multi-Scale Integrated Assessment of Society and Ecosystem Metabolism (MuSIASEM) is an innovative approach using accounting to integrate quantitative information generated by distinct types of conventional models based on different dimensions and scales of analysis (Giampietro et al., 2009). MuSIASEM has proven extremely useful in the characterization of the metabolic pattern of social systems. In MuSIASEM, fund elements are those in the observed system that are transformative agents expressing the functions required by society. To bridge the socioeconomic and the ecological view, MuSIASEM simultaneously uses two complementing but nonequivalent definitions of fund elements, one relevant for socioeconomic analysis (human activity and power capacity/technology) and one relevant for ecological analysis (land uses/land covers, water funds), at all levels and scales considered (e.g., local crop field, watershed, the whole country). In this way, it provides an integrated characterization of society's metabolic pattern and its effect on the metabolism of the embedding ecosystems by combining nonequivalent systems of accounting. MuSIASEM strengths are in the analysis of processes inside the economy to integrate different scales (Gerber and Scheidel, 2018). Its key applications are diagnosis and/or simulation of economies across space and scales with a strong focus on biophysical dimensions of societies about their functional subsectors (Gerber and Scheidel, 2018). Similar methods are used in the circular economy with a multiscale framework at organizational levels of regional, sectoral, national, etc. (Pauliuk, 2018).

3.2.5 WEF nexus framework NexSym

This approach provides a software tool for technoecological simulation of local food—energy—water systems (Martinez-Hernandez et al., 2017). NexSym is a state-of-the-art tool using explicit dynamic modeling of local technoecological interactions relevant to WEF operations. It is a modular tool that integrates models for ecosystems, WEF production, and consumption components and allows the user to build, simulate, and analyze a "flowsheet" of a local system. The clarification of critical interactions enhances the knowledge and understanding of the interrelationships between the components, thus supporting innovative solutions by balancing resource supply and demand while increasing synergies between WEF components. NexSym allows for assessing the synergistic design and analysis of a local nexus system to be applied at a catchment level. NexSym has been particularly useful in exploring potential improvements and/or validating and analyzing specific optimization results within a wider context of a local system. Thus, being used as a tool for studying local systems, NexSym requires sufficient locale details to allow meaningful assessments to be carried out using local data sets. This provides decision-makers with a holistic approach to be better informed about the trade-offs and synergies between the various development and management options and to assist in identifying choices on how to manage and plan these resources in a sustainable manner Hoff (2011).

3.3 Data sources

WEF nexus models are used for different purposes or applications by different users, so a "nexus task or challenge level" is included, together with the geographic scale and model type information. Indeed, decisions based on nexus-wide considerations rather than individual elements are likely to produce better, if not more, informed outcomes (McCarl et al., 2017a,b). Yet, data covering the full nexus scope are needed to achieve and capitalize on a better understanding of the relationships among nexus elements. A further complication is that data need to reflect changes over time to be useful for decision-making. Desired types of nexus data include information across each of the three sectors and at specific levels and for geographic areas so they can be related to each other in a meaningful way.

Data collected for this project across the Inkomati-Usuthu catchment have been a continuous consultative process with stakeholders in all three WEF nexus sectors. Several private and public organizations were consulted to request access to their valuable data sets. Data already obtained from the agricultural sector are directly linked to macadamia nut and sugarcane production and processing.

3.3.1 The water sector

The data required about the water sector stretches across all the natural resources—land, water, climate, and natural vegetation. A large amount of water resources information is available at IUCMA (IUCMA, 2015, 2017, 2019), including daily rainfall, water supplied to agriculture, municipalities, and industry from 2000 to 2020. Catchment level water discharge data from major dams in the catchment (Nooitgedacht Dam, Vygeboom Dam, DaGama Dam, Witklip Dam, and Kwena Dam) are also available. IUCMA stores water allocation information for selected farms and industries for the period 2010—2020. These will all be used for the input into the models. The information from the South African government department dealing with water will also be included (DWA, 2013a,b).

As the WEF nexus parameters are closely linked to the natural resources, they will play a vital role in the WEF nexus modeling activities. However, they are probably the most readily available and retrievable from many reliable national archives. Some of the scientific relationships between the water and production parameters are well documented. For example, the amount of water needed for crop and livestock production and the production efficiencies are well known and available from numerous sources from the "food" perspective. For this, many of the past WRC project reports can be used to secure the necessary relationships from evidence-based research results and verified sources, where the amounts and efficiencies of the water used by a range of industries are available. There is the possibility of triangulating the results using calculations from different databases and using a range of algorithms and internationally accepted equations. Considering the light industry (like consumer-oriented manufacturing) within the catchment, this information should be readily available to calculate water use for manufacturing and the cultivation of all crops and forests grown in Mpumalanga. For the water used for the energy generation by Eskom, again, these values and historical information should be available and be linked to the flow in the rivers and the changes in the groundwater levels over the years.

Data will also be sourced from many South African government departments—such as Agriculture, Land Reform and Rural Development (DALRRD); Communications and Digital Technologies (DCDT); Cooperative Governance (DCG); Forestry, Fisheries and the Environment (DFFE); Health (DH); Human Settlements (DHS); Mineral Resources and Energy (DMRE); Planning, Monitoring and Evaluation (DPME); Public Works and Infrastructure (DPWI); Science, Technology and Innovation (DSTI); Social Development (DSD); Tourism (DT); Transport (DTr); Water and Sanitation (DWS); Government Communication and Information System (GCIS); and Statistics South Africa (StatsSA) or FAO water users database (2020). Other information will be sourced from the parastatals, such as land type maps, climate, soils, and geological information.

There are several repositories of climate information in RSA, namely South African Weather Services (SAWS); Agricultural Research Council—Natural Resources and Engineering (ARC-NRE); Council for Scientific and Industrial Research (CSIR); DWS; University of KwaZulu-Natal—Centre for Water Resources Research (UKZN-CWRR) that each have a climate data set across the selected catchment either as observations or integrated surfaces or in GIS format. The climate data from the ARC-NRE Agrometeorology Climate Data Bank include daily precipitation values, maximum and minimum temperature, evaporation, and total solar radiation values. There are 10 available stations varying in length from 10 to 95 years and located across the catchment. For example, the spatial distribution of evapotranspiration (mm/year) (Fig. 7.3) across the catchment can be used in a GIS system to generate the water demand by agricultural systems.

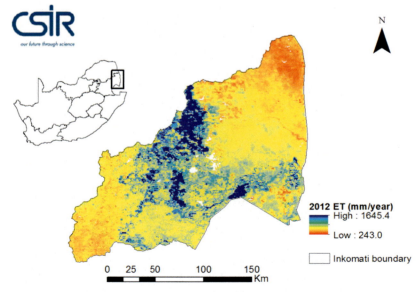

FIGURE 7.3
Spatial distribution of the annual evapotranspiration for the Inkomati-Usuthu catchment (https://csirwateruse.firebaseapp.com/inkomati).

3.3.2 The energy sector

The available energy sector data are from the three selected power stations, namely, Arnot, Hendrina, and Komati. Information related to power generation from 2000 to 2020 is captured. This information includes turbine water use per day (ML), monthly electricity production, and the total number of staff employed. Coal plays a major role in the energy sector in Mpumalanga and across South Africa. Water and employment information about coal mining companies supplying the three selected power stations has been captured. Information such as total monthly water use for washing and drilling during coal mining, total monthly electricity at the selected mines, and the total number of staff employed is captured.

As water is indispensable for the production, distribution, and use of energy, water use for energy generation must be documented. Chang et al. (2016) explain how the water footprint (WF) of power plants can be used to determine water needed to generate energy (m^3/GWh) by their thermal efficiency, their heat sink accessibility, and the cooling systems adopted. Electricity generation is defined as electricity generated from fossil fuels, nuclear power plants, hydropower plants (excluding pumped storage), geothermal systems, solar panels, biofuels, wind, etc. It includes electricity produced in electricity-only plants and combined heat and power plants. Both main activity producer and autoproducer plants are included, where data are available. Main activity producers generate electricity for sale to third parties as their primary activity (OECD, 2021). Energy production attracts a good deal of attention in current energy–water nexus research, mostly as water is used in coal and natural gas extraction and as a cooling liquid in thermal power generation (Chang et al., 2016), together with water for biomass cultivation and processing (Hoff, 2011). Extracting and refining fuel (coal, crude oil, natural gas) or transforming renewable energy sources (solar, hydropower, tidal energy, biocrops) are often water-intensive processes.

The database of water usage in energy production at three power stations in the catchment will include information about their capacity (megawatts), water use (ML/d), coal use (T/d), and the number of employees. Other details about the efficiency of the turbines and ramp rates will assist in the comparison of different energy generating systems. Throughout the catchment, the main energy source is Eskom providing electrical power, which primarily uses coal at the power stations situated in the highveld of Mpumalanga. These power stations are generally on the western border of IUCMA. Still, as the water is a shared resource, details of water use per kilowatt generated are included in this study. The amount of electric power coming into the catchment to meet the requirement is needed and included in the WEF nexus modeling exercise. The necessary information will be obtained from Eskom at the required scales (time and spatial scales), for some historical periods and a link for the ongoing future management activities. Some local industries also have private energy-generating plants at manufacturing factories or tourism operation sites. Therefore, they also need to be considered by collecting information about the amount and type of energy generated and its source. The required information from Eskom will need to include information on net power generation and cooling system information, as well as the grade of coal and the efficiencies of conversion to electricity, had data collected directly from facilities (USGS, 2020). Then one can develop the power plant–specific estimates of withdrawals and consumptive use of water.

As the Mpumalanga highveld and coal region of Emalahleni have been included in the renewable energy development zones (REDZs), probably there will be some shifts during the project term (Creamer, 2020). So the inclusion of green energy production and the transition from coal-powered energy generation to clusters of solar and wind-generating plants will need to be included in the scenarios used in the WEF nexus models. This would also be an ideal opportunity to test the operational status of the WEF nexus tools for practical management and compare power generation efficiency. One can imagine there will be opportunities for practical test runs of the models with alternative energy sources, as long as the linkages between sectors are included. As wind and solar power generation will not require high water consumption, there will be direct benefits for the water sector. Water currently used for energy generation could become available to other sectors. The interlinkages with the food and livelihoods, specifically the labor and employment levels, must also be considered. Theoretical calculations will also be made by using the basic information about coal-burning electricity generations. For example, steam coal typically contains 25–28 MJ/kg but can be variable, so the Mpumalanga coal values must be obtained. A steam boiler produces steam from the water to turn a turbine that spins a generator that is only 38% efficient. So a single kW of electrical power requires 2.63 kW of thermal power (=1/0.38). Then one uses the amount of energy contained in the coal (per kg) to calculate how much energy can be generated from the coal supply. As an example, the calculation is as follows: 2.63 kW/28 MJ/kg * 60 s/h * 24 h/d * 365 d/y = 2965 kg per year. Therefore, at 38% efficiency, 8766 kWh of electricity can be generated (Mook, 2020). So the necessary information about the coal that Eskom uses will need to be obtained.

3.3.3 The food sector

As most of this catchment is in the lowveld, there is little commercial large-scale production of the main staple foods of maize and wheat grown in other parts of RSA. The land cover map shows the distribution and range of agricultural commodities versus natural vegetation across the Inkomati-Usuthu

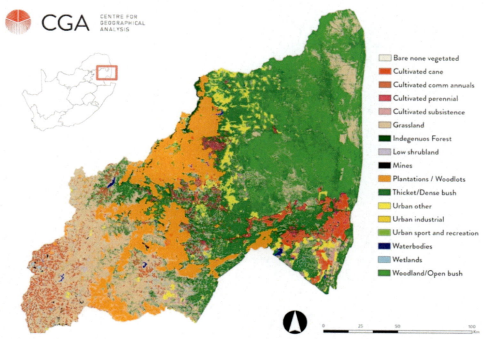

FIGURE 7.4
Land cover across the Inkomati-Usuthu catchment showing the distribution of cultivated lands and plantations, urban areas, and natural vegetation (including grassland, indigenous forest, low shrubland, woodlands/open bush, thickets/dense bush) (https://csirwateruse.firebaseapp.com/inkomati).

catchment (Fig. 7.4). Most cultivated lands are irrigated, requiring a water supply, including those marked as cane, annuals, perennials, and subsistence farms. The coal mines are shown on the western border, where the energy power stations are also located. The plantations and natural vegetation are generally nonirrigated and only use the water available from rainfall.

As the continuously increasing world population drives the increasing food demand, the water–food nexus must account for green, blue, and graywater consumption used during the cultivation, processing, marketing, and value chain to the garbage disposal of food waste and other food losses. The water–food nexus mainly refers to the water used for the production of food or agricultural produce (e.g., cereals, vegetables, fruits, edible oils etc.), animal products (e.g., meat, eggs, fish), and food and beverage (e.g., soft drinks, tea, coffee). The necessary information must be collected across the agricultural and food processing sectors (Schull et al., 2020). As there are several different agricultural industries in the catchment area of Mpumalanga, it will not be possible to do all the detailed calculations and analysis on all the food–agricultural products.

For this WEF analysis, the food sector will include the primary production of agricultural products at the farm level, particularly focusing on the main crops—sugarcane and subtropical fruit and nut trees under irrigated systems. So sugar and macadamia nuts have been selected to concentrate on for the calculations. The necessary data will be collected from several sources for each of these two commodities to characterize the water and energy use for the production and processing from inputs to the consumer.

For the water used in irrigated agriculture, a period of observation is set from 2000 to 2020. Monthly water use in the sugarcane and nuts industries from three of the top sugarcane growers (Crookes Brothers Ltd (https://www.cbl.co.za/), Kudu & Esperia Farms (https://ivorymacs.co.za/), and Elphick TF and Sons (PTY) LTD) has been collected together with the annual total area planted. The energy, water, and byproducts of sugarcane milling are compiled from one major milling company in Mpumalanga, RCL Foods Sugar and Milling (Pty) Ltd (https://rclfoods.com/), giving the following information—water and energy used in the milling and refining process, and the total number of employees. RCL Foods is a member of the South African Sugar Millers Association who is interested in all 14 of the sugar millers and refiners in South Africa. Information about the energy, water, and by-products of Macadamia nut production is compiled from three major farms in the catchment—Golden Macadamia (http://www.goldenmacadamias.com/), Sabie Valley Macadamia (Pty) Ltd (https://valleymacs.co.za/), and Kudu and Esperia Farms (https://ivorymacs.co.za/). The information about the total water used for growing and processing the nuts, the total energy for processing and irrigation, and total labor for growing and processing was compiled from 2000 to 2020.

In general, agriculture uses large amounts of water, but it is a low-energy intensity sector. Most farming activities use diesel for farm equipment and electricity for pumping groundwater or surface water for irrigation. The other energy that is indirectly used by agriculture is in the manufacturing of chemical fertilizers. When considering irrigation, reducing water losses by using pressurized pipes is an effective way to reduce water consumption in agriculture (Wakeel et al., 2016). However, in practice, a water-efficient pressurized delivery system consumes a large amount of energy (Siddiqi and Fletcher, 2015), resulting in a nexus point shift between the bulk water delivery system and pumping from groundwater. Agriculture can also pollute the water systems and causing environmental degradation through the high use of fertilizers and pesticides, forming long-term, nonpoint source pollution in downstream environments locally and nationally (Cai et al., 2018). However, these aspects will not be addressed in this project.

Although many agricultural value chains operate in the catchment, the two have been chosen as they both have a direct and dedicated processing system following the farm-level crop production system. The cane is processed in a sugar mill and then refined before being distributed to the retail market. The process is to cut the cane in the field and take it to the sugar mill, where it is crushed, and then the juice is heated and filtered before it is crystallized. Other processes include clarification, evaporation, crystallization, centrifugation, and refining where water is needed. The volumes of water and energy needed for both the milling/refining process will be considered together with the on-farm irrigation requirements. The macadamia nuts are mostly processes near the farming operations for human consumption, either as nuts or oil, both of which are destined for the local and export markets.

3.4 Spatial scale

In this project, the WEF will be analyzed at a catchment level; however, the whole large Inkomati-Usuthu catchment will not be used, but the study will focus on two smaller catchments. The one catchment, Lower Komati, stretches from the Eswatini border at the Driekoppies dam to the Mozambique border at Komatipoort. Lower Komati catchment is predominantly rural with settlements and two main roads, namely R570 and R571. The farming activities are dominated by sugarcane but include subtropical crops such as bananas, avocados, citrus, litchi, guavas, and mangoes, while there are also vegetables and small maize fields. The other catchment is the Crocodile River that stretches from the highveld (2100 masl) northwest of Dullstroom and flows through to the border with Mozambique at Komatipoort (200 masl). This catchment has a wider variety of activities, including similar farming activities in the lowveld section and Savanah grasslands in the highveld with livestock and game farming. The major city is Mbombela (formerly Nelspruit). Other towns include Malelane and Komatipoort in the lower stretches, Dullstroom, Belfast, and Middelburg in the highveld and Emalahleni (Witbank) on the western border. There are opencast coal mines and power stations in the upper reaches. These both use water and influence the water quality; hence, a need for coal mining and power stations is to be considered in the WEF nexus catchment analysis. The other industrial operations that need to be considered in this catchment include the paper mill at Ngodwana and the light industry in and around Mbombela.

3.5 Time/temporal scale

There are various options for the temporal scale to be used for the WEF nexus applications in these catchments according to the stakeholders' decision-making requirements. Therefore, this decision needs to be made together with the stakeholders as they will be the ones who will use the outputs from the tool. As a first snapshot picture of the application, it will be done in quarterly segments to sum to an annual basis to accommodate the different seasons.

Then using available historical data, decadal interval calculations can be compared to analyze the long-term trends. Later, the tools will also be formulated using the scenarios developed during the facilitative engagement process with the stakeholders in the catchment and with climate projections to show the projected changes under different assumptions.

3.6 Application of models

The tool selection depends on the proposed use and the stakeholders anticipated routine or operational requirements or questions. The applications can be for short-term decision-making at an operational time scale on a monthly or quarterly basis. IUCMA currently makes water allocation adjustments quarterly so that the WEF output will feed into this decision-making process. Alternatively, it can be used for either tactical decision-making on an annual basis or for strategic planning when the output could be used for strategic decision-making by policymakers with outputs based on future scenarios. The combined outputs over several of these time and spatial scales can also inform the current policy by varying stepwise inputs to provide a sensitivity analysis for future scenarios. Scenarios will be codeveloped with the stakeholders in the catchment during this project and reported elsewhere (Fakudze et al., 2021). The scenarios will include a business as usual; a hotter, drier future climate; a change in the socioeconomic status; and a change in the political and/or policy framework. The overall scenarios for the agricultural water in the future for the whole of South Africa developed during other WRC projects will also be considered as inputs to evaluate other possibilities or extremes in this catchment (Jordaan et al., 2020; Nyam et al., 2020a,b,c,d).

One of the main stakeholders and partners is Inkomati-Usuthu Catchment Management Agency (IUCMA). They are most interested in having an operational tool that can assist in the quarterly decision-making about water allocation in the catchment. As already stated, there are many water users in these catchments. They all have specified water requirements; however, often, these become conflicting demands for water when there is reduced flow. The IUCMA quarterly water allocations are currently based on the measured water flow and predictions for the upcoming 3- to 6-month period, according to the year's season, current usage and demand, and the seasonal climate forecast. The output from the WEF nexus tool will provide more useful information about the recent past and the balance between the different sectors, and a prediction for the upcoming period. Therefore, the inputs for the WEF nexus tool need to use currently available information according to the time of the year, and the measured water flow from particular weirs quarterly. In this way, a better informed decision can be made for the allocation for the upcoming period.

The inputs for the WEF nexus tool will include the calculations about each of the factors in each of the sectors. From the energy sector, one can obtain the amount of water used by the coal mining operations and that used at the thermal power stations. Although much of the latter is recycled, the monthly requirements must be considered in the water allocation budgeting. One of the ways of rationally calculating the effectiveness of the use of the water is by considering the knock-on benefits of the water used. This can be assessed by considering the several factors related to the population—for example, the size of the population served by the power station and the number of families befitting from those employed by such a power station. In this way, indices can be calculated to represent the efficiency of the water used in generating electricity related to the influence it has on livelihoods within the catchment. Probably, some of these factors can be considered dependent on the season of the year that is a cyclic pattern according to the higher demand for electricity during winter.

For the water use by the food sector, a wider range of aspects will be considered along the agricultural value chain. Initially, generalized indices could be used, such as the proportion of the population in the catchment below the breadline or food insecure and the productivity of the main agricultural systems. However, as there are many different farming systems in these catchments, some critical ones have been selected as described previously. Another matter that needs to be considered is that many small-scale farmers produce much of their produce for household use—especially maize and vegetables. In contrast, commercial farming systems are mostly not concerned with the main staple foods such as maize, wheat, and dairy products, so it is more difficult to relate them to the overall food security of the catchment. Most of the staple foods are produced outside of these catchments and brought in to be sold here. Therefore, the obvious choice was to use the sugar cane farming systems, as the sugar is also milled and processed within the catchment near Komatipoort. Another way to assess the "food" contribution to the WEF nexus is to select the farming systems that make the highest contribution to the country's GDP and, therefore, the livelihoods in the area. This is why the subtropical fruits and nuts will be investigated as they are often grown for export and contribute to the GDP.

4. Way forward and conclusion

As has been described, much work on the WEF nexus has been done at various large-scale levels, namely at regional and country levels. However, few studies focus on the catchment level to assist the catchment management agencies with their main decision-making tasks concerning resource allocation. This project tackles the two Mpumalanga catchments of the Crocodile River and the part of the lower Komati River in South Africa between the Kingdom of Eswatini and Mozambique. The many contrasting situations in these two subcatchments will be compared, for example, the presence of Mbombela city versus rural settlements, the presence of coal mining, and thermal electricity generation in the upper reaches of Crocodile River compared with a few small light industrial operations in the Lower Komati. The catchments are more similar from the food and agricultural side—both having sugarcane production and subtropical fruit and nut orchards. The other contrast is that the Crocodile River does not have a major dam impounding the water. In contrast, the Lower Komati is fed from the Driekoppies Dam on the border of Eswatini and supplying irrigation water for the sugar farms. Both catchments have international borders and must meet the requirements for transboundary water allocation to Mozambique.

One of the major challenges to implementing the WEF nexus frameworks is the availability of the necessary data at a spatial scale with sufficient detail and routine monitoring over at least 10 years, but preferable a 30-year period. This is necessary to perform the trend analysis and construct the time lines needed to establish the trends of the various indices that describe the interactions between the water, agriculture, and energy sectors. Another challenge is to select useful and meaningful indices that can represent the interactive relationship between the three sectors. At a larger scale, such as country scale, it is possible to use generic indices for food security. However, as there is a wide variation between the urban and rural populations in these subcatchments, such information could be misleading. Another challenge is how to compare the different food systems across a wide range of agricultural activities when most of the staple food is transported from other areas in South Africa, and many of the agricultural products are exported to Europe and northern hemisphere countries. One of the solutions to be explored is to consider the productivity according to the influence of the local population livelihoods by using alternative indices related to the population.

Therefore, to achieve and capitalize on a better understanding of the relationships among nexus elements, good data coverage of the full nexus scope is needed on the best spatial and temporal scales. Complete WEF nexus-wide catchment data have seldom been complied with, although separate component data have been gathered for many years. However, such data are frequently not comprehensive or compatible (in either space or time) across all WEF sectors, and they are seldom (if ever) integrated into a WEF-wide system database (McCarl et al., 2017b). In addition, data systems covering the full set of WEF domains are evolving as the models and frameworks develop. This presents emerging challenges that must be faced to better support WEF system-wide operational analyses. These challenges pertain to a wide scope of data development, retrieval, analysis, and use, as well as nexus issues such as scope, complexity, availability (both spatial and temporal), use, and assembly

(Dargin et al., 2019). In most cases, data required for nexus modeling and analysis usually exceed the data available, and thus, compromises are required to satisfy the tools and frameworks input requirements.

Moreover, nexus tools and frameworks should facilitate stakeholder and user understanding of the nexus scope, location, production, consumption, diversion, return flows, conveyance possibilities, and economic and policy implications. Data are needed to support efforts to understand system boundaries (that may traditionally be different for the different sectors), spatial dimensions, and the origin and fate of WEF commodities, including cross-sector interactions and interfaces. This is no simple task. A wide, rather comprehensive set of data are necessary across the full WEF scope as the nexus approach is still an active research field about widening perspectives to unexplored levels. In addition, WEF nexus data systems need to reflect the uniqueness of the study region, incorporating appropriate activities, meaning that the contents vary as it is applied from location to location or, in this case, different subcatchments. Challenges are apparent in representing an appropriate mix of enterprises (particularly to represent food); stochastic variation in water supplies (according to seasonal variations); WEF commodity production practices and market prices; and cost and return from ongoing and/or potential future technological choices. Data challenges further arise due to proprietary interests, scale differences in analysis, model requirements, representation of unexplored possibilities, assembly cost, projections of the future, representation of stochastic variation, quick query and retrieval systems, and integration with visualization. Therefore, comprehensive, innovative procedures and approaches are needed for data collection, storage/retrieval, and inference to support high-quality WEF nexus analyses and application into the selected framework.

Acknowledgments

The authors thank Water Research Commission in South Africa for funding through WRC Project No. C2019/2020-00326.

References

Aboelnga, H.T., Khalifa, M., McNamara, I., Sycz, J., 2018. Water-Energy-Food Nexus Literature Review. A Review of Nexus Literature and Ongoing Nexus Initiatives for Policymakers. Nexus Regional Dialogue Programme (NRD) and German Development Agency for International Cooperation (GIZ), Bonn.

Albrecht, T.R., Crootof, A., Scott, C.A., 2018. The Water-Energy-Food nexus: a systematic review of methods for nexus assessment. Environ. Res. Lett. 13 (4), 043002.

Amey, J., 2010. Think Global, Act Local. WMG, International Manufacturing Centre, University of Warwick, Coventry, United Kingdom. https://warwick.ac.uk/newsandevents/knowledge-archive/socialscience/thinklocal. (Accessed 27 March 2021).

Arnold, J.G., Srinivasan, R., Muttiah, R.S., Williams, J.R., 1998. Large area hydrologic modeling and assessment part I: model development. J. Am. Water Resour. Assoc. 34, 73–89.

Bhaduri, A., Ringler, C., Dombrowski, I., Mohtar, R., Scheumann, W., 2015. Sustainability in the water-energy-food nexus. Water Int. 40, 723–732.

Brouwer, F., Vamvakeridou-Lyroudia, L., Alexandri, E., Bremere, I., Griffey, M., Linderhof, V., 2018. The nexus concept integrating energy and resource efficiency for policy assessments: a comparative approach from three cases. Sustainability 10 (12), 4860.

Cai, X., Wallington, K., Shafiee-Jood, M., Marston, L., 2018. Understanding and managing the food-energy-water nexus: opportunities for water resources research. Adv. Water Res. 111, 259–273.

Carney, D., 2003. Sustainable Livelihoods Approaches: Progress and Possibilities for Change. Department for International Development (DFID), London.

Cash, D.W., Adger, W., Berkes, F., Garden, P., Lebel, L., Olsson, P., Pritchard, L., Young, O., 2006. Scale and cross-scale dynamics: governance and information in a multilevel world. Ecol. Soc. 11 (2), 8.

Cervigni, R., Liden, R., Neumann, J.E., Strzepek, K.M., 2015. Enhancing the Climate Resilience of Africa's Infrastructure: The Power and Water Sectors. The World Bank.

Chang, Y., Li, G., Yao, Y., Zhang, L., Yu, C., 2016. Quantifying the water-energy-food nexus: current status and trends. Energies 9 (2), 1–17.

Conway, D., van Garderen, E.A., Deryng, D., Dorling, S., Krueger, T., Landman, W., Lankford, B., Lebek, K., Osborn, T., Ringler, C., Thurlow, J., Zhu, T., Dalin, C., 2015. Climate and southern Africa's water-energy- food nexus. Nat. Clim. Change 5 (9), 837–846.

Conway, T.M., Almas, A.D., Coore, D., 2019. Ecosystem services, ecological integrity, and native species planting: how to balance these ideas in urban forest management? Urban For. Urban Green. 41, 1–5.

Creamer, T., 2020. 'Great Opportunity' to Tap Mpumalanga's Grid Capacity for Just Transition. IPP Office. https://www.engineeringnews.co.za/article/great-opportunity-to-tap-mpumalangas-grid-capacity-for-just-transition-ipp-office-2020-10-26 (Accessed 11 January 2021).

Daher, B.T., Mohtar, R.H., 2015. Water–energy–food (WEF) nexus tool 2.0: guiding integrative resource planning and decision-making. Water Int. 40 (5–6), 748–771.

Daher, B., Lee, S.-H., Assi, A., Mohtar, R.H., Mohtar, R., 2017. Modeling the water-energy-food nexus: a 7-Question Guideline. In: Abdul Salam, P., Shrestha, S., Vishnu Prasad Pandey, V.P., Anil, K.A. (Eds.), Water-Energy-Food Nexus: Principles and Practices 229. American Geophysical Union. John Wiley & Sons, pp. 57–66.

Dai, J., Wu, S., Han, G., Weinberg, J., Xie, X., Wu, X., Song, X., Jia, B., Xue, W., Yang, Q., 2018. Water-energy nexus: a review of methods and tools for macro-assessment. Appl. Energy 210, 393–408.

Dargin, J., Daher, B., Mohtar, R.H., 2019. Complexity versus simplicity in water energy food nexus (WEF) assessment tools. Sci. Total Environ. 650, 1566–1575.

de Vito, R., Portoghese, I., Pagano, A., Fratino, U., Vurro, M., 2017. An index-based approach for the sustainability assessment of irrigation practice based on the water-energy-food nexus framework. Adv. Water Resour. 110, 423–436.

DWA (Department of Water Affairs), 2013a. Business Case for the Inkomati-Usuthu Catchment Management Agency. Pretoria.

DWA (Department of Water Affairs), 2013b. National Water Resources Strategy 2. Government of South Africa. DWA, Pretoria.

El-Nasr, A.A., Arnold, J.G., Feyen, J., Berlamont, J., 2005. Modelling the hydrology of a catchment using a distributed and a semi-distributed model. Hydrol. Process. 19, 573–587.

Endo, A., Tsurita, I., Burnett, K., Orencio, P.M., 2017. A review of the current state of research on the water, energy and food nexus. J. Hydrol. 11, 20–30.

Endo, A., Yamada, M., Miyashita, Y., Sugimoto, R., Ishii, A., Nishijima, J., Fujii, M., Kato, T., Hamamoto, H., Kimura, M., Kumazawa, T., Qi, J., 2020. Dynamics of water–energy–food nexus methodology, methods, and tools. Curr. Opin. Environ. Sci. Health 13, 46–60.

Engström, R.E., Howells, M., Destouni, G., Bhatt, V., Baziliana, M., Rogner, H.-H., 2017. Connecting the resource nexus to basic urban service provision — with a focus on water-energy interactions in New York City. Sustain. Cities Soc. 31, 83–94.

Fakudze, B., Jacobs-Mata, I., Phahlane, M.O., Walker, S., 2021. Application of the Water Energy Food (WEF) Nexus as a Framework in the Inkomati-Usuthu Catchment. Submitted to Frontiers in Water.

FAO, 2014. Walking the Nexus Talk: Assessing the Water–Energy–Food Nexus in the Context of the Sustainable Energy for All Initiative. http://www.fao.org/3/a-i3959e.pdf (Accessed 20 July 2020).

FAO, 2020. Water Uses Database. http://www.fao.org/nr/water/aquastat/water_use/index.stm (Accessed 24 July 2020).

Ferroukhi, R., Nagpal, D., López-Peña, A., Hodges, T., Mohtar, R., Daher, B., Mohtar, S., Keulertz, M., 2015. Renewable Energy in the Water, Energy and Food Nexus, IRENA Technical Report.

Gelsdorf, K., 2010. Global Challenges and Their Impact on International Humanitarian Action. Occasional Policy Briefing Series (OCHA), Policy Development and Studies Branch (PDSB).

Gerber, J.-F., Scheidel, A., 2018. In search of substantive economics: comparing today's two major sociometabolic approaches to the economy — MEFA and MuSIASEM. Ecol. Econ. 144, 186–194.

Giampietro, M., Mayumi, K., Ramos-Martin, J., 2009. Multi-scale integrated analysis of societal and ecosystem metabolism (MuSIASEM): theoretical concepts and basic rationale. Energy 34 (3), 313–322.

Hang, D., Li, G., Sun, C., Liu, Q., 2020. Exploring interactions in the local water-energy-food nexus (WEF-Nexus) using a simultaneous equations model. Sci. Total Environ. 703, 135035.

Hermann, S., Rogner, H.H., Howells, M., Young, C., Fischer, G., Welsch, M., 2011. In the CLEW model — developing an integrated tool for modelling the interrelated effects of climate, land use, energy, and water (CLEW). In: 6th Dubrovnik Conference on Sustainable Development of Energy, Water and Environment Systems.

Hoff, H., 2011. Understanding the Nexus. Background Paper for the Bonn 2011 Conference: The Water Energy and Food Security Nexus. Environment Institute, Stockholm.

Howells, M., Hermann, S., Welsch, M., Bazilian, M., Segerström, R., Alfstad, T., Gielen, D., Rogner, H., Fischer, G., Van Velthuizen, H., 2013. Integrated analysis of climate change, land-use, energy and water strategies. Nat. Clim. Change 3, 621.

Howells, M., Rogner, H., Strachan, N., Heaps, C., Huntington, H., Kypreos, S., Hughes, A., Silveria, S., De Carolis, J., Bazillian, M., Rhoehl, A., 2011. OSeMOSYS: the open source energy modeling system: an introduction to its ethos, structure and development. Energy Pol. 39, 5850–5870.

IRENA (International Renewable Energy Agency), 2015. Renewable Energy in the Water, Energy & Food Nexus. http://www.irena.org/DocumentDownloads/Publications/IRENA_Water_Energy_Food_Nexus_2015.pdf (Accessed 20 July 2020).

IUCMA (Inkomati-Usuthu Catchment Management Agency), 2015. Annual Report 2015/2016 Financial Year. Inkomati-Usuthu Catchment Management Agency, Nelspruit.

IUCMA (Inkomati-Usuthu Catchment Management Agency), 2017. Five-year Strategic Plan and Budget for the Fiscal Years 2015/16-2020/21. Inkomati-Usuthu Catchment Management Agency, Nelspruit.

IUCMA (Inkomati-Usuthu Catchment Management Agency), 2019. Annual Performance Plan 1 April 2018 - 31 March 2019. Inkomati-Usuthu Catchment Management Agency, Nelspruit.

Jordaan, A.J., Ogundeji, A.A., Nyam, Y.S., Ilbury, C., Turton, A., Walker, S., van Coppenhagen, G., 2020. Development of Scenarios for Future Agricultural Water Management in South Africa. Report to the Water Research Commission, Project Number: K5/2711/4. Water Research Commission, Pretoria.

Krantz, L., 2001. The Sustainable Livelihood Approach to Poverty Reduction: An Introduction. Swedish International Development Cooperation Agency (SIDA), Stockholm.

Kurian, M., 2017. The water-energy-food nexus: trade-offs, thresholds and transdisciplinary approaches to sustainable development. Environ. Sci. Pol. 68, 97–106.

Li, G., Huang, D., Sun, C., Li, Y., 2019. Developing interpretive structural modelling based on factor analysis for the water-energy-food nexus conundrum. Sci. Total Environ. 651, 309–322.

Mabhaudhi, T., Mpandeli, S., Madhlopa, A., Modi, A.T., Backeberg, G., Nhamo, L., 2016. Southern Africa's water–energy nexus: towards regional integration and development. Water 8, 235.

Mabhaudhi, T., Mpandeli, S., Nhamo, L., Chimonyo, V., Nhemachena, C., Senzanje, A., Naidoo, D., Liphadzi, S., Modi, A.T., 2018. Prospects for improving irrigated agriculture in Southern Africa: linking water, energy and food. Water 15, 1881.

Mabhaudhi, T., Chimonyo, V.G.P., Hlahla, S., Massawe, F., Mayes, S., Nhamo, L., Modi, A.T., 2019a. Prospects of orphan crops in climate change. Planta 250, 695–708.

Mabhaudhi, T., Nhamo, L., Mpandeli, S., Nhemachena, C., Senzanje, A., Sobratee, N., Chivenge, P.P., Slotow, R., Naidoo, D., Liphadzi, S., Modi, A.T., 2019b. The water–energy–food nexus as a tool to transform rural livelihoods and wellbeing in Southern Africa. Int. J. Environ. Res. Publ. Health 16, 2970.

Mannan, M., Al-Ansari, T., Mackey, H.R., Al-Ghamdi, S.G., 2018. Quantifying the energy, water and food nexus: a review of the latest developments based on life-cycle assessment. J. Clean. Prod. 193, 300e314.

Martinez-Hernandez, E., Leach, M., Yang, A., 2017. Understanding water-energy-food and ecosystem interactions using the nexus simulation tool NexSym. Appl. Energy 206, 1009–1021.

Mohtar, R.H., Daher, B., 2016. Water-Energy-Food Nexus Framework for facilitating 441 multi-stakeholder dialogue. Water Int. 41 (5), 655–661.

McCarl, B.A., Yang, Y., Schwabe, K., Bernard, A.E., Alam, H.M., Ringler, C., Pistikopoulos, E.N., 2017a. Model use in WEF nexus analysis: a review of issues. Curr. Sustain. Renew. Energy Rep. 4 (3), 144–152.

McCarl, B., Yang, Y., Srinivasan, R., Pistikopoulos, E., Mohtar, R., 2017b. Data for WEF nexus analysis: a review of issues. Curr. Sustain. Renew. Energy Rep. 4, 137–143.

Mook, W., 2020. Energy Production. https://www.quora.com/profile/William-Mook (Accessed 12 November 2020).

Mpandeli, S., Nesamvuni, E., Maponya, P., 2015. Adapting to the impacts of drought by smallholder farmers in Sekhukhune District in Limpopo province, South Africa. J. Agric. Sci. 7 (2), 115–124.

Mpandeli, S., Naidoo, D., Mabhaudhi, T., Nhemachena, C., Nhamo, L., Liphadzi, S., Hlahla, S., Modi, A., 2018. Climate change adaptation through the water-energy-food nexus in Southern Africa. Int. J. Environ. Res. Publ. Health 15, 2306.

Ness, B., Urbel-Piirsalu, E., Anderberg, S., Olsson, L., 2007. Categorising tools for sustainability assessment. Ecol. Econ. 60, 498–508.

Nhamo, L., Mabhaudhi, T., Modi, A.T., 2019a. Preparedness or repeated short-term relief aid? Building drought resilience through early warning in Southern Africa. WaterSA 45 (1), 75–85.

Nhamo, L., Mabhaudhi, T., Mpandeli, S., Nhemachena, C., Senzanje, A., Naidoo, D., Liphadzi, S., Modi, A.T., 2019b. Sustainability Indicators and Indices for the Water-Energy-Food Nexus for Performance Assessment: WEF Nexus in Practice - South Africa Case Study. Preprint, 2019050359.

Nhamo, L., Mabhaudhi, T., Mpandeli, S., Dickens, C., Nhemachena, C., Senzanje, A., Naidoo, D., Liphadzi, S., Modi, A., 2020a. An integrative analytical model for the water-energy-food nexus: South Africa case study. Environ. Sci. Pol. 109, 15–24.

Nhamo, L., Ndlela, B., Nhemachena, C., Mabhaudhi, T., Mpandeli, S., Matchaya, G., 2018. The water-energy-food nexus: climate risks and opportunities in southern Africa. Water 10, 567.

Nhamo, L., Ndlela, B., Mpandeli, S., Mabhaudhi, T., 2020b. The water-energy-food nexus as an adaptation strategy for achieving sustainable livelihoods at a local level. Sustainability 12, 8582.

Nyam, Y.S., Kotir, J.H., Jordaan, A.J., Ogundeji, A.A., Adetoro, A.A., Orimoloye, I.R., 2020a. Towards understanding and sustaining natural resource systems through the systems perspective: a systematic evaluation. Sustainability 12 (23), 9871.

Nyam, Y.S., Kotir, J.H., Jordaan, A.J., Ogundeji, A.A., Turton, A.R., 2020b. Drivers of change in sustainable water management and agricultural development in South Africa: a participatory approach. Sustain. Water Res. Manag. 6 (4), 1–20.

Nyam, Y.S., Kotir, J.H., Jordaan, A.J., Ogundeji, A.A., 2020c. Developing a conceptual model for sustainable water resource management and agricultural development: the case of the Breede River catchment area, South Africa. J. Environ. Manag. 67 (4), 632–647.

Nyam, Y.S., Kotir, J.H., Jordaan, A., Ogundeji, A.A., 2020d. Identifying behavioural patterns of coupled water agriculture systems using system archetypes. Syst. Res. Behav. Sci. 1–19.

OECD, 2021. Electricity Generation (Indicator). https://doi.org/10.1787/c6e6caa2-en (Accessed 11 January 2021).

Pauliuk, S., 2018. Critical appraisal of the circular economy standard BS 8001:2017 and a dashboard of quantitative system indicators for its implementation in organizations. Resour. Conserv. Recycl. 129, 81–92.

Rasul, G., Neupane, N., 2021. Improving policy coordination across the water, energy, and food, sectors in South Asia: a Framework. Front. Sustain. Food Syst. 5, 602475.

Rich, P.M., Weintraub, L.H.Z., Ewers, M.E., Riggs, T.L., Wilson, C.J., 2005. Decision support for water planning: the ZeroNet water-energy initiative. In: Proceedings of the American Society of Civil Engineers - Environmental & Water Resources Institute (ASCE-EWRI) World Water and Environmental Resources Congress 2005: Impacts of Global Climate Change, 15-19 May, Anchorage, AK. LA-UR-05-1068.

Schull, V.Z., Daher, B., Gitau, M.W., Mehan, S., Flanagan, D.C., 2020. Analyzing FEW nexus modeling tools for water resources decision-making and management applications. Food Bioprod. Process. 119, 108–124.

Siddiqi, A., Fletcher, S., 2015. Energy intensity of water end-uses. Curr. Sustain. Renew. Energ. Rep. 2 (1), 25–31. https://doi.org/10.1007/s40518-014-0024-3.

Simpson, G.B., Jewitt, G.P.W., 2019. The development of the Water-Energy-Food nexus as a framework for achieving resource security: a review. Front. Environ. Sci. 7, 8. https://doi.org/10.3389/fenvs.2019.00008.

Strasser, L.D., Lipponen, A., Howells, M., Stec, S., Brethaut, C., 2016. A methodology to assess the water energy food ecosystems nexus in transboundary river basins. Water 8, 1–28.

Sušnik, J., Chew, C., Domingo, X., Mereu, S., Trabucco, A., Evans, B., Vamvakeridou-Lyroudia, L., Savić, D.A., Laspidou, C., Brouwer, F., 2018. Multi-stakeholder development of a serious game to explore the water-energy-food-land-climate nexus: the SIM4NEXUS approach. Water 10, 139.

USGS, 2020. Thermoelectric Power Water Use. https://www.usgs.gov/mission-areas/water-resources/science/thermoelectric-power-water-use?qt-science_center_objects=0#qt-science_center_objects.A (Accessed 12 November 2020).

Villamayor-Tomas, S., Grundmann, P., Epstein, G., Evans, T., Kimmich, C., 2015. The water-energy-food security nexus through the lenses of the value chain and the Institutional Analysis and Development frameworks. Water Altern. (WaA) 8 (1), 735–755.

Wakeel, M., Chen, B., Hayat, T., Alsaedi, A., Ahmad, B., 2016. Energy consumption for water use cycles in different countries: a review. Appl. Energy 178, 868–885.

Yu, L., Xiao, Y., Zeng, X.T., Li, Y.P., Fan, Y.R., 2020. Planning water-energy-food nexus system management under multilevel and uncertainty. J. Clean. Prod. 251, 119658.

Zhang, C., Chen, X., Li, Y., Ding, W., Fu, G., 2018. Water-energy-food nexus: concepts, questions and methodologies. J. Clean. Prod. 195, 625–639.

Zhang, X., Vesselinov, V.V., 2017. Integrated modeling approach for optimal management of water, energy and food security nexus. Adv. Water Resour. 101, 1–10.

Zhang, P., Xu, Z., Fan, W., Ren, J., Liu, R., Dong, X., 2019. Structure dynamics and risk assessment of water-energy-food nexus: a water footprint approach. J. Sustain. 11, 1187–1205.

CHAPTER 8

A regional approach to implementing the WEF nexus: a case study of the Southern African Development Community

Patrice Kandolo Kabeya[1], Dumisani Mndzebele[1], Moses Ntlamelle[1], Duncan Samikwa[1], Alex Simalabwi[2], Andrew Takawira[2], Kidane Jembere[2] and Shamiso Kumbirai[2]

[1]*Southern African Development Community Secretariat, Gaborone, Botswana;* [2]*Global Water Partnership Southern Africa, Pretoria, South Africa*

1. Introduction

Nhamo et al. (2018) explain that water, energy, and food are vital for human well-being, poverty reduction, and sustainable development. The SADC Regional WEF Nexus Framework report (2019) highlights that the Southern African region represents a wide range of resource and climate contexts with varied supplies of water, food, and energy. About 60% of the population of the Southern African Development Community (SADC) live in rural areas relying on rain-fed agriculture, lacking basic services of energy, water, and sanitation, yet the region is endowed with vast natural resources. Ensuring water, energy, and food security has dominated the development agenda of Southern African countries, centered on improving livelihoods and building resilience and regional integration (Nhamo et al., 2018).

Increasing demands for water, land, and energy resources due to population growth, increasing urbanization, and increasing economic growth are the major challenges in the region. These challenges are further exacerbated by climate change making this particularly concerning for the SADC region due to dependence on climate-sensitive sectors of agriculture and energy, which heavily depend on water resources (Kusangaya et al., 2014).

Mabhaudhi et al. (2016) state the water—energy—food (WEF) nexus approach has potential application in the region for ensuring security of water, energy, and food and for bringing resource use efficiency. It provides opportunity to stabilize competing demands and promote regional integration, particularly in the SADC where resources are mostly transboundary. The WEF nexus approach can help to ensure that development of one of the sectors has minimum impacts on the other. Sectoral collaboration is particularly relevant in SADC, as watercourses and electricity grids are shared among countries (Southern African Development Community, 2019).

1.1 Status on water, energy, and food security in the SADC region

Water resources availability in the region is generally good, estimated at 2300 km^3/year of the renewable freshwater resources against an extractive demand of 2% of the available resources (SADC, 2012). However, its distribution in the region is a major concern with mean annual rainfall ranging from as low as approximately 300 mm/year (Namibia) to 1530 mm/year (D.R. Congo) (Southern Africa Development Community, 2016). Total regional water storage is only 14% of the available annual renewable water resources. Efficacy in the harnessing and utilization of the region's water resources is important in attaining climate-resilient economic development in the region. Access to drinking water and sanitation in the region is still very low, estimated at 60% for water and 40% for sanitation (Southern Africa Development Community, 2016).

Energy is a critical input for the implementation of the SADC Industrialisation Agenda. The region is endowed with vast energy resources, namely hydropower, coal, biomass, and solar although availability across the region varies from country to country. However, the SADC Centre for Renewable Energy and Energy Efficiency (SACREEE) reports that only 48% (75% urban and 32% rural) of the total population in the region have access to electricity (SADC Centre for Renewable Energy Efficiency, 2018). It is worth noting that 62% of electricity generation is from coal followed by hydropower resources with 21% and the remainder is from medium-scale renewable (solar and wind) energy technologies and gas-fired power plants as well as one nuclear power plant. The overall hydropower potential of the SADC region is estimated at approximately 1080 terawatt hours per year (TWh/year), yet the current utilized capacity is less than 31 TWh/year (Southern African Development Community, 2018).

More than 90% of agricultural activities in the SADC region depend on rain-fed farming although there is an abundance of water. In the region, 77% of freshwater resources are available for agricultural activities; however, only about 7% of the region's irrigation potential has been developed (Southern Africa Development Community, 2016). Water security for the agriculture sector is the main challenge of the region in attaining food security since the agriculture sector is highly vulnerable to the impacts of climate change (drought and flood). Energy is less utilized in the agricultural sector especially in the smallholder setup.

1.1.1 Need for integrated effort to deliver on the regional development agenda

Harnessing the interdependency, interconnection, and complementarity among the water, energy, and agriculture sectors is key to enhancing efficiency in the utilization of the regions' limited resources, needed to uplift lives of SADC citizens. The demand for a more integrated approach to resolving the regional challenges has been noticeably rising over the years. Increasingly, each sector has realized that the developmental challenges they are facing are intertwined and go beyond their sectoral space. Therefore, there is a need for collective responsibility for joint management and utilization of resources to achieve the regional development agenda.

Noting this challenge, SADC leadership has in the recent past started to hold cross-sectoral dialogues to deal with shared developmental challenges. The first event was the Joint Energy and Water Ministerial Workshop in Gaborone, Botswana, in 2016. The workshop focused on finding solutions to respond to the 2015/16 drought, which had regional impacts on food security, water availability, and hydropower generation.

The second event was a Joint Investment Conference for Energy and Water Infrastructure held in 2017 in Ezulwini, Eswatini. The Conference facilitated the exchange of ideas with potential funders and forged practical solutions to the water and energy challenges confronting the region. The Joint Energy and Water Ministerial Meetings have been held since 2016. However, the agriculture sector as the largest user of the region's abstracted water resources (77%) and needing energy is yet to participate in these joint sessions (Southern Africa Development Community, 2016).

The SADC multisector dialogues have been instrumental in sensitizing stakeholders from various sectors and promoting integrated approaches. Regional experience has also shown that when infrastructure investment projects of a multisectoral nature are promoted by all involved sectors, they stand a better chance of securing funding than when promoted by only one sector. For example, better traction was realized when the Regional Infrastructure Development Master Plan (RIDMP) projects such as the Inga Dams, Batoka Gorge Hydroelectric Power Station, and Lesotho Highlands Water Project were jointly promoted by the energy and water sectors. The potential socioeconomic benefits and beneficiaries are broadened in such joint efforts.

2. Fostering water, energy, and food security nexus dialogue and multi-sector investment in the SADC region project

The SADC Nexus Dialogue Project on "Fostering Water, Energy and Food Security Nexus Dialogue and Multi-Sector Investment in the SADC Region" is a project supported by the European Commission as part of the global "Nexus Dialogues Programme."

The overall objective of the project is to support the transformation required to meet increasing water, energy, and food security demand in a context of climate change in the SADC region through the development of a truly integrated nexus approach. The specific objective of the project is to create an enabling environment that will drive cross-sectoral engagement and implementation of nexus investment projects that contribute to enhancing water, food, and energy security in the SADC region.

The Project began in 2017 and has been implemented in two phases. Phase I of the Nexus Dialogues Programme ran from 2017 to 2019 and aimed at helping regional organizations, and their Member States (MS) applied a nexus approach in the formulation of multisector policy recommendations, strategies, action plans, and investment programs. Phase I also aimed to identifying concrete investment projects—with a focus on multipurpose water infrastructure. The key results achieved from phase I involved:

- Establishing an SADC Regional WEF Nexus Governance Framework;
- The development of an SADC WEF nexus investment project screening and appraisal tool; and
- Developing a prioritized list of nexus investment projects.

Phase II of the project has a duration of 3 years, from 2020 until 2023. This phase will build on the achievements of the first phase, with the overall objective being to institutionalize the WEF nexus approach at regional and national governance structures and investment decisions for water, energy, and food security in the SADC region. The specific objectives of phase II are to

- Increase application of the nexus approach in planning, policymaking, and implementation, and
- Increase interest from public and private investors for investment projects applying the WEF nexus approach.

2.1 SADC WEF nexus conceptualization

The WEF nexus in the SADC region is understood as an approach to facilitate better interactions and synergies between the water, food, and energy sectors to unlock and optimize development potential for economic growth and transformation in the region. The two expected contributions of the WEF nexus approach in the SADC region include the following:

- Facilitating the simultaneous achievement of water, energy, and food securities; and
- Improving water, energy, and land resource use efficiencies.

The WEF nexus approach also presents opportunities for greater resource coordination, management, and policy convergence across sectors. SADC expects the WEF nexus approach to enhance investments in the region. Fig. 8.1 shows the conceptual understanding by SADC for the WEF nexus approach in the region.

3. Key planning, policy, and legal documents that are relevant for water, energy, and food security in the SADC region

3.1 Regional development context and sustainable development

The Treaty of the Southern African Development Community is the founding document for the establishment of the SADC and outlines the principles, objectives, and general undertakings of the Community. The SADC Treaty emphasizes sustainable and equitable growth and socioeconomic development for poverty alleviation, increased standard of living, and quality of life in the region (Southern African Development Community, 1992). It aims to realize the aforementioned, through sustainable utilization of the region's resources, deeper integration, and economic development.

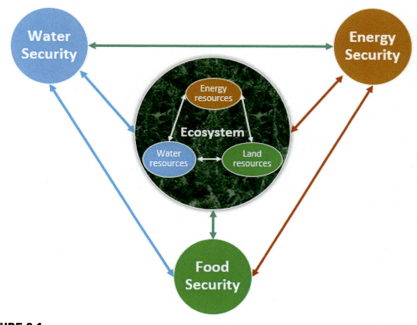

FIGURE 8.1
SADC WEF nexus conceptual understanding.

The Regional Indicative Strategic Development Plan (RISDP) elaborates the implementation of these aspirations by, among other things, prioritizing industrial development and market integration, infrastructure development for support of regional integration, and agriculture, food, and natural resources development. The energy, food, and water sectors are instrumental in the realization of the developmental priorities.

The SADC region also strongly embraces sustainable development, as provided in the protocols, policies, and strategies of the three sectors. An integrated approach toward meeting the Sustainable Development Goals (SDGs) for food, energy, and water security has been promoted in the region to increase the chances of meeting other related goals.

3.1.1 Key water sector planning, policy, and legal documents

SADC has put in place regional regulatory, policy, strategies, and planning instruments that provide the enabling environment for the implementation of the SADC water sector programs and plans. These are summarized in Table 8.1. The SADC Revised Protocol on Shared Watercourses (2000) provides the framework for the region that fosters closer cooperation for judicious,

Table 8.1 Main policy, planning, and legal documents in the SADC water sector. Courtesy, SADC 2019.

SADC document	Objective
The SADC declaration and treaty (1992)	Governs the regional activities of SADC and its MS aimed at achieving the SADC vision and agenda of regional integration, poverty eradication, industrialization, and economic development.
The SADC regional indicative strategic development plan (2003, 2007, 2015, and 2020–30)	Outlines the key interventions necessary to deepen the SADC vision over the period 2005 to 2020 and stipulates various targets for water and sanitation access.
The SADC revised protocol on shared watercourses (2000)	Fosters closer cooperation for judicious, sustainable and coordinated management, protection and utilization of the 15 SADC shared watercourses and advance the SADC vision and agenda.
The Southern African vision for water, life and the environment in the 21st century (2000)	Details the SADC vision for water, life, and the environment in the 21st century of *equitable and sustainable utilization of water for social, environmental justice, and economic benefit for present and future generations.*

3. Key planning, policy, and legal documents

Table 8.1 Main policy, planning, and legal documents in the SADC water sector. Courtesy, SADC 2019. *continued*

SADC document	Objective
The SADC regional water policy (RWP) (2005)	Aims at providing a framework for sustainable, integrated, and coordinated development, utilization, protection, and control of national and transboundary water resources in the SADC region, for the promotion of socioeconomic development and regional integration and the improvement in the quality of life of all people in the region.
The SADC regional water strategy (2006)	Provides for the framework for the implementation of the RWP. While the RWP deals with the "what" on regional water issues, the RWS deals with the "how," "who," and "when" in the implementation of the RWP.
The SADC regional awareness and communication strategy for the water sector (2009)	To improve awareness and understanding on water issues and initiatives in the SADC region contributing to poverty eradication and regional integration.
The SADC regional strategic action plans (RSAPs): I, II, III, IV, and V	The main objective of the RSAP I (1999—2004) was to create an enabling environment for joint management of regional water resources. The RSAP II (2005—10) put emphasis on infrastructure development. The goal of the RSAP III (2011—15) was to strengthen the enabling environment for regional water resources governance, management, and development through the application of IWRM at the regional, river basin, MS, and community levels. The RSAP IV (2016—20) was developed with a strong focus on unlocking the potential of water as a catalyst in socioeconomic development of the region. The RSAP V (2021—25) builds on the previous phases of RSAPs developed over the years as well as addressing the current challenges facing the regional water sector.
The SADC guidelines for strengthening river basin organisations (2010)	Gives guidelines on the establishment and development of RBOs, implementation of environmental management programs, procedures for RBOs to become financially sustainable, and procedures to assist RBOs with the implantation of participatory approaches.
Climate change adaptation in SADC: strategy for the water sector (2011)	To improve climate resilience in SADC region.

Continued

Table 8.1 Main policy, planning, and legal documents in the SADC water sector. Courtesy, SADC 2019. *continued*

SADC document	Objective
The SADC regional infrastructure development master plan (2012)	Defines the minimum but ultimate regional infrastructure development requirements and conditions to facilitate the implementation and realization by year 2027 of the key infrastructure in the water, energy, transport, tourism, meteorology, and telecommunication sectors that will move forward the SADC agenda and enable the SADC region realize its goal: *The attainment of an integrated regional economy on the basis of balance, equity, and mutual benefit for all MS.*
The SADC industrialisation strategy and road map (2015–2063)	Aims to increase competitiveness (at the firm/industry, country, and regional levels) with a quantitative goal to lift the regional growth rate of real GDP from 4% annually (since the year 2000) to a minimum of 7% a year.
UN sustainable development goal on water (SDG6)	Goal 6 addresses not only the issues relating to drinking water, sanitation, and hygiene but also the quality and sustainability of water resources worldwide.

MS, Member States; RBOs, River Basin Organizations; SADC, Southern African Development Community.
Southern Africa Development Community (SADC), 2016. Regional Strategic Action Plan on Integrated Water Resources Development and Management Phase IV. <https://www.sadc.int/files/9914/6823/9107/SADC_Water_4th_Regional_Strategic_Action_Plan_English_version.pdf> (accessed March 2021).

sustainable, and coordinated management, protection, and utilization of the 15 SADC shared watercourses and that advances the region's vision and agenda. Guidelines in the interpretation and application of the SADC Revised Protocol on Shared Watercourses (2000) have since been developed to assist MS and River Basin Organizations (RBOs) to develop their own appropriate rules and procedures, particularly with respect to data and information sharing and procedures for implementation of planned measures.

3.2 Key energy sector planning, policy, and legal documents

There are a number of legal documents, policies, and institutional frameworks aimed at facilitating availability of energy and energy security for the SADC region. These are summarized in Table 8.2. The SADC Energy Protocol of 1996 in particular clearly stipulates the region's desire to harmonize national and

Table 8.2 Main policy, planning, and legal documents in the SADC energy sector.

SADC document	Objective
The SADC declaration and treaty (1993)	Governs the regional activities of SADC and its MS aimed at achieving the SADC vision and agenda of regional integration, poverty eradication, industrialization, and economic development.
SADC energy protocol (1996)	To promote the harmonization of national and regional energy policies, strategies, and programs on matters of common interest based on equity, balance, and mutual benefit.
The SADC energy cooperation policy and strategy (1996)	To promote the sustainable, coordinated management, protection, and efficient utilization of energy resources.
The SADC energy action plan (1997) and (2000)	Programmes for creating the enabling environment and harmonization of policy/legal and regulatory frameworks.
The SADC regional indicative strategic development plan (2003, 2007, 2015)	Outlines the key interventions necessary to deepen the SADC vision over the period 2005 to 2020 and stipulates various targets for energy access.
The SADC regional energy access strategy and action plan (2010)	Details strategies and plans of halving the proportion of people without access to adequate, reliable, least-cost, environmentally sustainable energy services by 2020 for each end use and thereafter halving the proportion in successive 5 year periods until there is universal access for all end users.
The SADC regional infrastructure development master plan (2012)	Defines the minimum but ultimate regional infrastructure development requirements and conditions to facilitate the implementation and realization by year 2027 of the key infrastructure in the water, energy, transport, tourism, meteorology, and telecommunication sectors that will move forward the SADC Agenda and enable the SADC region realize its goal: *The attainment of an integrated regional economy on the basis of balance, equity, and mutual benefit for all MS.*
The SADC renewable energy and energy efficiency strategy and action plan (REEESAP) (2017)	To provide a framework for MS to develop their own renewable energy and energy efficiency strategies and action plans, leading to greater uptake of renewable energy resources as well as mobilization of financial resources for the sector.
SAPP plans	Regional electricity planning and trading; project development and implementation.
The SADC industrialisation strategy and road map (2015–2063)	To increase competitiveness (at the firm/industry, country, and regional level) with a quantitative goal to lift the regional growth rate of real GDP from 4% annually (since the year 2000) to a minimum of 7% a year.
RERA instruments	Electricity regulatory frameworks.

Continued

Table 8.2 Main policy, planning, and legal documents in the SADC energy sector. *continued*

SADC document	Objective
UN sustainable development goal on energy (SDG7)	Ensuring access to affordable, reliable, sustainable and modern energy for all (inspired formulation of the vision adopted for the REEESAP).

MS, *Member States;* SADC, *Southern African Development Community.*
Southern Africa Development Community (SADC), 2016. Regional Strategic Action Plan on Integrated Water Resources Development and Management Phase IV. <https://www.sadc.int/files/9914/6823/9107/SADC_Water_4th_Regional_Strategic_Action_Plan_English_version.pdf> (accessed March 2021).

regional energy policies, strategies, and programs on matters of common interest based on equity, balance, and mutual benefit. This Protocol and other legal and policy frameworks, as well as the strategic action plans, have created an enabling environment for investment and implementation of programs and projects in the energy sector and for inculcating economic cooperation among the SADC MS.

3.2.1 Key agricultural sector planning, policy, and legal documents

The agriculture sector contributes between 4% and 27% of GDP and approximately 13% of overall export earnings in the various SADC MS. About 70% of the region's population depends on agriculture for food, income, and employment; hence the performance of this sector has a strong influence on food security, economic growth, and social stability in the region (https://www.sadc.int/themes/agriculture-food-security/ accessed January 05, 2021).

The Food, Agriculture and Natural Resources Directorate (FANR) of the SADC Secretariat is responsible for programs in food security, crop and livestock production, and fisheries. Table 8.3 describes the main SADC legal documents, policies, and strategies for the agricultural sector.

Table 8.3 Main SADC policy, planning, and legal documents for food security.

SADC document	Objective
The SADC declaration and treaty (1993)	Governs the regional activities of SADC and its MS aimed at achieving the SADC vision and Agenda of regional integration, poverty eradication, industrialization, and economic development.
SADC protocol on wildlife conservation and law enforcement (1999)	To establish a common framework for conservation and sustainable use of wildlife in the region.
SADC protocol on fisheries (2001)	To promote and enhance food security and human health, the economic opportunities for nationals of the region are generated so as to alleviate poverty with the ultimate goal of poverty eradication.

Table 8.3 Main SADC policy, planning, and legal documents for food security. *continued*

SADC document	Objective
Dar-es-salaam declaration on agriculture and food security in the SADC region (2004)	Sets out SADC MS' commitment to enhancing agriculture as a means of improving access to food for people in the region. MS agreed to implement short-, medium-, and long-term objectives to advance the state of agriculture and food security in southern Africa. Short-term plans focus on raising the level of agriculture and food security through such means as ensuring small farmers access agricultural inputs, improving fertilizer usage in the region, and increasing production of drought-resistant crops and short-cycle livestock. Medium- to long-term approaches concentrate on maintaining sustainable agriculture and food security measures through environmental conservation, disaster preparation, and research into modern agricultural technologies. The declaration instructs the SADC integrated committee of ministers to implement the related plan of action, reviewing its progress every 2 years.
The SADC regional indicative strategic development plan (RISDP) (2003, 2007, 2015)	Outlines the key interventions necessary to deepen the SADC vision over the period 2005 to 2020 and stipulates various targets for food and nutrition security.
Charter establishing the centre for coordination of agricultural research and development (CCARDESA) (2010)	A charter establishing the centre for coordination of agricultural research and development for southern Africa.
The SADC regional infrastructure development master plan (RIDMP) (2012)	Defines the minimum but ultimate regional infrastructure development requirements and conditions to facilitate the implementation and realization by year 2027 of the key infrastructure in the water, energy, transport, tourism, meteorology, and telecommunication sectors that will move forward the SADC Agenda and enable the SADC region realize its goal: *The attainment of an integrated regional economy on the basis of balance, equity and mutual benefit for all MS.*
SADC regional agriculture policy (2014)	To enhance sustainable agricultural production, productivity, and competitiveness; improve regional and international trade and access to markets of agricultural products; improve private and public sector engagement and investment in the agricultural value chains; and reduce social and economic vulnerability of the region's population in the context of food and nutrition security and the changing economic and climatic environment.

Continued

Table 8.3 Main SADC policy, planning, and legal documents for food security. *continued*

SADC document	Objective
SADC food and nutrition security and vulnerability synthesis report (2020)	Provides an overview of vulnerability across the region, especially as it relates to food and nutrition security.
The SADC industrialisation strategy and road map (2015–2063)	Aims to increase competitiveness (at the firm/industry, country and regional level) with a quantitative goal to lift the regional growth rate of real GDP from 4% annually (since the year 2000) to a minimum of 7% a year.
SADC control strategy for peste de petit ruminants (PPR) (2010)	Peste de petit ruminants (PPR) is a serious viral disease of goats and sheep that causes high mortality in these two species with significant economic impact. Following an outbreak in two countries within the region in 2010, SADC developed a regional PPR control strategy in response to the outbreak of this disease.
The Global report on food crises (GRFC) (2020)	Describes the scale of acute hunger in the world. It provides an analysis of the drivers that are contributing to food crises across the globe and examines how the COVID-19 pandemic might contribute to their perpetuation or deterioration.
UN sustainable development goal on hunger (SDG2)	Zero hunger. The food and nutrition sector is central to hunger and poverty eradication and offers solutions for development.

MS, *Member States*; SADC, *Southern African Development Community.*

Of note is the SADC Protocol on Fisheries of 2001, which seeks to promote and enhance food security and human health, the generation of economic opportunities for nationals of the region are generated so as to alleviate poverty with the ultimate goal of poverty eradication.

4. Identified challenges related to the water−energy−food nexus approach in the SADC region

The major challenges related to the WEF nexus approach in the SADC region are summarized as follows.[1]

---

[1] Southern African Development Community, 2019. *SADC Water-Energy-Food (WEF) Nexus Framework.*

4. Identified challenges related to the water—energy—food nexus approach in the SADC region

4.1 Inadequate coordination of the three sectors at policy- and decision-making levels

Regional- and country-level assessments on WEF nexus indicated that one of the challenges of promoting the WEF nexus approach in the SADC region is the sector-focused policies and institutions with inadequate coordination mechanisms. The water, energy, and food policies in the SADC region are sector-focused with limited recognition of the interlinkages between the water—land—energy resources. Similarly, the institutions and governance arrangements are structured around the sectors. Sectors are working in silos (sectoral policy formulation, planning, budgeting, and implementation) with limited or no interaction between sectors, overlapping mandates and power dynamics among sectors.

The following table shows the level of WEF coherence of SADC polices, strategies, and plans (SADC, 2019). WEF coherence is greatest with respect to the policies, plans, and strategies for water, followed by agriculture, and finally energy. Policy statements and strategies such as the promotion of multipurpose reservoirs that serve drinking water, irrigation, and hydropower requirements are highly coherent with WEF nexus approach; for example, as advocated in the Regional Water Strategy (SADC, 2006). Energy plans are the relatively least coherent with regard to WEF, leaning toward a primarily focus on energy sector requirements and planning, particularly for economic growth, with relatively little attention to water and especially agriculture. The Regional Strategic Action Plan for IWRM (Southern African Development Community, 2016) is highly coherent with WEF, which can be expected owing to the holistic and intersector focused approach of Integrated Water Resource Management (IWRM) with respect to WEF. The Regional Indicative Strategic Development Plan (Southern African Development Community, 2015) scores relatively well, partial to high coherence, in part owing to its mandate acting as a guiding plan to direct future intersector activities for the SADC region. The Regional Agricultural Policy (Southern African Development Community, 2014) is also highly coherent with respect to WEF, including numerous cross-sector strategies and plans.

Generally, low WEF security mean score as compared with WEL resource use efficiency mean score indicates that the water, agriculture, and energy policies, strategies, and plans consider biodiversity, land resources management, and climate change issues as secondary importance relative to water, energy, and food security.

	WEF Security				WEL Resource Use Efficiency				Overall
	Water	Energy	Food	WEF mean score	Climate change	Biodiversity Ecosystems	Land	WEL mean score	Overall mean score WEF and WEL
Regional Water Policy (SADC, 2005)	-	3	3	3	1	2	1	1,33	2.12
Regional Water Strategy (SADC, 2006)	-	3	3	3	3	2	1	2	2.37
Guidelines for Strengthening River Basin Organisations (SADC, 2010a)	-	1	1	1	1	3	1	1,67	1.75
Climate Change Adaptation in SADC, Strategy for the Water Sector (2011)	-	2	2	2	-	1	1	1	1.71
Regional Strategic Action Plan. Integrated Water Resources Development and Management, 2016-2020 (SADC, 2016a)	-	3	3	3	3	3	2	2,67	2.85
Regional Infrastructure Development Master Plan, Energy Sector (2012)	3	-	1	2	3	1	1	1,67	1.5
Renewable Energy and Energy Efficiency Strategy and Action Plan, 2016-2030 (SADC, 2016b)	2	-	2	2	3	1	1	1,67	2
Regional Energy Access Strategy and Action Plan (SADC, 2010b)	1	-	1	1	0	0	1	0,33	0.77
Regional Agricultural Policy (SADC, 2014)	3	3	-	3	3	1	3	2,33	2.25
Regional Food and Nutrition Security Strategy (SADC, 2015a)	2	0	-	1	3	0	2	1,67	1.37
Regional Indicative Strategy Development Plan, 2015-2020 (2015b)	3	2	3	2.66	3	2	2	2,33	2.44
Regional Biodiversity Strategy (2008)	1	1	2	1.33	2	-	3	1,67	1.75
Subject mean coherence	2.14	2	2.1	-	2.27	1.45	1.58	-	-

Type of coherence	Description of coherence	Score
High coherence	The policy, strategy or plan aligns strongly across water, energy and food security. It also devotes specific attention to sub-components (climate change, ecosystems and biodiversity, land, integration, participation and gender, and livelihoods). It includes numerous, comprehensive and detailed complementary statement, strategies, activities and plans for both WEF and sub-components.	3
Partial coherence	Although the policy, strategy or plan supports water, energy and food security inter-sector alignment, as well as for sub-components, it is relatively less clear and distinct how it could be achieved. A few statements, strategies, activities and plans are included for WEF and sub-components, but lacks relative comprehensive details.	2
Limited coherence	The policy, strategy or plan supports water, energy and food security inter-sector alignment, as well as for sub-components, in the form of general statements. No statements, strategies, activities or plans are provided.	1
No coherence	No evidence that water, energy, food security or sub-components statements are aligned or coordinated. No relevant statements, strategies, activities or plans are provided.	0

Similarly, the SADC Governance structures are largely dominated by sectoral orientations. The Water, Energy and Agriculture Regional Technical Committees follow the lines of their respective sectoral senior officials and then the respective sectoral ministries. However, the Joint SADC Water and Energy Ministers meetings are establishing a very useful mechanism for coordinating the two sectors. There is a plan to bring the agriculture ministers into these meetings to make it an SADC WEF Ministers meeting.

4. Identified challenges related to the water–energy–food nexus approach in the SADC region

4.1.1 Inadequate coordination of the three sectors at the regional technical level

The SADC region generally has sector-focused and uncoordinated plans and targets. This can be demonstrated by looking at the SADC Regional Infrastructure Development Master Plan (RIDMP), which sets out the region's infrastructure development targets. The RIDMP targets were set for the different sectors (agriculture, water, energy) without adequately considering the available water, land, and energy resources of the region. Furthermore, the programs are more sector-based such as energy sector development, agricultural sector development, or water supply service programs. The focus is on attaining sector-specific targets rather than meeting comprehensive and integrated WEF targets.

4.1.2 Absence of a formal structure to facilitate coordination of the units responsible for the WEF sectors within the SADC secretariat

At the SADC Secretariat level, there are two directorates under the Deputy Executive Secretary (Regional Integration) that are dealing with water, energy, and food security issues. One is the Directorate for Infrastructure (responsible for water and energy), and another is the Directorate for Food, Agriculture and Natural Resources (FANR) (responsible for agriculture, food security, natural resources management, environment and climate change, and tourism). One weakness identified at the SADC Secretariat is that each Directorate generally works separately with relevant SADC subsidiarity organisations, partners, and stakeholders without adequate coordination. Each Directorate reports to their relevant cluster structures of the SADC main governance structure of technical committees rising up the decision-making channel to the Council of Ministers.

4.1.3 Weak coordination of programs by the regional implementing entities and other partners

SADC has regional subsidiary entities that support implementation of programs. However, coordination of their programs is through the respective sectoral units of the SADC secretariat. There has not been a clear mechanism for coordinating the programs of such entities from the WEF nexus perspective.

4.1.4 Inadequacy of the existing regional water multistakeholder platform to facilitate WEF nexus dialogues

One of the strengths of the SADC region with respect to the WEF nexus approach is availability of different regional platforms for multistakeholder consultation and dialogue. The SADC Regional Water Dialogue and the SADC River Basin Organizations' Meetings provided several opportunities for discussing WEF nexus issues at regional level. The scope for the SADC Water Multi-Stakeholder Dialogue has been broadened to include energy and food security sectors and has been renamed as the SADC Regional WEF Dialogue.

5. Operationalizing the WEF nexus in Southern Africa
5.1 SADC WEF nexus governance framework

The SADC Regional WEF Nexus Framework guides the institutionalization of the WEF nexus approach in the Community and was endorsed by the SADC Ministers responsible for water, energy, and agriculture in 2020. It facilitates integrated planning and development to support SADC's developmental agenda. The Framework is developed as an organizing mechanism for the coordination of institutions, policies, strategies, programs, and projects to achieve WEF security and ensure natural resource efficiency.[1]

The SADC Regional WEF Nexus Framework is expected to bring about alignment/coherence between the water, energy, and food policies; facilitate institutional coordination; align development strategies/targets/programs of the three sectors; and manage trade-offs and promote nexus investments in the region.

The objectives of the Framework are to

- Facilitate integrated planning and implementation of initiatives that will drive water, energy, and food security;
- Ensure simultaneous achievement of water, energy, and food security in the region through optimization of investments; and
- Improve sustainable use and management of natural resources underpinning development in the region (Fig. 8.2).

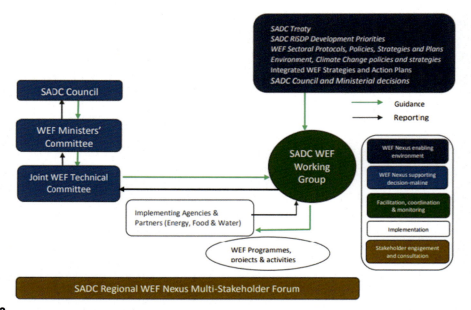

FIGURE 8.2
SADC WEF nexus framework. *Southern African Development Community (SADC), 2019. SADC Water-Energy-Food (WEF) Nexus Framework.*

The Framework is expected to contribute in addressing the main challenges related to the WEF nexus approach in the region by providing overall guidance for (1) coordinating the three sectors at policy- and decision-making levels, (2) coordinating the three sectors at the regional technical level, (3) coordinating the units responsible for the WEF sectors within the SADC Secretariat, (4) coordinating with regional implementing entities and other partners, (5) strengthening regional multistakeholder platforms. These points are elaborated further in the following paragraphs.

5.1.1 Coordinating the three sectors at policy- and decision-making levels

As described earlier, one of the challenges of promoting the WEF nexus approach in the SADC region is the sector-focused policies and institutions with inadequate coordination mechanism. The SADC Regional WEF Nexus Framework is expected to provide the guidance on how to strengthen coordination between the three sectors in developing and reviewing policies, plans, and strategies through establishing a Joint WEF Ministers' Committee in the SADC Governance Structure.

5.1.2 Coordinating the three sectors at the regional technical level

The SADC region generally has sector-focused and uncoordinated plans and targets. The SADC Regional WEF Nexus Framework is expected to provide the guidance on how to develop and implement integrated plans and programs to achieve water, energy, and food security targets while sustaining natural resources management. This will be achieved through establishing a Joint WEF Technical Committee with support from the SADC WEF Working Group at the SADC Secretariat.

5.1.3 Coordinating the Units responsible for the WEF sectors within the SADC Secretariat

One weakness identified at the SADC Secretariat is inadequate coordination between the directorates and between the units that are responsible for the water, energy, and food security sectors. The SADC Regional WEF Nexus Framework is expected to improve the coordination of the units responsible for WEF sectors at the SADC Secretariat through establishing an SADC WEF Working Group.

5.1.4 Coordinating with regional implementing entities and other partners

Another challenge identified is inadequate coordination of the programs and activities of the SADC implementing entities and other partners. The SADC Regional WEF Nexus Framework is expected to improve coordination of the programs and activities by SADC's subsidiary entities and other partners through establishing an SADC WEF Working Group. The SADC Secretariat will also provide guidance and technical support (including WEF Nexus project screening tool) to such entities and partners.

5.1.5 Strengthening regional multistakeholder platforms

Broadening the existing regional multistakeholder platforms to include the three sectors and other relevant stakeholders is an important requirement for promoting the WEF nexus approach. The SADC Regional WEF Nexus Framework is expected to provide a wider forum with a balanced representation of water, energy, and agriculture sectors through establishing an SADC regional WEF nexus multistakeholder forum.

In summary, the SADC regional policies and strategies provide the overall enabling environment in adopting the WEF nexus approach in the region. This will include the SADC regional overarching treaties, policies, strategies, programs, or the decisions of the SADC Council of Ministers. It also includes the water, energy, and agriculture sectoral policies, strategies, and programs.

The existing SADC governance structure provides the institutional framework for coordinating the development and implementation of policies related to water, energy, and food security sectors. The WEF Nexus Working Group under the SADC Secretariat provides an overall secretariat function and plays technical coordination role. The Joint WEF Technical Committee is not a new structure. It is mainly organizing joint meetings of the existing technical committees for water, energy, and agriculture sectors. Similarly, the Joint WEF Ministers' Committee is also organizing joint meetings of the Ministers responsible for water, agriculture, and energy. The Water and Energy Ministers meetings are already organized as joint meetings. It is important to note that the Ministers of water, energy, and agriculture may agree to form a joint Ministerial Committee to facilitate their work. The same approach may be followed at regional technical committee levels. The decisions of the SADC Council of Ministers provide the highest level of strategic or policy guidance on WEF nexus issues.

The SADC regional WEF nexus multistakeholder forum provides a platform for multisectoral engagement in the region. This is also not a new structure. It is mainly widening the existing SADC Water Sector multistakeholder dialogue to include stakeholders from the energy and agriculture sectors.

5.2 Implementing the SADC regional WEF nexus framework

Implementing the SADC Regional WEF Nexus Framework requires undertaking of some actions toward achieving the following.

5.2.1 Coordinating the three sectors at policy- and decision-making levels

Strengthening coordination between the three sectors in developing and reviewing policies, plans, and strategies requires the formation of a Joint WEF Ministers Committee in the SADC Governance Structure. The Meeting is expected to provide policy guidance regarding the WEF nexus approaches in the region and regarding the implementation of the Nexus Framework.

5.2.2 Coordinating the three sectors at the regional technical level

Developing and implementing integrated plans and programs to achieve water, energy, and food security targets while sustaining natural resources management requires the establishment of a regional-level technical coordinating structure. The Regional WEF Nexus Framework recognizes the formation of a Joint WEF Technical Committee to provide technical clearance service on WEF nexus related initiatives in the region and advise the WEF Ministers to take appropriate decisions.

5.2.3 Coordinating the units responsible for the WEF sectors within the SADC secretariat

The key coordination mechanism considered by the Regional WEF Nexus Framework at the SADC Secretariat level is the formation of a WEF Working Group at the SADC Secretariat. This arrangement is expected to improve the coordination of the units responsible for WEF sectors at the SADC Secretariat. The main functions of the Working Group are facilitation, coordination, and monitoring.

5.2.4 Coordinating with regional implementing entities and other partners

The establishment of a WEF Nexus Working Group at the SADC Secretariat is also expected to improve coordination of the programs and activities by SADC's subsidiary entities and other partners. The SADC Secretariat, through the Working Group, will provide guidance and technical support (including use of the WEF Nexus project screening tool) to the regional subsidiary entities and other partners. The guidance will cover how to integrate the WEF nexus approach in programs and align with the SADC development agenda. Other functions of the Secretariat will be providing MS with guidelines, tools, information, and strengthening their capacities. It will also facilitate experience and knowledge sharing on WEF nexus issues.

5.2.5 Strengthening regional multistakeholder platforms

The WEF nexus approach requires a multistakeholder and multisectoral dialogue platform. The SADC Regional WEF Nexus Framework considered the strengthening of an SADC regional WEF nexus multistakeholder forum.

5.3 SADC WEF nexus screening tool for guiding discourse in the region

A WEF Nexus Investment Projects Screening Tool was developed for SADC based on the conceptual framework that defined the WEF nexus parameters for the region. The tool was used in identifying and screening investment projects that have WEF nexus potential. A long list of projects was identified as projects where the nexus approach could be applied. The sources of the projects were from the SADC regional development Master Plans, Sectoral Programs, Basin Programs, or country prioritized projects. The Tool (web-based) was tested by the SADC MS and is expected to support SADC in identifying nexus opportunities of investment projects in the region. A list of 15 potential projects for applying WEF nexus approach was prioritized for further nexus analysis to improve the quality of the projects to attract funding for implementation.

5.4 Capacity development and guiding discourse in the region

Capacity building of regional stakeholders engaging closely with the WEF nexus approach is required, especially on conducting WEF nexus analyses and using the WEF nexus approach to inform decision-making. Furthermore, the documenting of regional case studies to facilitate the practical learning and regional knowledge sharing of WEF Nexus applications will be critical to achieve stakeholder buy-in.

Phase II of the SADC Nexus Dialogue Project on "Fostering Water, Energy and Food Security Nexus Dialogue and Multi-Sector Investment in the SADC Region" places a particular focus on strengthening capacity of stakeholders at both the regional and national levels for planning, policymaking, and implementation of the WEF nexus approach. From 2020 to 2023, the project will endeavor to organize regional trainings and 16 country WEF Nexus Dialogues to support the establishment of governance structures that will enable the integration of a nexus approach at national decision-making levels.

Furthermore, phase II of the project aims to develop Regional WEF Nexus Guidelines to support implementing agencies at the national, regional, and transboundary levels on how to integrate the WEF nexus in investment planning and project preparation. The Regional Guidelines will make use of case studies within the SADC region to practically demonstrate and provide technical guidance and support required to identify and prioritize WEF nexus investment projects and facilitate the operationalization of the WEF nexus in the SADC region.

6. Key lessons from the implementation of the SADC WEF nexus regional dialogues project

The results expected from implementing the EU supported SADC Regional WEF Nexus Dialogue Project were successfully achieved. The SADC Regional WEF Nexus Framework was developed and endorsed by the SADC Ministers responsible for water, energy, and food security in 2020. A list of 15 investment projects (with brief project profiles) that have potential for applying WEF nexus approach was also developed.

The WEF Nexus approach is gaining momentum in the SADC region. WEF nexus is understood as an approach to facilitate better interactions and synergies between the water, food, and energy sectors to unlock and optimize the development potential for economic growth and transformation in the region. Discussions on the WEF nexus are going on at higher SADC governance structures. In 2018 and 2019, WEF nexus was discussed at SADC Water, Energy, and Agriculture Ministerial levels. The SADC Council Decisions in August 2018 called for strengthened collaboration among water, energy, and food sectors.

The approach is also being embraced by the SADC RBOs. The 8th SADC RBOs workshop in 2018 recognized the WEF nexus as an approach to facilitate investment at basin level. Moreover, the 3rd Zambezi Basin Stakeholders' Forum in 2018 discussed the role of the WEF nexus approach at river basin level, specifically at Zambezi river basin.

The main lessons learned from implementing the SADC Nexus Dialogue Project is that adopting a WEF nexus approach is a long process that requires the following:

- Defining the scope and objectives of the nexus approach: SADC defined the conceptual understanding and objectives of the nexus approach
- Securing political support for the nexus approach: SADC provided high-level political support for the WEF nexus approach establishing and facilitating a continuous multistakeholder/sectoral dialogue/engagement: the SADC regional multistakeholder dialogues provided useful platforms
- Building on existing structures rather than creating new ones. The SADC regional framework was building on the existing SADC structures.
- Considering capacity building as part of the intervention.

7. Summary and conclusions

The WEF nexus in the SADC region is understood as an approach to facilitate better interactions and synergies between the water, food, and energy sectors to unlock and optimize development potential for economic growth and transformation in the region. The WEF nexus approach also presents opportunities for greater resource coordination, management, and policy convergence across sectors. SADC expects the WEF nexus approach to enhance investment in the region.

The SADC has developed a regional WEF Nexus Framework that will drive implementation of the WEF nexus approach toward integrated planning and development among the water, energy, and food security sectors in the SADC region.

It is strongly believed that the next WEF nexus initiatives will build on the achievements of the project. It is expected that the Regional Nexus Dialogue Project Phase II will focus around supporting MS in embracing the WEF nexus approach and operationalizing the SADC regional Nexus Framework. It is also expected that the other regions in Africa and the African Union will take similar initiatives with contribution from the SADC region through sharing its experience.

In conclusion, the implementation of the Regional Nexus Dialogue Project in the SADC region was very successful in achieving its goal of establishing an enabling framework for WEF nexus in the region, and in identifying investment projects with potential for nexus application. These results, together with the high-level political support, will provide a solid foundation for implementing the nexus approach in the region. The success of this project has attracted a lot of interest not only from the SADC MS but also from the other regions in Africa and beyond.

References

Kusangaya, S., Warburton, M., Archer van Garderen, E., Jewitt, G., 2014. Impacts of climate change on water resources in Southern Africa: a review. Phys. Chem. Earth, Parts A/B/C 67–69, 47–54.

Mabhaudhi, T., Mpandeli, S., Madhlopa, A., Modi, A., Backeberg, G., Nhamo, L., 2016. Southern Africa's water—energy nexus: towards regional integration and development. Water 8 (6), 235.

Nhamo, L., Ndlela, B., Nhemachena, C., Mabhaudhi, T., Mpandeli, S., Matchaya, G., 2018. The water-energy-food nexus: climate risks and opportunities in southern Africa. Water 10 (5), 567.

SADC Centre for Renewable Energy and Energy Efficiency(SACREEE), 2018. SADC Renewable Energy and Energy Efficiency Status Report 2018. https://www.ren21.net/2018-sadc-renewable-energy-and-energy-efficiency-status-report/. (Accessed March 2021).

SADC, 2000. SADC Revised Protocol on Shared Watercourses [online]. https://www.sadc.int/files/3413/6698/6218/Revised_Protocol_on_Shared_Watercourses_-_2000_-_English.pdf. (Accessed March 2021).

Southern African Development Community (SADC), 2006. Southern African Development Community Regional Water Strategy. SADC. https://www.sadc.int/documents-publications/show/Regional_Water_Strategy.pdf. (Accessed March 2021).

References

SADC, 2012. Regional Infrastructure Development Master Plan — Water Sector Plan. Available at. http://www.sadc.int/files/6313/5293/3538/Regional_Infrastructure_Development_Master_Plan_Water_Sector_Plan.pdf. (Accessed March 2021).

Southern Africa Development Community (SADC), 2016. Regional Strategic Action Plan on Integrated Water Resources Development and Management Phase IV. SADC. https://www.sadc.int/files/9914/6823/9107/SADC_Water_4th_Regional_Strategic_Action_Plan_English_version.pdf. (Accessed March 2021).

Southern African Development Community (SADC), 2015. Regional Indicative Strategic Development Plan. SADC. www.sadc.int/files/5713/5292/8372/Regional_Indicative_Strategic_Development_Plan.pdf. (Accessed March 2021).

Southern African Development Community (SADC), 2014. SADC Regional Agricultural Policy. SADC. www.nepad.org/publication/sadc-regional-agricultural-policy-0. (Accessed March 2021).

Southern African Development Community (SADC), 2018. SADC Energy Monitor 2018: Enabling Industrialisation and Regional Integration in SADC. SADC. www.sadc.int/files/5515/6837/8450/SADC_ENERGY_MONITOR_2018.pdf. (Accessed March 2021).

Southern African Development Community (SADC), 2019. SADC Water-Energy-Food (WEF) Nexus Framework.

Southern Africa Development Community (SADC), 1992. Treaty of the Southern African Development Community. SADC [online]. www.sadc.int/files/9113/5292/9434/SADC_Treaty.pdf. (Accessed March 2021).

CHAPTER 9

Exploring the contribution of Tugwi-Mukosi Dam toward water, energy, and food security

Never Mujere[1] and Nelson Chanza[2]
[1]Department of Geography, Geospatial Science and Earth Observation, University of Zimbabwe, Mt Pleasant, Harare, Zimbabwe; [2]Department of Urban and Regional Planning, University of Johannesburg, Johannesburg, South Africa

1. Introduction

About a billion of the world's population faces water insecurity, which is interconnected equally with devastating energy and food security (Wolde et al., 2019). Access to safe and adequate water, energy, and food is critical to meet the Sustainable Development Goals (SDGs) (UN, 2020). The SDGs are a global blueprint development agenda toward a pathway for a sustainable future by all countries by 2030. The goals indicate that ending poverty and other development challenges must be synchronized with strategies that improve health and education, reduce inequality, and promote economic growth while tackling climate change and preserving forests, ecosystems, and water resources (United Nations General Assembly, 2015). As such, the water—energy—food (WEF) nexus is prominently articulated in the global development action plan (Wolde et al., 2019). The "web" specifically addresses SDGs 2, 6, and 7. However, since 2014, the number of food-insecure people worldwide has gradually risen. The situation is likely to be aggravated by the COVID-19 pandemic. In Africa, the prevalence of undernutrition rose from 17.6% in 2014 to 19.1% in 2019 (FAO, IFAD, UNICEF, WFP, WHO, 2020). This suggests that Africa is not on track to achieve zero hunger by 2030. About 29% of the global population has no access to safe water, with the largest share being in Africa (UN, 2020). Notwithstanding the progress toward increasing access to electricity and improving energy efficiency witnessed globally, the world's energy deficit is still concentrated in sub-Saharan Africa, where 53% of the population lack access to electricity (UN, 2020).

There is now a growing acknowledgment of the intrinsic WEF nexus in development discourse (Albrecht et al., 2018; Chen et al., 2019; Lebel et al., 2020; Purwanto et al., 2021; Pueppke, 2021; Zarei et al., 2021). Joint treatment of WEF is understood as a robust tool to transform the livelihoods of communities that face poverty in Africa (Tarisayi, 2014; Mabhaudhi et al., 2018; Gebreyes et al., 2020). Interest to interlink these resources emanates from increased understanding that interventions to attain food security, access safe, and adequate water and obtain clean energy are symbiotic, and the limitation in one element has negative implications on other elements. For example, water and energy are important requirements in food production; energy is also important in water management; and energy generation needs water (Wolde et al., 2019). Given this strong nexus and how it influences the attainment of livelihoods security, there is a need to continue examining the WEF issues, particularly in sub-Saharan Africa where climate change and other stressors threaten the achievement of SDGs (Mpandeli et al., 2018; Pardoe et al., 2018).

Previous WEF nexus studies associated with dam projects have shown mixed impacts of such infrastructure. Gao et al. (2021) used a WEF nexus model to investigate the impacts of value preferences from riparian countries in the Lancang-Mekong River Basin, indicated water-use conflicts for energy generation and food production, and ecosystem maintenance between upstream and downstream countries. In Ethiopia, a study by Gebreyes et al. (2020) of households around two large-scale irrigation and hydropower dams in the Upper Blue Nile basin revealed that the impact of dams and the perception of communities around is socially diverse. In Zambia, the Itezhi-Tezhi dam, primarily designed for hydropower generation, has great irrigation potential to the farming community and enhances food security in the country. Similarly, Zimbabwe's dams have the potential to unleash numerous development prospects to the surrounding communities. Although Kariba Dam, which is shared between Zambia and Zimbabwe, was primarily commissioned for hydropower generation, its other major uses include aquaculture, urban water supply, tourism, wildlife support, and lake transportation. There are planned efforts to accessorize such dams for multipurpose use. In Zimbabwe, the Tugwi-Mukosi Dam has been earmarked to provide multisectoral development projects (Tarisayi, 2015; Hove, 2016).

This study provides a critical appraisal of the synergies between WEF interconnection and sustainable livelihoods, both of which aim to promote sustainable development. By adopting a case study review of Tugwi-Mukosi Dam in Zimbabwe, this study examines the WEF nexus and the opportunities for enhancing the livelihoods of surrounding rural communities. The study's research question was: How can the availability of water, energy, and food improve the capacities of communities to increase livelihood options? The results of this study are intended to inform evidence-based decision-making for transforming the livelihoods of riparian communities. This chapter is structured as follows: the next section articulates the WEF nexus framework and its connection to livelihoods, followed by an overview of the Tugwi-Mukosi Dam. The study then discusses the WEF interconnection and livelihoods

3. Tugwi-Mukosi Dam   171

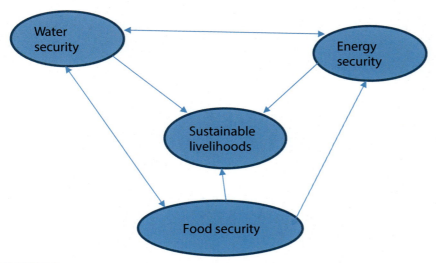

FIGURE 9.1
Conceptual model showing the water—energy—food nexus and its linkage to livelihood.

implications of the dam to the surrounding community. Lastly, the research offers suggestions on policy directions for addressing the complex WEF nexus to ensure that local communities would benefit from the opportunities associated with dam infrastructure.

2. The WEF linkage conceptual framework

A framework that describes WEF security was chosen to guide this research (Fig. 9.1). It focuses on the interlinkages between WEF resources and the livelihoods of local communities. This framework depicts water security, energy security, and food security as connected and affects livelihoods (Mabhaudhi et al., 2018). For example, water is essential for food and energy production. Energy is required to deliver water to crops and to process food. The indicators of assessing water, energy, and food security are shown in Table 9.1.

3. Tugwi-Mukosi Dam

Tugwi-Mukosi (formerly Tokwe-Mukosi) Dam (Fig. 9.2) was constructed across the confluence of Tugwi and Mukosi Rivers in the Chivi district of Masvingo Province in Zimbabwe.

It is the country's largest inland dam with a storage capacity of 1.8 billion m^3, a catchment area of 7120 km^2 and a dam wall measuring 90 m high. A foreign investor built the dam at the cost of nearly US$270 million. The completion

Table 9.1 Indicators of the WEF security.

1. Water	2. Energy	3. Food
Proportion of crops produced per unit of water used (water productivity)	Proportion of population with access to electricity (accessibility)	Absent prevalence of moderate or severe food insecurity in the population (self-sufficiency)
Proportion of available freshwater resources per capita (availability)	Energy intensity measured in terms of primary energy and GDP (productivity)	Proportion of sustainable agricultural production per unit area (cereal productivity)
Proportion of available safe water resources (safety)	Amount of energy available to meet demand (sufficiency)	Proportion of available sufficient food per capita (availability)
Proportion of the time to which water supply remains stable and adequate (stability)	Energy mix in terms of clean, renewable, and nonrenewable (energy type)	Proportion of population with access to food (accessibility)
Willingness to pay for water by customers measured in terms of pricing (affordability)	Proportion of time when energy is sufficient (reliability)	The duration to which food prices remain stable and whether the customers can afford to pay (affordability, stability)

3. Tugwi-Mukosi Dam

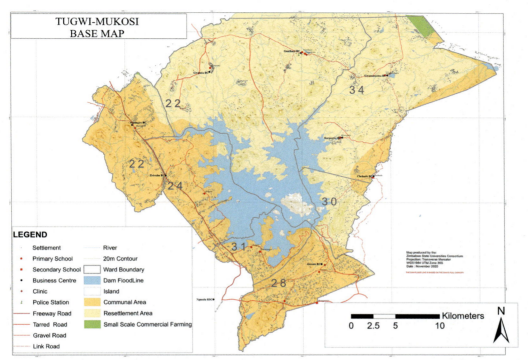

FIGURE 9.2
Layout of Tugwi-Mukosi master plan.

of Tugwi-Mukosi in May 2017 was one of key deliverables of the country's medium-term economic blueprint, which considers full utilization of critical resources as a springboard toward national development.

The dam's main purpose is to provide irrigation water and hydropower electricity to communities in the semiarid southern part of Masvingo province. It is intended to mitigate the adverse effects of climate change and water shortages in this drought-prone area. The dam is expected to act as a catalyst for socioeconomic transformation in the province due to opportunities in agriculture, fisheries, and tourism. Significant benefits of mass employment and increased agricultural production are expected. Controlled water for irrigation will allow the expansion of commercial sugar plantation to sustain irrigation schemes for small-scale and large-scale commercial farmers. Also, the dam is believed to boost sugar production in Triangle, Hippo Valley, and Mkwasine estates downstream (Hove, 2016). The development of irrigation plots will also benefit the relocated families.

The government promised to relocate 6393 families living around the dam site to Nuanetsi Ranch, Masangula, Chisase, and Chingwizi resettlement sites in

Masvingo province. Almost 400 families (approximately 2500 individuals) were displaced from their localities in Chivi district to make room for dam construction and relocated about 100 km away to Nuanetsi Ranch, in Mwenezi district. Chisase site in Masvingo district has 126 families serviced by 1 clinic, 1 school, and 42 boreholes. Chingwizi site in Mwenezi district accommodates 5782 families with 2 clinics, 5 schools and 63 boreholes. Masangula site in Mwenezi district is inhabited by 485 families, with 2 clinics, 3 schools, and 42 boreholes. Each displaced family was promised a 17-ha arable land and some monetary compensation. However, by January 2014, the relocated families were only allocated 4 ha each as compensation. Also, the government had only resettled 712 of 6393 households because most families resisted being relocated before compensation (Mutangi and Mutari, 2014; Tarisayi, 2018).

The resettlement location is said to be unsuitable for anything except livestock farming. There are reports that resettled people often engage in clearing up trees to sell firewood instead of productively utilizing the land (Mutangi and Mutari, 2014). The relocation negatively affected school children whose education was disturbed during movements. Schools are far away present from homesteads. The dam also displaced wild animals such as hyenas, jackals, and foxes, from the hills engulfed by its water. The animals are destroying crops and livestock, escalating human-wildlife conflicts.

4. Contribution of Tugwi-Mukosi toward water, energy, and food security

Tugwi-Mukosi Dam is viewed as the catalyst of socioeconomic development owing to its vast irrigation potential (Tarisayi, 2018). The project full master plan (Fig. 9.2), which has been a work in progress since 2016, is expected to enhance livelihoods by improving water, energy, and food security of the local community and nation at large in various ways. Both existing and new irrigation schemes, mini hydropower plant, hospitality and recreational facilities, a game park, fisheries, and crocodile farms are expected to be developed. Many families are set to benefit directly and indirectly from economic activities centered on Tugwi-Mukosi Dam. Table 9.2 evaluates the proposed investment opportunity at Tugwi-Mukosi and progress to date.

4.1 Water security

At a volumetric capacity of 364,000 megaliters per annum, Tugwi-Mukosi will have sufficient water to irrigate up to 25,000 ha of land stretching from the Hippo Valley Estates to the eastern parts of the province and down the Runde River to the south (Rusvingo, 2014; Hove, 2016). This will cover both commercial farming and communal farmland. The dam will supply water to the nearby city of Masvingo, where water shortages have become frequent in recent years (Tarisayi, 2015; Hove, 2016).

4. Contribution of Tugwi-Mukosi toward water, energy, and food security

Table 9.2 Socioeconomic expectations from the dam.

Sector	Potential	Status
Agriculture	To irrigate more than 25,000 ha of land	Water has started flowing from the Tugwi-Mukosi Dam to Chiredzi and triangle to irrigate sugarcane fields. The locals have not started benefiting. 6000 ha of land at Masangula has already been cleared to establish irrigation schemes for the displaced families.
Tourism and hospitality industry	An international tourist attraction	Apart from some local schools that have been taking children on tours of the dam, there is little activity that is of immediate benefit to the locals.
Hydropower generation	15 MW power generation	Work in progress.
Biofuel	Ethanol plants from sugarcane	No progress as yet.
Fisheries	To breed 1.5 million kapenta fingerlings	Stocked with 200,000 tilapia fingerlings. The major activity so far has been the harvesting of fish, which benefits a few cooperatives (especially the youths) that have been established.
Other socioeconomic opportunities	To unlock an array of business opportunities in the tourism, hospitality, retail, and transport sectors	No progress as yet.

4.2 Energy security

The Tugwi-Mukosi Dam is earmarked for increased energy generation with a design capacity of 15 MW of hydropower. The power generated will be fed onto the national grid, thus easing power outages in the southern parts of the country (Tarisayi, 2018). It will be supplied to local fishery projects and tourism-related activities such as lodges and hotels. With adequate supply of water from the dam to sugar plantations, renewable energy (biofuel) production is expected to increase. For example, at Chingwizi area in Mwenezi district where 3000 families were resettled, an $80 million ethanol plant by a private investor is set to be developed (Hove, 2016). The availability of electricity and water will help in setting up industries at service centers around the dam. Community members will also benefit as some will be incorporated into the maintenance works. A public institute will be electrified, hence improving health, education services, and quality of life in the region.

4.3 Food security

Tugwi-Mukosi Dam was constructed primarily to boost agriculture production in Masvingo province. The dam is expected to make Chivi South, a green belt with the potential to contribute to national food security and to provide water to irrigate 25,000 ha of land in the local area and downstream plantations. Irrigation will improve production of food crops and cash crops for local consumption and for export. The dam is expected to address food shortages in the district as families will be able to cultivate crops throughout the year under irrigation. There has been indication to engage stakeholders in planning for agricultural development in the area because it is generally affected by drought, which has resulted in some families relying on food handouts from government almost every season (Tarisayi, 2018).

Tugwi-Mukosi Dam is expected to improve food security through fishery projects. The fisheries industry at the dam is expected to support about 1.2 million people at the primary production level. This contributes toward the provision of food, thus reducing poverty and improving the nutritional status of the beneficiaries. The Command Fishery Programme has brought kapenta fish to Tugwi-Mukosi from Lake Kariba. The dam is favorable for fish population growth, game fish angling, and crucial fish tournaments. Although registered fishermen use commercial nets, poachers use poison prepared from a local herb known locally as *mutsvatsva* to kill fish. Illegal fish harvesting and predation have increased due to lax security (Mutengi and Mutari, 2014; Chazireni and Chigonda, 2018).

4.4 Promotion of tourism

Tugwi-Mukosi Dam serves as a tourist attraction for nationals and internationals. Ecotourism projects around the dam are expected to include self-catering chalets, hotels, cable car, conference center and casino, game park, boating facilities, boat cruising, and angling (Chazireni and Chagonda, 2018). Such projects will improve local communities' living standards, recreation activities and employment opportunities and bring foreign currency, hence raising the livelihoods of local communities.

The dam is surrounded by majestic indigenous trees such as *Bjalbertii* (*mutuputupu in the local Shona language*). The tree is distinct, self-pruning, and borer proof and can grow up to 30 m. It is categorized under International Union for Conservation Nature (IUCN) red data list as near threatened. Some of the overlapping benefits include upgrading the nearby Buffalo Range Airport, which will subsequently improve the livelihoods of the local communities through job creation (Chazireni and Chagonda, 2018).

Nevertheless, the dam construction is not devoid of challenges. It displaced people from the homesteads and famlands. New disease ecologies such as schistosomiasis, malaria, and water-borne diseases emerged in the area. The proposed game reserve is likely to trigger human-wildlife conflicts. The dam and its environs can experience seismic earth tremors due impounding of water and sediments (Rusvingo, 2014; Tarisayi, 2018).

5. Discussion

This study has revealed several outcomes associated with the development of Tugwi-Mukosi dam. These impacts permeate the WEF nexus, including job creation and other economic impacts from the envisaged tourism development. Tugwi-Mukosi Dam is expected to improve access and availability of scarce water resources in the semiarid region of Masvingo Province. If managed well, the dam has great potential to promote agricultural development, food security, and local economic growth by harnessing water for irrigation development (Tarisayi, 2015). These opportunities can immensely contribute toward poverty alleviation.

Farming communities surrounding the dam are expected to benefit from established irrigation schemes. Apparently, the province has been a net importer of food owing to overreliance on rain-fed agriculture, a practice that has been seriously affected by climate change. Existing irrigation schemes have largely been on small scale, and their potential has also been limited by governance and technical operational challenges (Mabhaudhi et al., 2018; Gadzirayi et al., 2020). There are also opportunities to be realized from the growth of the fisheries. The fishery industry is likely to involve local people who have been experienced poverty mainly because of limited development of rural industries in Zimbabwe (Hove, 2016). However, there is need to put strong mechanisms to address the problem of poaching that as highlighted by Mutengi and Mutari (2014) and Chazireni and Chigonda (2018).

The major beneficiaries of water from the dam are downstream lowveld sugar estates, whose irrigation potential has been boosted by additional water. This is likely to enhance the adaptation of the agriculture sector to the growing risks brought about by climate change in this region (Malanco et al., 2018; Gadzirayi et al., 2020). More water is expected to be pumped to Mwenezi district when the development of Kilimanjaro sugarcane estate is finalized. Vast arable land surrounding the dam is expected to be put under irrigation purposes. The dam is expected to attract domestic and international tourists. This will improve businesses to support the tourism industry in the area. It is also evident that the dam will contribute toward addressing the perennial water supply challenges in the City of Masvingo. Such benefits have also been reported in other places following the development of dam projects (Albrecht et al., 2018; Gebreyes et al., 2020; Gao et al., 2021). However, there is need for public education and to strengthen early warning systems to avert the risk of floods on downstream communities. Part of the public awareness interventions should also be around educating the communities about crocodile attacks, drowning, and water-related diseases.

Since the dam was finally commissioned in 2017 (after taking 19 years to complete), delay in the development master plan, which was only availed in July 2021, was a major stumbling block in realizing the water body's full potential. Benefits have taken long to be realized by riparian communities (Hove, 2016; Tarisayi, 2018). Participation of the local communities who are directly affected by the dam projects is crucial to avoid resistance during implementation of the master plan. Delays in the master plan meant that planned developments were put on halt before a proper development plan that guides the physical changes to the area was in place. The development master plan ensures sustainable use of the land around the dam and a successful contribution to the country's development. Investors interested in setting up projects around the water reservoir require clear roadmap on land use to establish permanent businesses. It is evident that the development master plan was supposed to be availed during project initiation to ensure strategic planning and avoid haphazard land-use activities and settlements. This would also ensure that stakeholders and development partners are clear on proposed land uses around the dam.

6. Summary and conclusion and policy implications

Dam construction has social, economic, political, cultural, and environmental impacts on society. The current study sought to identify and examine the contribution of the Tugwi-Mukosi Dam towards enhancing WEF security in communities in Zimbabwe. The research taps on information available in published literature. It was observed that the dam has positive and negative impacts on riparian and surrounding communities. These included increased availability of fish; water provision for livestock production, irrigation, and domestic purposes gainsedenhanced However, expected negative impacts of dam construction comprise drowning of people; attack of people and livestock by crocodiles; displacement of people; and increased incidences of water-borne and vector-borne diseases.

Various measures are suggested to minimize the negative impacts of Tugwi-Mukosi Dam to local communities and the environment. These include awareness programs to minimize dangers of the drowning of people and attacks by crocodiles; healthcare interventions by the government and other stakeholders to prevent or minimize water-borne diseases and vectors; development of tourist facilities by locals in partnership with other entities to enhance benefits from tourism; training of locals in aquaculture to increase benefits from fish resources; and the setting up of an integrated environmental management plan to enhance the long-term conservation and sustainable utilization of the dam and its resources.

When implementing mega projects such as Tugwi-Mukosi Dam, which took long to complete, emphasis should be placed on tapping into private capital through joint ventures, public—private partnerships and public-private-community partnerships. These can raise funds for speeding up completion of projects. Such development models would ensure project sustainability and community ownership. There is need to deal with biodiversity and water protection challenges beyond the agricultural and industrial sectors to promote environmental sustainability.

Conflict of interest

The authors declare no conflict of interest.

Funding

This research did not receive direct funding from any institution.

References

Albrecht, T.R., Crootof, A., Scott, C.A., 2018. The water-energy-food nexus: a systematic review of methods for nexus assessment. Environ. Res. Lett. 13 (4), 043002.

Chazireni, E., Chigonda, T., 2018. The socio-economic impacts of dam construction: case of Tugwi Mukosi in Masvingo Province, Zimbabwe. Euro. J. Soc. Sci. Stud. 3 (2), 209—218.

Chen, D., Zhang, P., Luo, Z., Zhang, D., Bi, B., Cao, X., 2019. Recent progress on the water-energy-food nexus using bibliometric analysis. Curr. Sci. 117 (4), 577—586. https://doi.org/10.18520/cs/v117/i4/577-586.

FAO, IFAD, UNICEF, WFP and WHO, 2020. The State of Food Security and Nutrition in the World 2020.Transforming Food Systems for Affordable Healthy Diets. FAO, Rome.

Gadzirayi, T.C., Manatsa, D., Mutandwa, E., 2020. Tailoring conservation farming to climate change in the smallholder farming sector: case of lowveld areas in Zimbabwe. Af. J. Sci., Technol., Innov. Develop. 12 (5), 581—590.

Gao, J., Zhao, J., Wang, H., 2021. Dam-impacted water-energy-food nexus in Lancang-Mekong River basin. J. Water Resour. Plann. Manag. 147 (4), 04021010.

Gebreyes, M., Bazzana, D., Simonetto, A., Müller-Mahn, D., Zaitchik, B., Gilioli, G., Simane, B., 2020. Local perceptions of water-energy-food security: livelihood consequences of dam construction in Ethiopia. Sustainability 12 (6), 2161.

Hove, M., 2016. When flood victims became state victims: Tugwi-Mukosi, Zimbabwe. Democr. Secur. 12 (3), 135–161.

Lebel, L., Haefner, A., Pahl-Wostl, C., Baduri, A., 2020. Governance of the water-energy-food nexus: insights from four infrastructure projects in the Lower Mekong Basin. Sustain. Sci. 15 (3), 885–900.

Mabhaudhi, T., Mpandeli, S., Nhamo, L., Chimonyo, V.G.P., Nhemachena, C., Senzanje, A., Naidoo, D., Modi, A.T., 2018. Prospects for improving irrigated agriculture in Southern Africa: linking water, energy and food. Water 10 (12), 1881.

Malanco, J.A., Makurira, H., Kaseke, E., Gumindoga, W., 2018. Water management challenges at Mushandike irrigation scheme in Runde catchment, Zimbabwe. Proc. Int. Assoc. Hydrol. Sci. 378, 73–78.

Mutangi, G.T., Mutari, W., 2014. Social-cultural implications and livelihoods displacement of the moved communities as a result of the construction of Tokwe Mukosi dam. Masvingo. Greener J. 4 (2), 4–12.

Mpandeli, S., Naidoo, D., Mabhaudhi, T., Nhemachena, C., Nhamo, L., Liphadzi, S., Hlahla, S., Modi, A.T., 2018. Climate change adaptation through the water-energy-food nexus in Southern Africa. Int. J. Environ. Res. Publ. Health 15 (10), 2306.

Pardoe, J., Conway, D., Namaganda, E., Vincent, K., Dougill, A.J., Kashaigili, J.J., 2018. Climate change and the water–energy–food nexus: insights from policy and practice in Tanzania. Clim. Pol. 18 (7), 863–877.

Pueppke, S.G., 2021. Ancient wef: water-energy-food nexus in the distant past. Water 13 (7), 925.

Purwanto, A., Sušnik, J., Suryadi, F.X., de Fraiture, C., 2021. Water-energy-food nexus: critical review, practical applications, and prospects for future research. Sustainability 13 (4), 1–18.

Rusvingo, S., 2014. An assessment of the disaster preparedness of the Zimbabwe government: a case study of Tokwe Mukosi flood victims now temporarily camped at Chingwizi transit camp Mwenezi District Masvingo. Int. J. Polit. Law Res. 2 (2), 1–7.

Tarisayi, K.S., 2014. Ramifications of flooding on livelihoods: a case of two communal areas in Chivi District in Zimbabwe. Int. J. Human. Soc. Stud. 2 (2), 165–167.

Tarisayi, K.S., 2015. A Bourdieu perspective of the Chingwizi transit camp, Zimbabwe. J. Stud. Managem. Plann. 1 (9), 297–303.

Tarisayi, K.S., 2018. Traditional leadership and the Tugwi-Mukosi induced displacements: finding the missing link. Jàmbá: J. Disas. Risk Stud. 10 (1), a592.

United Nations General Assembly, 2015. Transforming Our World: The 2030 Agenda for Sustainable Development. http://www.un.org/ga/search/view_doc.asp?symbol=A/RES/70/1&Lang=E. (Accessed 30 April 2021).

United Nations (UN), 2020. Progress towards the Sustainable Development Goals. United Nations Economic and Social Council, New York. https://undocs.org/en/E/2020/57.

Wolde, Z., Wei, W., Kunpeng, W., Ketema, H., 2019. Local community perceptions toward livelihood and water-energy-food nexus: a perspective on food security. Food Ener. Secur. https://doi.org/10.1002/fes3.207.

Zarei, S., Bozorg-Haddad, O., Kheirinejad, S., Loáiciga, H.A., 2021. Environmental sustainability: a review of the water-energy-food nexus. Aqua Water Infrastruct., Ecosyst. Soci. 70 (2), 138–154. https://doi.org/10.2166/aqua.2020.058.

CHAPTER 10

The water—energy—food nexus as an approach for achieving sustainable development goals 2 (food), 6 (water), and 7 (energy)

Aidan Senzanje[1], M. Mudhara[2] and L. Tirivamwe[2]

[1]*Bioresources Engineering Programme, School of Engineering, University of KwaZulu-Natal (UKZN), Pietermaritzburg, South Africa;* [2]*School of Agriculture, Earth and Environmental Sciences, University of KwaZulu-Natal, Pietermaritzburg, South Africa*

1. Introduction

Water, energy, and food are vital resources for human well-being, poverty reduction, and sustainable development (Rasul and Sharma, 2016). With rapid population growth and migration, economic development, urbanization, cultural and technological changes, deepening poverty levels, and unprecedented climate change in the recent age, projections are that the global demand for resources, such as water, food, and energy, will rise (Corvalan et al., 2005). The approach recognizes the tension and complementarity in the use and availability of the three resources. Demand for food, water, and energy is expected to grow by 35%, 40%, and 50% by 2030, respectively. The poor also lack basic amenities such as access to electricity and clean water. Therefore, there is an urgent need to enhance the livelihoods of the poor as they tend to be undernourished, and lack access to power and clean water (Ferroukhi et al., 2015).

1.1 The WEF nexus—past to present discourse

The WEF nexus approach has attracted growing attention within international politics, academia, and other areas of society. Initially, the concept originated within the realms of international politics as a result of the World Economic Forum and other policymakers in 2011 (Terrapon-Pfaff et al., 2018). For instance, the work of Cairns and Krzywoszynska (2016) dated the nexus back to the year 2008, when business leaders at the World Economic Forum issued a request to deal with nexus concerns between economic growth and water, energy, and food resource systems. One could argue that an integrated approach to water, energy, and food could improve resource security, efficiency, poverty reduction and achieve better resource governance across sectors, and this has set the tone and premises for future discourses.

The global financial crisis of 2008–09, and its aftermath gave further impetus to the notion of a nexus encompassing water, energy, and food (Allouche et al., 2014). The increasing awareness of the impacts of biofuel production on food prices and availability, particularly in developing countries, also stimulated public and scholarly concern regarding trade-offs between food and energy (Pittock et al., 2015). The competition for land and water for crop production of biofuels and food sparked interest pertaining to the water node of the nexus.

As the water–energy–food nexus debate gains traction, it is increasingly becoming topical such that it progressively influences international development and resource governance approaches. The WEF nexus concept is gaining recognition as a conceptual framework that can be used to achieve sustainable utilization of resources across the different spheres, i.e., international organizations, academics, policy analysts, and other developmental stakeholders. The WEF nexus is regarded as a strong pillar in the successful implementation of the United Nations (UN) Sustainable Development Goals (SDGs). The United Nations and European Union (EU) Commission, for instance, seek to adopt a nexus perspective to implement the SDGs. The analysis of SDGs from the WEF nexus perspective has proved to be an important tool for establishing a holistic approach toward achieving sustainability and meeting the SDG targets.

The increasing interest in the WEF nexus discourse is partly spurred by the increasing awareness among scholars and public officials that processes influencing the sustainability of natural resources are dynamic, complex, and uncertain. The uncertainty associated with climate change has added a new dimension requiring consideration. Much of the discussion of the nexus reflects an increasing interest of the impacts of climate change–induced uncertainty on interactions involving water, energy, and food production (Scott et al., 2015; Kurian and Ardakanian, 2015). Yet the increasing interest, alone, might not be sufficient to overcome the challenges that have thwarted earlier efforts to establish integrated policy programs.

Although WEF nexus challenges are significantly evident at local scales, literature shows that the major focus of the WEF nexus has been at national, regional, and global scales (Biggs et al., 2015). Much focus has been placed on macrolevel drivers, material flows, and large infrastructure development. The discussion of WEF nexus challenges at local and household levels, especially in rural areas, is still missing (Terrapon-Pfaff et al., 2018). There is need for the creation of appropriate strategies that address local development strategies, especially in developing countries. The role of renewable energy in the WEF nexus has not been adequately explored and has been discussed independently of the nexus system. Owing to the linkages between the WEF components, it is important to address water, food, and energy issues jointly since they affect each other positively and negatively. Due to the complexity of the interdependencies, the trade-offs and synergies of water, food, and energy are described as the WEF nexus.

1.2 The WEF nexus as a tool for natural resources management

Water, energy, and food are vital resources for human well-being, poverty reduction, and sustainable development (Rasul and Sharma, 2016; Khacheba et al., 2018). The current societal megatrends in urbanization, migration, consumerism, coupled with environmental, technological, economic, and demographic changes, continue to exert pressure on already scarce and deplete natural resources, threatening their sustainability, and, thereby, undermining the resilience of communities (Gelsdorf, 2011). The resultant challenges require transdisciplinary and transformative approaches in resource management, development, and utilization, using integrated approaches such as the WEF nexus, which allow for inclusive and equitable development, as well as coordinated resource planning and management (Nhamo et al., 2018).

The WEF nexus approach suggests that the three sectors are not only interdependent, but they also have impacts on each other (WWF, 2017). For instance, the generation of energy causes water pollution, while the pumping and distribution of water for food production and processing consume significant amounts of energy. The sustainable management of natural resources requires a holistic approach that recognizes the interdependencies and relationships that exist between them. In the WEF nexus, water, food, and energy are not treated as independent systems but as subsystems of the nexus.

Water, energy, and food are all complex sectors on their own, but they become even more complex upon studying their interactions increasing the complexity (Cosgrove and Loucks, 2015). In reality, the WEF nexus can be viewed in the following complex interactive relationships: (1) water for food (in excess of 70% of global freshwater withdrawal goes to food production), (2) water for energy (water is needed for energy extraction, electricity generation, refining, and processing in the energy sector), (3) water for energy and food (hydropower generation exhibits energy–water–food–environment connectivity), (4) agriculture and land for energy and water (agriculture has a dual role as an energy user and supplier in the form of bioenergy, and furthermore, agriculture production impacts the water sector through its effects on land condition, runoff, groundwater discharge, water quality, and land/water availability for other purposes), (5) agriculture, water, and the environment (overabstraction from surface water affects the minimum environmental flow that is required to maintain ecosystem services), (6) energy is required for food and water (directly or indirectly, for transportation, processing, packaging, etc.), and (7) energy for water supply and sanitation services (including activities such water pumping, water distribution networks, water and wastewater treatment, etc.). Rather than viewing the interrelatedness of the water, energy, and food security sectors as a constraint, their relationships should be used as an opportunity to tackle development issues using a multisectoral approach. The WEF nexus approach aims to understand how each of these three sectors relates to the other two and seeks to use this understanding to make policy decisions that promote sustainable development and poverty reduction (Bizikova et al., 2013). The recognition of the complex interrelationships between the water, energy, and food sectors creates a basis for a new approach to integrated management and governance across sectors and ultimately across scales.

Participation in the WEF nexus brings together all relevant stakeholders, including local and national governments, development banks and agencies, civil society, research institutes and universities, NGOs, the private sector, and international and regional organizations (Daher et al., 2018). The global partnership allows nations to share knowledge, technical expertise, and resources while also encouraging intraorganizational collaboration.

Policy decisions and investments are still confined to sectorial boundaries, hence limiting the full operationalization of the WEF nexus in natural resource management. The existing policies across many countries still handle the three sectors separately as it is increasingly regarded as a conceptual framework for sustainable development (Biggs et al., 2015). The WEF nexus is intended to enhance policy to achieve sustainable development goals. The WEF nexus offers an integrated approach for analyzing the synergies and trade-offs between the different sectors and maximizes the efficiency of using the resources for adapting optimum policies and institutional arrangements. In other words, the WEF nexus guarantees the sustainable use of natural resources while reducing waste and loss (Daher et al., 2018).

A frequent critique of the WEF nexus is that it appears to add relatively little to already existing integrated resource management approaches, such as the integrated landscape approach or integrated water resources management (IWRM). However, in IWRM, the conceptual framework arguably pursues the integrated and coordinated management of water and land as a means of balancing different water uses, while meeting social and ecological needs and promoting economic development, it has a fatal weakness when compared with the WEF nexus. By explicitly focusing on water, IWRM has a risk of prioritizing water-related development goals over others, thereby reinforcing traditional sectoral approaches. By contrast, the WEF nexus approach considers the different dimensions of water, energy, and food equally and acknowledges the interdependencies of different resource uses in sustainable development.

Some critics of the WEF nexus argue that analyzing one resource sector is sufficiently complex, suggesting that integrating multiple resource sectors simultaneously is a substantial challenge (de Loë and Patterson, 2017). Wichelns (2017) concurs, contending that given the lack of success in implementing integrated natural resource management (INRM) and IWRM in practice, another call for integration should be questioned. It has, however, been suggested that the criticism of IWRM is well founded because it underestimates the importance of administrative boundaries, with its focus being hydrological catchments (Kurian, 2017). Belinskij (2015) argues for utilizing a nexus approach since it removes the institutional "silos" that are prevalent in governance and policy circles. Leck et al. (2015) argues that although the nexus concept is attractive, it is challenging to implement. Al-Saidi et al. (2017), while acknowledging the complexity of modeling the nexus (i.e., computer-based modeling), emphasize that there is no one-size-fits-all model to address WEF-related issues. However, common grounds have been found in the description of how localizing and contextualizing a nexus assessment will be vital to addressing trade-offs.

2. The SDGs dimensions and the WEF nexus

The UN developed 17 SDGs, three of them, i.e., SDG 2, 6, and 7 pertain to the key components of WEF, i.e., food, water, and energy, respectively. Therefore, a discussion of WEF should first develop an understanding of the three SDGs.

2.1 SDG 2—zero hunger

The SDG 2 aims to end hunger, accomplish food security and improved nutrition, and encourage sustainable agriculture by 2030. As with all SDGs, the zero hunger goal has targets, metrics, and indicators that are measurable and time bound (Gil et al., 2019). Globally, there has been a lack of universally accepted and simple definition of hunger, which has hampered progress in successfully addressing the problem. Generally, hunger has been defined by various scholars as an unpleasant feeling that is caused by lack of eating (e.g., Riches, 2016; Espel-Huynh et al., 2018). In both the developed and developing nations, hunger is often linked to poverty (Siddiqui et al., 2020), meaning that as the level of poverty increases globally, the number of people experiencing hunger also increases. FAO (2021) indicates that hunger is also directly linked to resource access to resources, implying that those with more access to resources are less likely to be affected by hunger compared with their resource-poor counterparts. This could explain the findings reported by Haque et al. (2017) who noted that women and children are more vulnerable to hunger, malnutrition, and starvation compared with men. Although the fight against hunger has been debated at various global forums, the number of people affected by hunger is increasing, especially in developing countries. Additionally, the efforts to curb hunger have been crippled by the COVID 19 pandemic as it is directly and indirectly intensifying the weaknesses and shortcomings of global food systems.

The total eradication of hunger requires a holistic approach that views the problem from a system perspective (Gil et al., 2019). FAO (2021) defined hunger as the experience of having inadequate food required for meeting dietary requirements. The establishment of the linkages between water and energy has been identified as a prerequisite for the attainment of the zero hunger target (FAO, 2016). Recently, the WEF nexus emerged as a system that can aid in the fight against hunger by employing a multisectoral approach (Rasul and Sharma, 2016). Specifically, the approach considers the impacts of water and energy as the most influential sectors, which have a direct impact on local, regional, and global food systems. Although widely discussed in literature, the operationalization of the WEF nexus for achieving zero hunger at various scales remains a challenge. The nexus approach requires political commitment and knowledge on how it can effectively be utilized to formulate policies that reduce various forms of poverty and positively impact food security (Gödecke et al., 2018). Under the nexus approach, the implementation and monitoring of the models to eradicate hunger requires key indicators with accompanying threshold values. The main indicators help in measuring the success of approaches and identifying trade-offs with other sectors.

2.2 SDG 6—clean water and sanitation

The SDG 6 mandate is to ensure the availability and sustainable management of water resources globally to give access to clean water and sanitation to the world's population (Tortajada and Biswas, 2018). Previously, the Millennium Development Goals (MDGs) focused more on the global population access to water, with less attention given to water quality. According to the World Bank (2020) statistics, 68% of the world's population has reliable access to water and basic sanitation. However, only 39% of the people who were reported to have access to water and sanitation had access to safely managed sanitation, which comprises appropriate collection treatment and disposal of waste (Shandra et al., 2011). Poor management of wastewater systems has led to contamination and pollution of underground water sources causing serious health risks to people (Blackett, 2015). Unlike the MDGs, the SDG 6 included and emphasized water quality as part of the 2030 goals. However, achieving clean water and sanitization is largely dependent on understanding the trade-offs and synergies that exist across the water, food, and energy sectors.

Stephan et al. (2018) pointed that the WEF nexus is a platform that can be used to effectively implement the SDG goals. Therefore, the improvement of the access to clean water and sanitation at global level requires the nexus approach. Water is transboundary resource, and it requires systems approach that goes beyond single sectors or individual territory.

2.3 SDG 7—affordable and clean energy

The primary objective of the SDG 7 is to ensure equal access to affordable, reliable, sustainable, and modern energy for all. Ahuja and Tatsutani (2009) highlighted that as the global population grows so will the demand for inexpensive energy. There has been a new drive to encourage alternative energy sources, and in 2011 renewable energy accounted for more than 20% of global power generated (Franco et al., 2020). However, one in five people lack access to electricity, and as the demand continues to rise, there is a need for a matching increase in the production of renewable energy across the world (Křížková, 2019). The SDG programs promote the use of renewable and sustainable sources of energy. At the same time, they promote construction techniques that are more energy efficient. For example, in Mozambique, it is supporting, through UNIDO and national partners, a technology exchange with South African National Cleaner Production Center (SDGF, 2020).

The WEF nexus is critical to the achievement of affordable and clean energy for all (Saladini et al., 2018). Ensuring universal access to affordable electricity by 2030 means investing in clean energy sources such as solar, wind, and thermal. Adopting cost-effective standards for a wider range of technologies could also reduce the global electricity consumption in buildings and industry by 14%, preventing the construction of around 1300 medium-sized power plants. Expanding infrastructure and upgrading technology to provide clean energy sources in all developing countries is a crucial goal that can both encourage growth and benefit the environment.

3. Food and nutrition security

The concept of food and nutrition security has been explicitly defined in literature as a stage when a person has physical, social, and economic access to food all the times (El Bilali et al., 2019). To fulfill the nutrition aspect, the food should be consumed in adequate quantities and acceptable quality to effectively meet the dietary needs and food preferences (Pangaribowo et al., 2013).

As the world population continues to experience food shortages, the importance of food security in various facets of society has been emphasized. SDG 2's primary goals are to eliminate hunger while increasing food security and nutrition (Griggs et al., 2017). Under the SDG 2, the United Nations outlines eight targets and 13 indicators, which all links to the improvement of global food and nutrition security (Fonseca et al., 2020). Several authors mentioned climate change as one of the major factors negatively impacting the global water and food systems and leading to food insecurity (Fanzo et al., 2018; Myers et al., 2017; Loboguerrero et al., 2019; Clapp et al., 2018). As part of curbing the devastating impacts of climate change on food and nutrition security, SDG 2 outlines practical solutions that countries can adopt as part of their efforts to achieve zero hunger. For example, agricultural systems, forestry, and fisheries were identified as sources that can provide nutritious food as well as facilitate income generating projects that can improve livelihoods while protecting the environment (FAO 2018). However, in this example, agricultural systems, forestry and fisheries are not only interconnected but also directly or indirectly influenced and impacted by water and energy resources. Therefore, it is necessary to take a multisectoral approach (e.g., the WEF nexus) to address food and nutrition security challenges.

Improvement in food security and nutrition can be effectively achieved through better coordination between the water, food, and energy sectors (Hoff, 2011), other factors being constant. The work of Mahlknecht et al. (2020) postulated that for sustainability to be achieved in the process of reducing global hunger, the interconnections of the three sectors have to be recognized at local, regional, and global scales. The practical execution of SDG 2 objectives has considerably improved at various scales since the formation of the WEF nexus at the Bonn 2011 Nexus Conference; hence, the WEF nexus strategy is essential in achieving food and nutrition security.

4. Synergies and trade-offs in the WEF nexus

Galafassi et al. (2017) defined trade-offs as a decision-making process that benefits one or a few components of an interconnected system at the expense of the other components, whereas synergy can be taken as the interaction or cooperation of two or more components to produce a greater combined effect. Scott (2017), however, indicated that trade-offs are the sources of contestations or conflicts that emanates from resource allocation and utilization within the WEF nexus system.

4.1 Synergies and trade-offs in the WEF nexus toward achieving food and nutrition security (SDG 2)

The recognition of key synergies between water, energy, and food is essential in resource management decisions. Synergies are defined as the close connections of two or more groups, sectors, or institutions to generate a collective effect that would not be produced if they would have acted independently (Li et al., 2019). The complex dynamic interactions that exist between water, energy, and food sectors form synergies within the WEF system. In simple terms, Scott (2017) defined synergies as "win—win" situations, whereby the needs of all sectors are considered in the resource allocation and utilization. For the attainment of food and nutrition security, it is essential to quantify and analyze the synergies that occur in water, energy, and food systems (Xu et al., 2019). A balance in the use and generation of the WEF components must be achieved; otherwise, unsustainability may result. Wu et al. (2021) evaluated the synergies and trade-offs in the WEF nexus in Canada. They noted that sources of synergies in the WEF nexus primarily emanate from the production and demand sides of the water, energy, and food sectors. For instance, the agricultural sector provides an opportunity to develop and generate bioenergy using food crops such as wheat and canola. On the other hand, the production of crops requires energy and water. If quantified and well balanced, these synergies can improve food and nutrition security.

Martinez-Hernandez et al. (2017) added that the close linkages that exist in the WEF nexus enable the recycling of resources, thus minimizing unnecessary losses especially in food production. For instance, wastewater from agriculture can be channeled toward energy generation. Evidence from literature shows that the effective assessment of WEF nexus synergies toward achieving food and nutrition security requires the participation of all relevant stakeholders especially at a local scale. The development of more synergies within the WEF nexus is encouraged since it can improve the capacity of nations to achieve the objective of the nexus, as well as moving toward achieving food and nutrition security globally (Bhaduri et al., 2015).

The WEF nexus was established to specifically challenge and address the conflicts that emanates from resource utilization among the water, energy, and food sectors. The attainment of food security and nutrition has been affected by the trade-offs within the nexus system. For example, in many cases where one sector loses out, the decision about trade-offs becomes political and not sustainable. Several WEF nexus modeling studies have explored the sustainable utilization of scarce resources by means of balancing trade-offs in the nexus, thus improving food and nutrition to the world population (Wu et al., 2021; Zhang and Vesselinov, 2017; Yuan et al., 2018).

An example of WEF trade-offs that affects food and nutrition security is the scarcity of water resources within the system. If water resources are limited, there are contestations with regards to its allocation to food production versus hydroelectric power generating projects. The works of Biemans et al. (2016) and Rasul (2014) confirmed that it is challenging to ensure that hydroelectric power generation and food production sectors are allocated adequate water, especially under the increasing water stress globally. Climate change is the primary source of water stress around the world, and exacerbates the WEF nexus trade-offs and undermines the efforts toward achieving food and nutrition security. To improve the resource flow in the nexus, Dhaubanjar et al. (2017) proposes that the holistic nexus approaches should promote evaluation of the benefits, impacts, and risks of projects across the three sectors to minimize trade-offs.

The following sections provide a couple of case studies on the synergies and trade-offs for achieving SDGs 6 and 7 and the link to food and nutrition security, SDG 2.

4.2 Synergies and trade-offs for achieving clean water and sanitation (SDG 6)

4.2.1 Case study: Guatemala agricultural extension project

Having noted that SDG 6 has the primary objective to provide access to clean water and sanitation (Tortajada and Biswas, 2018), it is worth noting that there are synergies and trade-offs that either promote or restrict the successful implementation and operationalization of the WEF nexus toward this goal. Synergies result in mutual benefits between the water sector and the adjacent sectors, while trade-offs increase contestations in resource utilization with the system. Therefore, if there are more synergies than trade-offs in the system, it will be a positive step toward achieving clean water and sanitization. Banerjee et al. (2019) assessed the synergies and trade-offs in achieving zero hunger and clean sanitation in Guatemala. Their study reported that investments in water and sanitation and agriculture lead to trade-offs. Specifically, the expansion in commercial irrigation schemes affects the amount of water allocated for domestic use. Additionally, the excessive use of chemicals to improve agricultural productivity also resulted in pollution of water. While improving food security, agricultural expansion projects have denied certain communities of their right to clean water.

Positively, agricultural extension projects have also resulted in the creation of synergies that favor the achievement of access to clean water and sanitation for all. For example, Kurian (2017) noted that the recycling of agricultural water and possibly channeling it toward energy generation is a sustainable approach that minimizes resource waste of resources while achieving several SDGs at the same time.

4.3 Synergies and trade-offs for achieving affordable and clean energy (SDG 7)

4.3.1 Case study: off-grid solar energy in Rwanda

The SDG 7 deals with ensuring access to affordable, reliable, sustainable, and modern energy for all by 2030. However, in the process of achieving the targets, there are unavoidable synergies and trade-offs that are created. Bisaga et al. (2021) mapped synergies and trade-offs between the energy sector and sustainable development goals in Rwanda. The off-grid solar energy project created synergies that substantially added value toward achieving clean energy in the country. Unlike hydroelectric power generation that consumes large amounts of water, the off-grid solar system is sustainable and puts less pressure on water resources (Sutthivirode et al., 2009; Kalogirou, 2004). Eventually, more water resources can be allocated for agriculture, while accessibility to clean water is still achieved. Additionally, the off-grid solar energy also significantly contributes to the attainment of zero hunger goal through solar-powered irrigation, which provides clean energy for perennial food crops production.

The off-grid solar systems case also presents trade-offs that bring several challenges in the achievement of the access to clean energy goal. Bisaga et al. (2021) identified several trade-offs including potential competition between the need for agricultural land and solar grid development, as well as the depletion of groundwater because of increased solar-driven irrigation.

5. Drivers of the WEF nexus toward achievement of SDGs 2, 6, and 7

The discussion in the preceding sections showed the relevance of the WEF nexus approach to understanding the opportunities for achieving SDGs 2, 6, and 7 and food and nutrition security. The sections argued that the possibilities of using the WEF nexus as an approach for achieving the highlighted SDGs and food and nutrition security is conditional on several drivers prevailing across varying contexts. An understanding of the context and drivers in WEF nexus can be achieved through the sustainable livelihoods approach (SLA). The SLA was instrumental in highlighting the importance of the context within which development occurs and to pointing to the drivers that are critical to achieving desired livelihood outcomes (Scoones, 2015). Relevant drivers area varied and included climate change, population and economic growth rates, urbanization, policies and institutions, and technological advances. The extent to which the different drivers express themselves under varying contexts is all relevant to the expression of the WEF nexus toward achievement of food security, e.g., how climate change manifests itself, the level and growth of the population and the economy, the percentage of the population that is urbanized, the prevailing policies and institutional settings, the pace of technological changes, etc. Policies, for example, can stifle or stimulate the use of various resources, such as water, as well as the production, availability, and pricing of energy. Therefore, the suggestion that WEF nexus does not exist in a vacuum, and other factors should be considered in understanding or devising ways in which it is relevant to food and nutrition security outcomes.

6. Upscaling and outscaling the WEF nexus as a natural resources management tool for attaining SDGs 2, 6, and 7

Upscaling and outscaling an innovation, practice, or process require a defined communications plan and pathway. Technically speaking, upscaling of an innovation or practice means improving on its quality and making it better, whereas outscaling implies getting an innovation or practice out to a wider audience or to cover a larger scale. Therefore, upscaling can mean use of an innovation or practice could increase, starting from a lower level, e.g., local scale, and then be taken to cover a larger scale at the district, provincial, or national scales. Admittedly, the concepts of upscaling and outscaling lend themselves to different interpretations and applications, but both require clear communications pathways to enhance adoption and hence success in either direction. In this chapter, we adopt the definition where upscaling is related to quality, while outscaling is related to higher quantum in coverage. In a way, the WEF nexus as a resource planning and use tool can be classified as an innovation or practice that requires to be clearly communicated to various stakeholder within the development levels. This can range from policy makers at the top of government, through implementers in the provinces and municipalities down to the target beneficiaries right on the ground who benefit from properly planned resource utilization. Similarly, WEF nexus tools require to be outscaled for wider applicability. This, too, needs a communication strategy to enhance adoption of the practice. This point is further buttressed by the fact that there has been an outcry that the WEF nexus has remained in the realm of academic research alone, with no progress toward wider adoption and implementation by the target stakeholders (users) for sustainable natural resources management, both locally and globally.

A communication plan is the complete set of methods and procedures that will be used to convey and share information about an innovation, practice, or lesson. It assumes that there are stakeholders who need to know about the innovation or lessons or outputs, and that the innovation, lessons, or outputs must be packaged accordingly to suit the needs and be desirable to specific group of stakeholders, and then how to actually carry out the communication activities, and finally monitor the communication plan. At each and every stage or step, the following questions should be asked; what, why, by who, by when, and how? These steps include (1) identifying WEF nexus communication stakeholders, (2) designing and producing WEF nexus communication products, (3) undertaking WEF nexus communication activities, and (4) monitoring and evaluating (M&E) the WEF nexus communication plan.

In upscaling the WEF nexus tools, it means they need to be improved (if that is required) and also they need to be applied at different levels from the local level to the national level. Upscaling can be a natural process that happens with minimal input depending on how refined an innovation is. If one takes the example of the integrative analytical WEF nexus (IAWN) model, as of now the model is spreadsheet based and can benefit from further development and enhancement such as having a friendly graphical user interface (GUI), having a database of development scenarios applicable to South Africa, and taking inputs from policy makers. Incorporating such will help in upscaling the model. The question then is what is the communication pathway for this to happen? As discussed before, the steps are by and large the same, who are the concerned stakeholders, what are the communication products and how can they be packaged, how best to then communicate these upscaling products, and finally the required M&E on this communication plan to ensure that it succeeds. The same concept applies if the upscaling implies moving from a lower level of application (local) to higher level (national). Of course, the reverse would be downscaling the technology, from national level to local level.

Outscaling of WEF nexus practices means that they progress from being practiced at a smaller scale to a wide scale. Scale can be considered from a catchment perspective, i.e., moving from a subcatchment or catchment to several catchments, or from an administrative area perspective, i.e., moving from one district or municipality to many. As previously alluded to, the communication plan and pathway needs to be flexible since conditions in one catchment can be significantly different from those in the next for outscaling to occur. Consequently, the communication plans must be able to adapt and be tailored to those conditions. For example, currently the WEF nexus is being tried out in a number of spatial scale environments in South Africa and the region, from urban settings through catchment scale and finally at the national scale, and even at the Southern African Development Community (SADC) regional scale and continental level. For the practice to progress from local scale to national scale, there is need for a clear communication pathway for the WEF nexus models and tools—from who are the concerned stakeholders at all these levels through to how to M&E evaluate what is going on in that pathway.

7. Conclusion

Achieving SDGs 2, 6, and 7 is one of the thrusts of the whole world, in addition to the other 13 SDGs. The three SDGs are interrelated as they collectively address issues around the water, energy, and food, suggesting the need for their joint consideration. The WEF nexus approach considers the three resources and has advantages for achieving the three SDGs. However, the WEF approach has highlighted the need to consider synergies and trade-offs between the three resources.

The development and implementation of the WEF nexus should include all relevant stakeholders that relate to the three resource systems. The WEF nexus identifies interlinkages, synergies, and trade-offs from the interactions that exist between water, energy, and food sectors. The knowledge of the connections that exist between WEF resources is critical in making resource management decision and can eventually lead to sustainable utilization of the resources and indeed the fulfillment of food and nutrition security and the SDGs 2, 6, and 7. Recently, WEF nexus research publications have been increasing exponentially and have demonstrated the capability of investigating the complex relationships between the three subsystems. Emphasis should be placed on the development of quantitative assessment tools that can enable the operationalization of the WEF nexus in both developed and developing nations and their contribution to food and nutrition security and the SDGs 2, 6, and 7 outcomes. Governments should ensure that the enabling environment and facilitating drivers exists. Lastly the WEF nexus as a practice needs to be upscaled and outscaled for the benefits to reach a wider audience at different spatial scales.

References

Ahuja, D., Tatsutani, M., 2009. Sustainable energy for developing countries. Surv. Perspect. Integrat. Environ. Soci. 2 (1).

Al-saidi, M., Elagib, N.A., Ribbe, L., Schellenberg, T., Roach, E., OEZHAN, D., 2017. Water-energy-food security nexus in the eastern Nile Basin: assessing the potential of transboundary regional cooperation. Water-Energy-Food Nexus: Princ. Pract., Geophys. Monogr. 229, 103—116.

Allouche, J., Middleton, C., Gyawali, D., 2014. Nexus Nirvana or Nexus Nullity? A Dynamic Approach to Security and Sustainability in the Water-Energy-Food Nexus.

Banerjee, O., Cicowiez, M., Horridge, M., Vargas, R., 2019. Evaluating synergies and trade-offs in achieving the SDGs of zero hunger and clean water and sanitation: an application of the IEEM Platform to Guatemala. Ecol. Econ. 161, 280—291.

BelinskiJ, A., 2015. Water-energy-food nexus within the framework of international water law. Water 7, 5396—5415.

BhadurI, A., Ringler, C., Dombrowski, I., Mohtar, R., Scheumann, W., 2015. Sustainability in the Water—Energy—Food Nexus. Taylor & Francis.

Biemans, H., Siderius, C., Mishra, A., Ahmad, B., 2016. Crop-specific seasonal estimates of irrigation-water demand in South Asia. Hydrol. Earth Syst. Sci. 20, 1971—1982.

Biggs, E.M., Bruce, E., Boruff, B., Duncan, J.M., Horsley, J., Pauli, N., Mcneill, K., Neef, A., Van ogtrop, F., Curnow, J., 2015. Sustainable development and the water—energy—food nexus: a perspective on livelihoods. Environ. Sci. Pol. 54, 389—397.

Bisaga, I., Parikh, P., Tomei, J., TO, L.S., 2021. Mapping synergies and trade-offs between energy and the sustainable development goals: a case study of off-grid solar energy in Rwanda. Energy Pol. 149, 112028.

Bizikova, L., Roy, D., Swanson, D., Venema, H.D., McCandless, M., 2013. The Water-Energy-Food Security Nexus: Towards a Practical Planning and Decision-Support Framework for Landscape Investment and Risk Management. International Institute for Sustainable Development, Winnipeg.

Blackett, I., 2015. Improving On-Site Sanitation and Connections to Sewers in Southeast Asia: Insights from Indonesia and Vietnam. Water and Sanitation Program: Research Brief. The World Bank.

Cairns, R., Krzywoszynska, A., 2016. Anatomy of a buzzword: the emergence of 'the water-energy-food nexus' in UK natural resource debates. Environ. Sci. Pol. 64, 164—170.

Clapp, J., Newell, P., Brent, Z.W., 2018. The global political economy of climate change, agriculture and food systems. J. Peasant Stud. 45, 80—88.

Corvalan, C., Hales, S., Mcmichael, A.J., Butler, C., McMichael, A., 2005. Ecosystems and Human Well-Being: Health Synthesis. World Health Organization.

Cosgrove, W.J., Loucks, D.P., 2015. Water management: current and future challenges and research directions. Water Resour. Res. 51, 4823—4839.

Daher, B., Mohtar, R.H., Pistikopoulos, E.N., Portney, K.E., Kaiser, R., Saad, W., 2018. Developing socio-techno-economic-political (STEP) solutions for addressing resource nexus hotspots. Sustainability 10, 512.

De Loë, R.C., Patterson, J.J., 2017. Rethinking water governance: moving beyond water-centric perspectives in a connected and changing world. Nat. Resour. J. 57, 75—100.

Dhaubanjar, S., Davidsen, C., Bauer-Gottwein, P., 2017. Multi-objective optimization for analysis of changing trade-offs in the Nepalese water—energy—food nexus with hydropower development. Water 9, 162.

El Bilall, H., Callenius, C., STrassner, C., Probst, L., 2019. Food and nutrition security and sustainability transitions in food systems. Food Energy Secu. 8, e00154.

Espel-Huynh, H., Muratore, A., Lowe, M., 2018. A narrative review of the construct of hedonic hunger and its measurement by the Power of Food Scale. Obes. Sci. Pract. 4, 238—249.

Fanzo, J., Davis, C., Mclaren, R., Choufani, J., 2018. The effect of climate change across food systems: implications for nutrition outcomes. Global Food Secur. 18, 12—19.

FAO, 2016. Agriculture: Key to Achieving the 2030 Agenda for Sustainable Development. Food and Agriculture Organization of the United Nations, Rome.

FAO, 2021. The State of Food Security and Nutrition in the World. Food and Agriculture Organization of the United Nations, Rome.

Ferroukhi, R., NagpaL, D., Lopez-Peña, A., Hodges, T., Mohtar, R.H., Daher, B., Mohtar, S., Keulertz, M., 2015. Renewable Energy in the Water, Energy & Food Nexus. IRENA, Abu Dhabi.

Fonseca, L.M., Domingues, J.P., Dima, A.M., 2020. Mapping the sustainable development goals relationships. Sustainability 12, 3359.

Franco, I.B., Power, C., Whereat, J., 2020. SDG 7 Affordable and Clean Energy. Actioning the Global Goals for Local Impact. Springer.

Galafassi, D., Daw, T.M., Munyi, L., Brown, K., Barnaud, C., Fazey, I., 2017. Learning about social-ecological trade-offs. Ecol. Soc. 22.

Gelsdorf, K., 2011. Global Challenges and Their Impact on International Humanitarian Action, Office for the Coordination of Humanitarian Affairs (OCHA).

Gil, J.D.B., Reidsma, P., Giller, K., Todman, L., Whitmore, A., Van ittersum, M., 2019. Sustainable development goal 2: improved targets and indicators for agriculture and food security. Ambio 48, 685—698.

Gödecke, T., Stein, A.J., Qaim, M., 2018. The global burden of chronic and hidden hunger: trends and determinants. Global Food Secur. 17, 21—29.

Griggs, D., Nilsson, M., Stevance, A., McCollum, D., 2017. A Guide to SDG Interactions: From Science to Implementation. International Council for Science, Paris.

Haque, M.A., Farzana, F.D., Sultana, S., Raihan, M.J., Rahman, A.S., Waid, J.L., Choudhury, N., Ahmed, T., 2017. Factors associated with child hunger among food insecure households in Bangladesh. BMC Pub. Health 17, 1–8.

Hoff, H., 2011. Understanding the Nexus. Background Paper for the Bonn2011 Nexus Conference: The Water, Energy and Food Security Nexus. Stockholm Environment Institute (SEI), Stockholm.

Kalogirou, S.A., 2004. Environmental benefits of domestic solar energy systems. Energy Convers. Manage. 45, 3075–3092.

Khacheba, R., Cherfaoui, M., Hartani, T., Drouiche, N., 2018. The nexus approach to water-energy-food security: an option for adaptation to climate change in Algeria. Desalination Water Treat. 131, 30–33.

Křížková, A., 2019. Sustainable Development Goal 7: Affordable and Clean Energy Panel Data Analysis.

Kurian, M., 2017. The water-energy-food nexus: trade-offs, thresholds and transdisciplinary approaches to sustainable development. Environ. Sci. Pol. 68, 97–106.

Kurian, M., Ardakanian, R., 2015. Governing the Nexus. Springer.

Leck, H., Conway, D., Bradshaw, M., Rees, J., 2015. Tracing the water–energy–food nexus: description, theory and practice. Geograph. Compass 9, 445–460.

Li, G., Wang, Y., Li, Y., 2019. Synergies within the water-energy-food nexus to support the integrated urban resources governance. Water 11, 2365.

Loboguerrero, A.M., Campbell, B.M., Cooper, P.J., Hansen, J.W., Rosenstock, T., Wollenberg, E., 2019. Food and earth systems: priorities for climate change adaptation and mitigation for agriculture and food systems. Sustainability 11, 1372.

Mahlknecht, J., GonzáleZ-Bravo, R., Loge, F.J., 2020. Water-energy-food security: a nexus perspective of the current situation in Latin America and the Caribbean. Energy 194, 116824.

Martinez-Hernandez, E., Leach, M., Yang, A., 2017. Understanding water-energy-food and ecosystem interactions using the nexus simulation tool NexSym. Appl. Energy 206, 1009–1021.

Myers, S.S., Smith, M.R., Guth, S., Golden, C.D., Vaitla, B., Mueller, N.D., Dangour, A.D., Huybers, P., 2017. Climate change and global food systems: potential impacts on food security and undernutrition. Annu. Rev. Publ. Health 38, 259–277.

Nhamo, L., Ndlela, B., Nhemachena, C., Mabhaudhi, T., Mpandeli, S., Matchaya, G., 2018. The water-energy-food nexus: climate risks and opportunities in southern Africa. Water 10, 567.

Pangaribowo, E.H., Gerber, N., Torero, M., 2013. Food and nutrition security indicators: a review. ZEF Working. Paper No. 108. Bonn, Germany: Center for Development Research (ZEF), University of Bonn. http://www.zef.de/fileadmin/webfiles/downloads/zef_wp/wp108.pdf.

Pittock, J., Orr, S., Stevens, L., Aheeyar, M., Smith, M., 2015. Tackling trade-offs in the nexus of water, energy and food. Aquat. Proc. 5, 58–68.

Rasul, G., 2014. Food, water, and energy security in South Asia: a nexus perspective from the Hindu Kush Himalayan region. Environ. Sci. Pol. 39, 35–48.

Rasul, G., Sharma, B., 2016. The nexus approach to water–energy–food security: an option for adaptation to climate change. Clim. Pol. 16, 682–702.

Riches, G., 2016. First World Hunger: Food Security and Welfare Politics. Springer.

Saladini, F., Betti, G., Ferragina, E., Bouraoui, F., Cupertino, S., Canitano, G., Gigliotti, M., Autino, A., Pulselli, F., Riccaboni, A., 2018. Linking the water-energy-food nexus and sustainable development indicators for the Mediterranean region. Ecol. Indicat. 91, 689–697.

Scoones, I., 2015. Sustainable Rural Livelihoods and Rural Development. Practical Action Publishing, Warwickshire.

Scott, A., 2017. Making Governance Work for Water−Energy−Food Nexus Approaches. https://odi.org/en/publications/making-governance-work-for-water-energy-food-nexus-approaches/. (Accessed 30 September 2021).

Scott, C.A., Kurian, M., Wescoat, J.L., 2015. The Water-Energy-Food Nexus: Enhancing Adaptive Capacity to Complex Global Challenges. Governing the Nexus. Springer.

SDGF, 2020. Goal 7: Affordable and Clean Energy. www.sdgfund.org/es/node/233?. (Accessed 30 September 2021).

Shandra, C.L., Shandra, J.M., London, B., 2011. World bank structural adjustment, water, and sanitation: a cross-national analysis of child mortality in Sub-Saharan Africa. Organ. Environ. 24, 107−129.

Siddiqui, F., Salam, R.A., Lassi, Z.S., Das, J.K., 2020. The intertwined relationship between malnutrition and poverty. Front. Pub. Health 8, 1−5. https://doi.org/10.3389/fpubh.2020.00453. Article 453.

Stephan, R.M., Mohtar, R.H., Daher, B., Embid irujo, A., Hillers, A., Ganter, J.C., Karlberg, L., Martin, L., Nairizi, S., Rodriguez, D.J., 2018. Water−energy−food nexus: a platform for implementing the Sustainable Development Goals. Water Int. 43, 472−479.

Sutthivirode, K., Namprakai, P., Roonprasang, N., 2009. A new version of a solar water heating system coupled with a solar water pump. Appl. Energy 86, 1423−1430.

Terrapon-Pfaff, J., Ortiz, W., Dienst, C., Gröne, M.-C., 2018. Energising the WEF nexus to enhance sustainable development at local level. J. Environ. Manag. 223, 409−416.

Tortajada, C., Biswas, A.K., 2018. Achieving universal access to clean water and sanitation in an era of water scarcity: strengthening contributions from academia. Curr. Opin. Environ. Sustain. 34, 21−25.

Wichelns, D., 2017. The water-energy-food nexus: is the increasing attention warranted, from either a research or policy perspective? Environ. Sci. Pol. 69, 113−123.

World Bank, 2020. Sanitation. www.worldbank.org/en/topic/sanitation. (Accessed 30 September 2021).

Wu, L., Elshorbagy, A., Pande, S., Zhuo, L., 2021. Trade-offs and synergies in the water-energy-food nexus: the case of Saskatchewan, Canada. Resources. Conserv. Recycl. 164, 105192.

WWF, 2017. The Food-Energy-Nexus as a lens for delivering the UN's Sustainable Development Goals in Southern Africa. Cape Town, South Africa: World Wide Fund for Nature.

Xu, S., He, W., Shen, J., Degefu, D.M., Yuan, L., Kong, Y., 2019. Coupling and coordination degrees of the core water−energy−food nexus in China. Int. J. Environ. Res. Publ. Health 16, 1648.

Yuan, K.-Y., Lin, Y.-C., Chiueh, P.-T., Lo, S.-L., 2018. Spatial optimization of the food, energy, and water nexus: a life cycle assessment-based approach. Energy Pol. 119, 502−514.

Zhang, X., Vesselinov, V.V., 2017. Integrated modeling approach for optimal management of water, energy and food security nexus. Adv. Water Resour. 101, 1−10.

Further reading

Botai, J.O., Botai, C.M., Ncongwane, P.N., Mpandeli, S., Nhamo, L., Masinde, M., Adeola, A.M., Mengistu, M.G., Tazvinga, H., Murambadoro, M.D., Lottering, S., Motochi, I., Hayombe, P., Zwane, N.N., WamitI, E.K., Mabhaudhi, T., 2021. A review of the water−energy−food nexus research in Africa. Sustainability 13 (4), 1762.

Mabhaudhi, T., Simpson, G., Badenhorst, J., Senzanje, A., Jewitt, G.P.W., Chimonyo, V.G.P., Mpandeli, S., Nhamo, L., 2021. Developing a framework for the water-energy-food nexus in South Africa. In: Salif Diop, S., Peter Scheren, P., Awa Niang, A. (Eds.), Climate Change and Water Resources in Africa — Perspectives and Solutions towards an Imminent Water Crisis. y Springer Future Earth, Switzerland.

World Health Organization, 2018. The State of Food Security and Nutrition in the World 2018: Building Climate Resilience for Food Security and Nutrition. Food & Agriculture Org.

CHAPTER 11

Enhancing sustainable human and environmental health through nexus planning

Luxon Nhamo[1], Sylvester Mpandeli[1,4], Shamiso P. Nhamo[2], Stanley Liphadzi[1,4] and Tafadzwanashe Mabhaudhi[3,5]

[1]*Water Research Commission of South Africa (WRC), Pretoria, South Africa;* [2]*Department of Pharmacology, Faculty of Health Sciences, University of Pretoria, Hatfield, Pretoria, South Africa;* [3]*Centre for Transformative Agricultural and Food Systems (CTAFS), School of Agricultural, Earth and Environmental Sciences, University of KwaZulu-Natal, Pietermaritzburg, South Africa;* [4]*School of Environmental Sciences, University of Venda, Thohoyandou, Limpopo, South Africa;* [5]*International Water Management Institute (IWMI-GH), West Africa Office, Accra, Ghana*

1. Introduction

Extreme events, such as droughts, floods, cyclones, and novel pathogens, continue manifesting with increased intensity and frequency, impacting heavily on economies, human wellbeing, livelihoods, and human health (Abdallah et al., 2013; Naidoo et al., 2021; Nhamo et al., 2021; Patz et al., 2014). In particular, the emergence of novel infectious diseases has increased the vulnerability of humans to new zoonotic health threats, adding to already existing stressors bedeviling humankind. In 2019, at the very end, the world was bedeviled by a novel severe acute respiratory syndrome coronavirus 2 (SARS-CoV-2), a virus responsible for the COVID-19, which originated from China and spread to the rest of world causing hundreds of thousands of infections and mortality (Dong et al., 2020; Xu et al., 2020). Technological advancements, globalization, and easy means of transport eased its spread worldwide within a short space of time (Boulos and Geraghty, 2020; WHO, 2020). The threat of the COVID-19 pandemic was so huge that most governments declared total lockdowns for prolonged periods of time to reduce contact transmission, but this reactive response triggered economic recessions, company closures, and widespread job losses (Adhikari et al., 2020). The threat of novel zoonoses (Ebola, HIV/AIDS, and COVID-19) is so huge that more resources need to be directed toward preparedness and readiness that future novel pathogens may not end in pandemics.

Continued alterations on wildlife habitats due to demographic, climatic, and environmental changes are instigating wild animals to invade human habitats, particularly in urban areas (Wong and Candolin, 2015), posing a serious threat on human health (Naicker, 2011; Nava et al., 2017; Smith and Wang, 2013). Currently, over 55% of the world population reside in cities, and the number is estimated to go up to about 68% by 2050 (Leeson, 2018). The abundance of food and poor waste management in urban areas (and the scarcity of food in the wild) have attracted wildlife into urban areas in search of food, including rats, rabbits, mice, baboons, raccoons, squirrels, hares, foxes, birds, and jackals, among others, risking human health (Cox and Gaston, 2018; Newsome and Van Eeden, 2017). Urban areas have, thus, become hot spots for evolving infectious diseases with origins from wildlife (Lindahl and Grace, 2015). This is identified as the main reason for a sudden surge of novel zoonoses, as over 60% of about 400 novel infectious diseases that have inflicted humankind since 1940 have been traced to wildlife (Morse et al., 2012; WHO, 2018).

Previous trends of zoonotic diseases had given rise to the notion that these novel infectious diseases are "once off" events, but, although this could be true to some extent, the impacts of the COVID-19 highlighted the grave risk imposed by wildlife to humans, particularly in an era of globalization and technological advancements (WHO, 2020). Some of these pandemics like HIV/AIDS have existed for longer periods. Just like the severe acute respiratory syndrome (SARS) and Middle East respiratory syndrome (MERS) outbreaks, initial research on the SARS-CoV-2 indicates that the virus originated from wildlife (Andersen et al., 2020). The trend in the emergence of zoonotic diseases highlights the increasing risk posed by close human–wildlife interactions on human health. The main drivers of these novel human–wildlife interactions include loss of wildlife habitat, population increase, urbanization, globalization, and climate change (Galvani et al., 2016; Hassell et al., 2017; Saker et al., 2004).

The spread of novel infectious diseases has been rapid, reaching continents within short spaces of time, after making an evolutionary change of host from wildlife to humans (Lindahl and Grace, 2015; Morse, 2001). Examples of infectious diseases with global impact include the HIV/AIDS crisis that started in the 1980s, which was traced from apes (Sharp and Hahn, 2011); the 2004/7 Avian flu pandemic, with origins from birds (Lycett et al., 2019); the 2009 Swine flu pandemic, which was traced from pigs (Gibbs et al., 2009); the SARS, which came from bats through civet cats (Wang and Eaton, 2007); the Middle East respiratory syndrome coronavirus (MERS-CoV), which was traced from camels (El-Kafrawy et al., 2019); the Ebola pandemic, which was also traced to bats (De Nys et al., 2018); and the current COVID-19 that has also been traced from wildlife (WHO, 2020). The trend indicates a harmful human–wildlife relationship as worst pandemics that have afflicted humankind have originated from wildlife (Hassell et al., 2017), and some of them like HIV/AIDS are still without a cure (Kallings, 2008).

The challenge is complex, involving multiple socioecological factors that require transformative interventions to comprehend the intricately interlinked socioecological interactions and how they are impacting human health and well-being. A water—health—ecosystem—nutrition (WHEN) nexus is established to simplify these intricate interactions and provide the lens to comprehend naturally connected components, providing evidence on risk reduction initiatives, ensuring sustainable ecosystems and human health, and enhancing preparedness. Current challenges call for the mainstreaming of the health sector into the transformative analyses that allow holistic and cross-sectoral interventions. The complexity in today's cross-cutting challenges has witnessed the emergence of the WHEN nexus that integrates health-related challenges to inform policy on pathways for timely interventions and preparedness. This chapter addresses the complexity and interconnectedness of current challenges and discusses how nexus planning quantitively simplifies these cross-sectoral relationships. Thus, this chapter integrates the health component into nexus planning for cross-sectoral transformational change.

2. Linking socioecological interactions with nexus planning

This chapter uses a set chronological sequence to establish the intricate interlinkages between human health and socioecological interactions. Initially, a systematic review of literature was conducted to better comprehend the trends of novel infectious diseases, focusing mainly on viruses whose origins are traced from wildlife. These included novel pathogens that have grappled humankind, including the COVID-19 pandemic. The review facilitated the understanding of the impacts of environmental and societal changes on ecosystems, changes that are constantly altering wildlife habitats, forcing wild animals to live in proximity with human beings. This knowledge is vital for comprehending the evolution of zoonotic viruses, and their rapid transmission and spread at global scale. The analysis facilitated a better understanding of the risk posed by wildlife on human health and well-being.

A database of over 120 relevant published articles was developed using search engines such as Web of Science, DOAJ, PubMed, Scopus, and Google Scholar. Keywords and search terms such as infectious diseases, virus from wildlife, disease pandemics, human—wildlife interactions, origins of infectious diseases, zoonotic infectious diseases, coronaviruses, climate change and infectious diseases, trends of infectious diseases, urbanization, and the spread of infectious diseases, among others, were used to retrieve the relevant publications. To cover a wide range of relevant research, we supplemented papers pertaining to individual pandemics such as HIV/AIDS, SARS, Ebola, MERS, Spanish Influenza, and COVID-19. Besides the knowledge on the dynamics involved in human–wildlife interactions, the literature review also found out that little has been done on understanding the intricate interlinkages between climatic and environmental changes with socioecological systems from a nexus planning perspective. Apart from closing this gap through nexus planning, the study also endeavored to provide pathways toward preparedness and readiness that future infectious diseases may not end in pandemics.

2.1 Defining the water–health–environment–nutrition nexus

The literature search identified four thematic areas related to socioecological changes: (1) factors that are driving change, (2) risk and exposure to novel infectious diseases, (3) preparedness and resilience to future pandemics, and (4) adaptation and resilience (Fig. 11.1). As a polycentric and transformative approach, nexus planning facilitates an understanding of the intricate interlinkages and interactions among the multisectoral socioecological components (Mabhaudhi et al., 2019; Nhamo and Ndlela, 2020). Nexus planning refers to a systematic and holistic assessment of different but interlinked attributes together, a cross-sectoral approach to comprehend and assess the intricately interconnected natural interactions (Bleischwitz et al., 2018; Naidoo et al., 2021). Each part of the nexus is equally evaluated without any prioritization of one particular component (Hoff, 2011; Nhamo et al., 2018). Important attributes of nexus planning include the capability to (1) identify trade-offs and synergies, (2) indicate areas needing immediate intervention, and (3) inform strategic formulations to reduce the risk of placing challenges from one sector to the others (Mpandeli et al., 2018; Nhamo et al., 2020a). This is the main difference between linear and circular models, which also include nexus planning (Nhamo and Ndlela, 2020).

Four interwoven components that make up a functional socioecological system include water, human health, environment, and nutrition (Nhamo and Ndlela, 2020). The four drive the processes and all the dynamics occurring a natural system. Because of the interconnectedness and close relationships, the four have been called the "WHEN nexus" (Mabhaudhi et al., 2021; Nhamo and Ndlela, 2020). It is a nexus representing this close and intricate connectedness between the WHEN components (Fig. 11.1). This nexus is built on the understanding that any variations made on one of the components will result in total change to the whole system, and this may include species extinction, migration, and invasion of alien species (Bellard et al., 2012; Nhamo and Ndlela, 2020; Nhamo et al., 2018; Wong and Candolin, 2015). The WHEN nexus is, therefore, meant to facilitate an understanding of these complexities and provide pathways for transformational change. The transformational change coming through nexus planning facilitates an understanding of novel interactions between animals and people, and it provides decision tools to reduce the risk

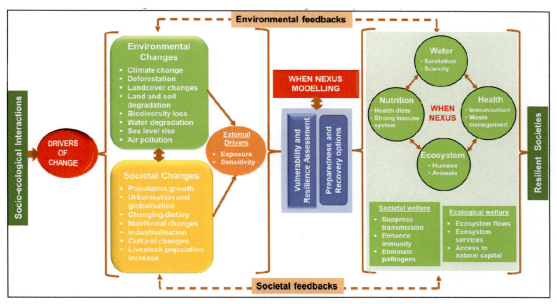

FIGURE 11.1
Dynamic processes and procedures of a socioecological system and the impacts on human and environmental health. Nexus planning simplifies these intricately interconnected relations and guide policy and decision-making on coherent and strategic policy formulations.

of novel infectious diseases and pathogens (Nhamo and Ndlela, 2020). An understanding of the complex interlinkages and interdependences among the socioecological components facilitates the plotting of a conceptual framework (Fig. 11.1), to illustrate and interpret the dynamic processes that are caused by prevailing human—wildlife interactions.

3. Understanding the risk posed by wildlife on human health

3.1 The role of nexus planning in simplifying socioecological systems

Nexus planning integrates distinct but interconnected components of the same system (Mabhaudhi et al., 2019; Nhamo et al., 2020b). In socioecological systems, these components include (1) the factors that drive change, (2) social and ecological responses, and (3) the numerous results (Fig. 11.1). This is based on that these ecosystems are controlled (directly or indirectly) to benefit people. Nevertheless, there is risk when the concentration is on a single ecosystem service (provision of food) and forgetting the importance of other services (flood attenuation and hygiene) habitually results in discord and risking human health (Purvis et al., 2019). However, anthropogenic activities continue to alter ecosystems and, at the same time, destroy wildlife habitats, increasing the risk of novel infectious diseases (Lindahl and Grace, 2015). As a transformative approach for cross-management of resources, the WHEN nexus provides informed interventions when faced with uncertainty and unknown outcomes. Adopting the WHEN nexus has the benefit of reducing risk of novel pathogens through informed preparedness and readiness.

The conceptual framework (Fig. 11.1) illustrates the major interactions within a natural system and the fundamental factors influencing socioecological interactions. The framework demonstrates the spatiotemporal flow and nature societal and environmental outcomes. Novel pathogens with origins from the wild have inflicted humankind for the past 100 years, and the trend significantly correlates with both accelerated population growth and the spread of biodiversity (Allen et al., 2017). Nexus planning facilitates an understanding of these intricacies, simplifying those complex socioecological interactions. Nexus planning also guides policy on improving health and sanitation, waste disposal, and pest control, pathways for reducing risk and vulnerability on human health (Nhamo and Ndlela, 2020).

3.2 WHEN nexus and sustainability indicators

Unlike linear or monocentric processes, nexus planning offers cross-sectoral solutions and provides pathways toward sustainable development (Mabhaudhi et al., 2019; Nhamo et al., 2020a). The WHEN nexus approach unpacks the complexities of complex interactions between the WHEN nexus components (water, human health, environment, and nutrition) and their spatiotemporal heterogeneity (Nhamo and Ndlela, 2020). Nexus planning is full of nonlinear socioecological feedbacks (Fig. 11.1). The approach unbundles multifaceted and cross-sectoral causal challenges within each component (Carpenter et al., 2009; Nhamo et al., 2020a). As a transformative approach, nexus planning provides the traits that lead to sustainable development and risk reduction. The WHEN nexus results in outcomes and interactions that minimize certain behaviors that compound human vulnerability to novel zoonotic pathogens.

A set of sustainability indicators (Table 11.1) for the WHEN nexus is critical for determining quantitative relationships among distinct WHEN nexus components. This is achieved through a multicriteria decision-method (MCDM) method (Nhamo et al., 2020a). The sustainability indicators provide an indication of resource security and sustainability (Hoff, 2011). Adopting the WHEN nexus in transformational changes is essential for achieving sustainable development and balancing the competing demands from different sectors on the background of the knowledge of human economic and socioecological

Table 11.1 WHEN nexus sustainability indicators.

Component	Subcomponent	Indicator	Units	SDG indicator
Water	Water security	Proportion of population using safely managed drinking water services	%	6.1.1
		Proportion of bodies of water with good ambient water quality	%	6.3.2
Human health	WASH	Mortality rate attributed to unsafe water, unsafe sanitation, and lack of hygiene	Per 100K of pop	3.9.2
Environment	Functional ecosystem	Forest area as a proportion of total land area	%	15.1.1
		Proportion of land that is degraded over total land area	%	15.3.1
Nutrition	Sustainable diets	Prevalence of moderate or severe food insecurity in the population	%	2.1.2
		Prevalence of malnutrition	%	2.2.2

Adapted from Nhamo, L., Ndlela, B., 2020. Nexus planning as a pathway towards sustainable environmental and human health post Covid-19. Environ. Res. 192, 110376.

boundaries (Meadows et al., 1972). This is the first step toward achieving sustainable development (Breslow et al., 2017; Shilling et al., 2013).

Sustainability indicators are, therefore, the basic decision support tools essential for simplifying complex interlinkages among distinct but related components and are capable of adapting those interactions into simplified declarations that unbundle the complexities (Ciegis et al., 2009; Mabhaudhi et al., 2021; Nhamo and Ndlela, 2020). Hence, sustainability indicators, being the basic unit of measurement, are central to WHEN nexus modeling and for comprehend complex relationships. The WHEN nexus indicators given in Table 11.1 are related to SDG indicators, qualifying the approach to assess progress toward the 2030 sustainable development goals (Mabhaudhi et al., 2021; Nhamo et al., 2020a). Indicators are a source of crucial information used to calculate indices in nexus planning. The process identifies priority areas for immediate intervention as a means to reduce risk on human and environmental health and ensuring sustainable natural interactions (Milner-Gulland, 2012).

3.3 Modeling vulnerability and resilience

Scenario planning and modeling of novel infectious diseases is possible through global- and large-scale computer simulations (Van den Broeck et al., 2011; Walters et al., 2018). In epidemics preparedness, nexus planning informs scenario planning, which is essential in building resilience, preparedness, and readiness (Khan et al., 2018; Nhamo and Ndlela, 2020). Resilience is better understood from experiences of previous exposure and the responses undertaken usually taken when there are shocks and stresses. This facilitates the formulation of coherent strategic policies that enhance resilience (Patel et al., 2017). These trajectory evaluations are vital for policy formulations and resilience building, and they offer insights into previous socioecological system dynamics that consider both previous and imminent conditions, as the current outlook is framed by historical events (Stringer et al., 2014, 2018). The vulnerability or resilience of natural transformations is negatively influenced by socioecological alterations (Folke et al., 2016) (Fig. 11.1). These global processes are driving change and are compounded by population increase, rapid urban development, globalization, deforestation, and environmental degradation.

There are three components that constitute vulnerability and resilience, and these are sensitivity, exposure, and resilience (Miller et al., 2010). Vulnerability (V) is, therefore, a dependent on the recovery potential (RP) of the components, as well as the potential impacts (PI). These are presented through exposure (E) and sensitivity (S) and expressed as (Allen and Prosperi, 2016)

$$V = f(PI, PR), \text{ with } PI = f(E, S) \tag{11.1}$$

Presenting vulnerability and resilience is essential for showing the distinct components of a system, as well as for establishing quantitative relationships and calculating integrated indices as demonstrated by Nhamo and Ndlela (2020). This is one of the approaches for assessing risk and exposure and formulating strategic policies that enhance resilience.

3.3.1 Mapping risky and vulnerable zoonotic hot spots

Tracing back the origins of novel infectious diseases from the time they first appeared in humans facilitates an understanding of unique patterns that could lead to risk reduction through disease control (Morse, 2001). The risk to human health is high as the frequency of emergence of novel pathogens is increasing, and some of them have ended in pandemics (Madhav et al., 2017; WHO, 2020). Vulnerability to novel pathogens is also increasing as exposure and sensitivity are increasing due to societal and environmental changes (Lindahl and Grace, 2015).

As novel pathogens are significantly correlated to human population density (Vanden Broecke et al., 2019), it is sufficient evidence that disease emergence and spread are driven by anthropogenic changes such as land-use changes, expansion of agriculture land, urbanization, globalization, and easy of travel and trade (Hassell et al., 2017). Geographic information system (GIS) is an important tool to model and identify hot spots of emerging zoonoses and identify regions with the most likelihood of the next emergence of a zoonotic infection (Jones et al., 2008). Fig. 11.2 is a map developed using data of known novel infections since 1940, based on the origins of each outbreak (Jones et al., 2008; Wang et al., 2019). The presence or absence of novel infections from wildlife was examined using logistic regression analysis, alongside a series of drivers that include population density, rapid urbanization, and mammalian species richness at 1 km^2 resolution (Allen et al., 2017; Morse et al., 2012).

Zoonoses hot spots are regions where anthropogenic activities are highly altering wildlife habitat, causing wildlife to invade human settlements, increasing the risk of zoonotic diseases (Morse et al., 2012). One of the

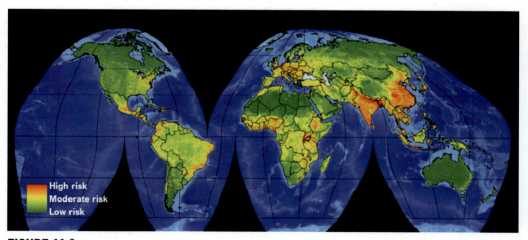

FIGURE 11.2
Global hot spots for novel zoonotic infectious diseases. Red areas are the high-risk regions with highest likelihood of the emergent of novel zoonotic infectious disease. *Reproduced from: Morse, S.S., Mazet, J.A., Woolhouse, M., Parrish, C.R., Carroll, D., Karesh, W.B., Zambrana-Torrelio, C., Lipkin, W.I., Daszak, P., 2012. Prediction and prevention of the next pandemic zoonosis. Lancet 380 (9857), 1956–1965.*

identified high risky hot spots is east and southeast Asia where the SARS-CoV-2 originated. Surveillance and resources should be targeted to zoonoses hot spots as a means of prevention of diseases.

Globalization and the increase in air travel have also facilitated global transit of novel infectious diseases on a daily basis (Fig. 11.3), causing accelerated spread and transmission in a short space of time, and even during the latent period for some infections (Hufnagel et al., 2004; Kilpatrick and Randolph, 2012; Tatem et al., 2006). The lines on the map (Fig. 11.3) indicate daily airports connectivity, and the colors represent daily passenger capacity in 1000 (red), 100 (yellow), and 10 (blue). The intensity of the choropleth routes agrees with the regions that are mostly affected by the SARS-CoV-2 infections (China, Europe, and North America). Air traffic routes are thus pathways for pathogens to move from one region to another, and the rate of infestation is represented by the intensity of air traffic route.

4. Modeling multisector and complex systems

A nexus analytical model simplifies the complex interlinkages and dynamics among the WHEN nexus components and to mathematically represent the interactions between the inputs and outputs occurring in a socioecological system. The principal outcomes include sustainable environments, provision of water and sanitation, enhanced immunity, efficient ecosystems, and sustainable diets (Chiabai et al., 2018; Martinez-Juarez et al., 2015). Nexus planning

FIGURE 11.3

Daily global aviation network. The network indicates how airplanes accelerate transmission and spread of novel infectious diseases in a short space of time. *Reproduced from: Kilpatrick, A.M., Randolph, S.E., 2012. Drivers, dynamics, and control of emerging vector-borne zoonotic diseases. Lancet 380 (9857), 1946–1955.*

was used to calculate these outcomes that include the composite indices for each component to demonstrate the susceptibility and vulnerability of a region to zoonotic diseases. The indices are classified into health hazard risk classes (highly or lowly risky) (Nhamo et al., 2020a). The indicators given in Table 11.1 are the decision support tools to determine the spatiotemporal indices (Mabhaudhi et al., 2021; Nhamo et al., 2020a).

Information on the state of a system at any given time is important as an input trajectory, as it provides the lens to determine the qualities and attributes of the system over time (Allen and Prosperi, 2016). The state of variables (x) of the system provides the least amount of information that explains its state at any time. A quantitative representation of variables (x) of the system, along with the understanding of the variables at an original time (t_0) and the system inputs for time (t), provides the crucial information needed to approximate future system changes and outputs for all time (t) (Åström and Murray, 2009).

Outcomes (y) are generally critical for describing input—output relationships. The dynamics of a system can also be represented through differential equations, where time becomes the independent variable. These transformational equations represent the spatiotemporal matrix structure of natural systems and are expressed as (Allen and Prosperi, 2016; Rowell, 2002)

$$y_t = h(x_t, u_t, e_t), \tag{11.2}$$

where y are outcomes and h is a vector with n elements for the n outputs y of concern. The variables differ with time t. This equation is transformational in that it provides an all-time status of a system (Rowell, 2002). It is expressed as

$$\ddot{x} = \frac{dy}{dt} = f(x_t, u_t, e_t), \tag{11.3}$$

where (f) is a vector function. The system status at any given time (t) is indicated as an m-dimensional state-space, and the active response x_t is described as a sequence tracked out in state space (Rowell, 2002).

Eqs. (11.2) and (11.3) are alternatively expanded to account for feedback to the inputs as Eqs. (11.4) and (11.5) (Åström and Murray, 2009). The main challenge with this kind of modeling is data availability. However, most of this information can be obtained from remotely sensed data or from open data platforms such as the World Bank Indicators and AQUASTAT.

$$\ddot{e} = \frac{de}{dt} = (x_t, u_t, e_t) \tag{11.4}$$

$$\ddot{u} = \frac{du}{dt} = \emptyset(x_t, e_t) \tag{11.5}$$

An alternative method to quantitatively represent these intricate but distinct variables is the through applying the analytic hierarchy process (AHP), which is a multicriteria decision method (MCDM) (Saaty, 1987). The AHP, numerically relates distinct components like different sustainability indicators through a process called pairwise comparison matrix (PCM) (Nhamo et al., 2020a). The PCM determines an integrated WHEN nexus sustainability index (Table 11.1) to convey relational information about preparedness at a given time. This information is necessary for indicating how susceptible or resilient a country or any region is to novel infectious diseases. This is represented in the form of priority weights (also known as indices) for each indicator (Mabhaudhi et al., 2021; Nhamo et al., 2020a, 2020b). The priority weights are symbolized as w (Saaty, 1990), with individual weights determined through a principal input of the matrix, A, of n criteria, in the order $(n \times n)$ (Rao et al., 1991). A is a sequence of elements a_{ij}. The matrix, which is reciprocal, is expressed as

$$a_{ij} = \frac{1}{a_{ij}} \tag{11.6}$$

After establishing the matrix, it is then standardized or normalized as pattern B, of a pattern A, with components b_{ij} which is expressed as

$$b_{ij} = \frac{a_{ij}}{\sum_{j=1}^{n} a_{ij}} \tag{11.7}$$

The indicator weight (w_i) is expressed as

$$w_i = \frac{\sum_{j=1}^{n} b_{ij}}{\sum_{i=1}^{n} \sum_{j=1}^{n} b_{ij}}, \quad i,j = 1, 2, 3\ldots, n \tag{11.8}$$

The integrated index is determined by weighting all the indices. The indices are then represented visually through a spider graph to provide a visual overview on the preparedness status to zoonotic diseases. The relationship among the indicators vividly demonstrates the degree of vulnerability of a country to infectious diseases and the risk on human health. The information is used to identify the areas needing priority intervention (Nhamo et al., 2020a). However, there is need to determine the consistency ratio (CR), of the comparison matrix (Nhamo et al., 2020a).

The choice between the two methods in establishing numerical relationships among distinct indicators depends on the objectives of the study. The MCDM was the preferred method as it establishes indices for each sustainability indicator and facilitates the graphical representation of the relationships among the indicators (Mabhaudhi et al., 2021; Nhamo et al., 2020a; Nhamo and Ndlela, 2020), whereas the differential method only establishes integrated composite indices, which are difficult to interpret or cannot be used to identify priority areas for intervention (Nhamo et al., 2020a; Nardo et al., 2005).

5. Calculating WHEN nexus indices for South Africa

In the PCM for South Africa (Table 11.3), the diagonal values of unity are always 1 (Saaty, 1987). The symmetrical matrix is in two parts, the shaded section and the nonshaded section. The nonshaded section is the one that needs to be filled, yet the shaded is reciprocal. The values represented by the bottom unshaded triangle are reciprocals of the shaded half. An established scale indicates a relational classification that ranges between 1/9 and 9 (Table 11.3) (Saaty, 1990). The weights are based on expert consultation and the country status on the indicators for a given year as shown in Table 11.2. Thus, the indicator values (Table 11.2) are important in establishing the numerical relationships and essential for providing the basis to classify whether a country is vulnerable or resilient to novel infectious diseases (Mabhaudhi et al., 2021; Nhamo and Ndlela, 2020).

The weights are then standardized using Eqs. (11.7) and (11.8). The standardized indices (Table 11.4) indicate that the indicators are now quantitatively linked, an important feature of the procedure that facilitates the determination of integrated quantitative analysis (Nardo et al., 2005; Saaty, 1990). The sum of the values in the column should always be 1. The calculated CR was 0.10, which is acceptable. The weighted average becomes the integrated index and is classified according to the categories shown in Table 11.5. The integrated index for South Africa in 2018 was 0.170.

Table 11.2 The status of WHEN nexus-related indicators in South Africa in 2018.

WHEN nexus	Indicator	Status 2018
Water	Proportion of population using safely managed drinking water services (water accessibility)	74%
	Proportion of bodies of water with good ambient water quality (water quality)	46.92%
Human health	Mortality rate attributed to unsafe water, unsafe sanitation, and lack of hygiene (WASH mortality)	13.7/100K pop
Environment	Forest area as a proportion of total land area (forested area)	7.6%
	Proportion of land that is degraded over total land area (degraded area)	60%
Nutrition	Prevalence of moderate or severe food insecurity in the population (food insecurity)	52%
	Prevalence of malnutrition (malnutrition)	6.2%

Courtesy: World Bank Indicators

CHAPTER 11: Enhancing sustainable human and environmental health through nexus planning

Table 11.3 Relational pairwise comparison matrix for WHEN nexus indicators in South Africa.

Indicator	Pairwise comparison matrix							
	Water accessibility	Water quality	WASH mortality	Forested area	Degraded area	Food insecurity	Malnutrition
Water accessibility	1	1	1	1/4	1/2	1/3	1/2
Water quality	1	1	1	1/3	1/3	1/2	1
WASH mortality	1	1	1	1/2	1/3	1	1/2
Forested area	4	3	2	1	1	1/3	1/3
Degraded area	2	3	3	1	1	1/3	1/3
Food insecurity	3	2	1	3	3	1	1
Malnutrition	2	1	2	3	3	1	1

Table 11.4 The normalized WHEN nexus pairwise comparison matrix and the integrated indices.

| Indicator | Normalized pairwise comparison matrix ||||||| Indices |
	Water accessibility	Water quality	WASH mortality	Forested area	Degraded area	Food insecurity	Malnutrition	
WASH mortality	0.071	0.083	0.091	0.028	0.055	0.074	0.107	0.073
Water quality	0.071	0.083	0.091	0.037	0.036	0.111	0.214	0.092
WASH mortality	0.071	0.083	0.091	0.055	0.036	0.222	0.107	0.095
Forested area	0.286	0.250	0.182	0.110	0.109	0.074	0.071	0.155
Degraded area	0.143	0.250	0.273	0.110	0.109	0.074	0.071	0.147
Food insecurity	0.214	0.167	0.091	0.330	0.327	0.222	0.214	0.224
Malnutrition	0.143	0.083	0.182	0.330	0.327	0.222	0.214	0.215
CR = 0.10								$\Sigma = 1$
Composite index (weighted average)								0.170

Table 11.5 Health risk classification categories.

	Severe risk	High risk	Moderate risk	Low risk
Category	0–0.9	0.1–0.2	0.3–0.6	0.7–1

The indices (Table 11.4) oscillate between 0 and 1. According to the classification categories given in Table 11.5, the indices are classified into health risk categories either as severe risk, high risk, moderate risk, or low risk. South Africa has a composite index of 0.170, which is in the high-risk category.

The classification categories given in Table 11.5 are also used to classify individual indicators. The classifications are indicative of the level of preparedness and readiness. These health risk classification categories used to interpret the health risk level and for informed policy formulations aimed at reducing vulnerability and building resilience.

6. Understanding the integrated health indices

As already alluded to, the overall health index ranking for South Africa (0.170) places the country into a high-risk category (Table 11.5), which indicates that the country is susceptible to novel infectious diseases. The classification concurs with the identified global zoonoses hot spots, which identified South Africa as a moderate risk country to emerging infectious diseases (Fig. 11.2). The risk for South Africa is worsened by the high rate of aviation traffic into the country, which is the highest in African, as its airports connect the continent to the rest of the world (Fig. 11.3). The high influx of visitors into the country on a daily basis poses the risk of rapid disease transit from high-risk regions.

The indices (Table 11.4) are represented by means of a spider web (Fig. 11.4), which vividly shows how the indicators are related and how each of them contributes to the health risk level. In relation with the other indicators, malnutrition and food insecurity indices are very high in South Africa at 0.224 and 0.215, respectively, posing a high risk of disease infection due to deficiencies in the immune system. Decision-makers should pay more attention to reducing the risk as indicated by the negative indices. The indices are an indication of the level of vulnerability of people living mainly in poor communities (Satterthwaite et al., 2020). The vulnerability level for South Africa to novel infectious diseases is compounded by the high-risk indices of important indicators related to water accessibility and water quality. In contrast, the food insecurity and malnutrition indices should be reduced, yet those for water accessibility and water quality should be improved. In South Africa, achieving the desired levels in water-related indicators is deterred by the water scarcity challenges that the country faces (Sershen et al., 2016). However, besides these

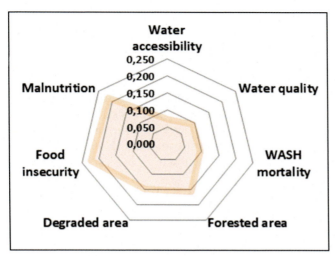

FIGURE 11.4
A quantitative relational socioecological interactions among WHEN nexus indicators in South Africa.

challenges, it has achieved commendable low levels in water, sanitation, and hygiene (WASH)—related deaths at 13.7 deaths/100 000 population (Fig. 11.4).

The environmental indices for South Africa are quite low. These refer to the proportion of forested area and land degradation, which should be getting priority attention as over 60% of the country's land area is degraded and only 7.6% is forested (Niedertscheider et al., 2012). These statistics indicate the high environmental changes that are altering wildlife habitats mainly due to rapid urbanization and expanding agricultural land. A good example is in Gauteng Province where wild animals are spotted in urban areas (Nhamo et al., 2021). These changes are instigating wildlife to encroach into human habitats, especially in urban areas as food is more available than in wild where it has become scarce (Nava et al., 2017; Wong and Candolin, 2015). These novel natural interactions are posing the greatest risk of zoonotic diseases to humans (Jones et al., 2008; WHO, 2018).

The generated knowledge can be linked to the epidemic preparedness index (EPI) (Madhav et al., 2017; Oppenheim et al., 2019) to provide the required information on formulating coherent strategies on enhancing resilience. In contract to sustainability studies where the spider graph is supposed to take a circular shape to achieve a certain level of sustainability, in the health sector, the weights of negative indicators such as food insecurity and malnutrition have to be drastically reduced to ensure resilience (Nhamo et al., 2020a).

7. Recommendations

Societal changes that include rapid urbanization, industrialization, increased population growth, and globalization, compounded by environmental and climatic changes, biodiversity loss, sea level rise, and water and land degradation, continue to modify ecological, biological, and social conditions, immensely giving rise to novel infectious disease. These factors are driving environmental changes, which has seen the emergence, prevalence, transmission, and novel infectious diseases (Saker et al., 2004). With the increasing vulnerability to novel pathogens and the risk they pose on human health due to novel socioecological interactions, we recommend the following:

- As poor communities (especially in the global south) are the most vulnerable due to the insecurity of essential resources and malnutrition, there is need for decision-making to focus on improving accessibility to clean and reliable water, and good sanitation and hygiene (WASH) in poor communities. There is also a need to improve on food and nutritional security to boost the immunity system, particularly in poor communities. In an attempt to reduce the risk of infectious disease originating from wildlife, there should be concerted efforts to curb air pollution and unsanitary conditions that prevail in poor communities such as informal settlements. Unsanitary conditions may include the consumption of urban wildlife, or bushmeat hunted from the surrounding area (Sarti et al., 2015).
- Anthropogenic activities that include agricultural extensification and increase in human settlement are altering environments, disrupting ecosystems in the process, and creating viable conditions for novel pathogens (WHO, 2005). While change could be irreversible, there is need to research more on viruses living in wildlife and develop vaccines against them in preparedness and readiness of their emergence and change of host from animals to humans. Presently, only 10% of the world's known pathogens are documented, leaving a huge gap for research to study the rest, including the host animals (WHO, 2018).
- There are research gaps on tracking those animals that are now dwelling in urban areas in search of food, and if urban dwellers are consuming them. Lack of such information increases the vulnerability of people to novel infectious diseases with origin from wildlife. Interventions to reduce risk should include the improvement in sanitation, adoption of the circular economy, waste disposal, and pest control (Bartram and Cairncross, 2010).
- The challenge of novel infectious diseases is of a global scale (HIV/AIDS and COVID-19), which calls for planetary strategies to reduce risk and vulnerability to these novel pathogens in a timely manner. Besides national efforts, and because the challenge is planetary in nature,

international organizations such as the World Health Organization (WHO) should spearhead the infectious diseases risk reduction initiatives and ring fence zoonoses hot spots and reduce worldwide transmission. To effectively respond to novel infectious diseases, policymakers need to acknowledge and respond with "planetary consciousness" by taking holistic and global measures on public health, including the health of the natural environments.

8. Conclusions

Comprehending the intricate interrelationships between the drivers of change (climate change, ecosystems, and human health) is critical for planning adaptive responses against novel zoonotic diseases. Nexus planning simplifies these intricate linkages and provides indicators for those areas needing intervention to inform strategic policy formulations that result in resilience and adaptation. Thus, the WHEN nexus assessment facilitated the identification of areas for priority intervention to ensure mutual socioecological cobenefits and reduce the risk of novel infectious diseases originating from wildlife. Linking nexus planning and epidemiology studies allows the documentation of novel pathogens and moves from circumstantial to investigative and analytical processes that facilitate predictive and transformational modeling. Further research is needed to document zoonotic viruses as a risk reduction measure through forward development of vaccines to combat potential harmful pathogens beforehand. This requires huge investments to keep pace with fats changing socioecological transformations. Gone are days of pursuing strategies based on a single challenge as the COVID-19 experiences exposed the vulnerability of sector-based solutions. Today's challenges are interlinked and cross-cutting, and they require polycentric and transformative approaches such as nexus planning, circular economy, and scenario planning, which provide integrated solutions.

References

Abdallah, K.B., Belloumi, M., De Wolf, D., 2013. Indicators for sustainable energy development: a multivariate cointegration and causality analysis from Tunisian road transport sector. Renew. Sustain. Energy Rev. 25, 34–43.

Adhikari, S.P., Meng, S., Wu, Y.-J., Mao, Y.-P., Ye, R.-X., Wang, Q.-Z., Sun, C., Sylvia, S., Rozelle, S., Raat, H., 2020. Epidemiology, causes, clinical manifestation and diagnosis, prevention and control of coronavirus disease (COVID-19) during the early outbreak period: a scoping review. Infecti. Dis. Poverty 9, 1–12.

Allen, T., Murray, K.A., Zambrana-Torrelio, C., Morse, S.S., Rondinini, C., Di Marco, M., Breit, N., Olival, K.J., Daszak, P., 2017. Global hotspots and correlates of emerging zoonotic diseases. Nat. Commun. 8, 1–10.

Allen, T., Prosperi, P., 2016. Modeling sustainable food systems. Environ. Manag. 57, 956–975.

Andersen, K.G., Rambaut, A., Lipkin, W.I., Holmes, E.C., Garry, R.F., 2020. The proximal origin of SARS-CoV-2. Nat. Med. 1–3.

Åström, K.J., Murray, R.M., 2009. Feedback Systems: An Introduction for Scientists and Engineers. Princeton University Press, Princeton and Oxford.

Bartram, J., Cairncross, S., 2010. Hygiene, sanitation, and water: forgotten foundations of health. PLoS Med. 7 (11).

Bellard, C., Bertelsmeier, C., Leadley, P., Thuiller, W., Courchamp, F., 2012. Impacts of climate change on the future of biodiversity. Ecol. Lett. 15, 365–377.

Bleischwitz, R., Spataru, C., VanDeveer, S.D., Obersteiner, M., van der Voet, E., Johnson, C., Andrews-Speed, P., Boersma, T., Hoff, H., van Vuuren, D.P., 2018. Resource nexus perspectives towards the United Nations sustainable development goals. Nat. Sustain. 1, 737–743.

Boulos, M.N.K., Geraghty, E.M., 2020. Geographical tracking and mapping of coronavirus disease COVID-19/severe acute respiratory syndrome coronavirus 2 (SARS-CoV-2) epidemic and associated events around the world: how 21st century GIS technologies are supporting the global fight against outbreaks and epidemics. Int. J. Health Geogr. 19 (8), 1–12.

Breslow, S.J., Allen, M., Holstein, D., Sojka, B., Barnea, R., Basurto, X., Carothers, C., Charnley, S., Coulthard, S., Dolšak, N., 2017. Evaluating indicators of human well-being for ecosystem-based management. Ecosys. Health Sustain. 3 (12), 1–18.

Carpenter, S.R., Mooney, H.A., Agard, J., Capistrano, D., DeFries, R.S., Díaz, S., Dietz, T., Duraiappah, A.K., Oteng-Yeboah, A., Pereira, H.M., 2009. Science for managing ecosystem services: beyond the Millennium Ecosystem Assessment. Proc. Natl. Acad. Sci. U.S.A. 106 (5), 1305–1312.

Chiabai, A., Quiroga, S., Martinez-Juarez, P., Higgins, S., Taylor, T., 2018. The nexus between climate change, ecosystem services and human health: towards a conceptual framework. Sci. Total Environ. 635, 1191–1204.

Ciegis, R., Ramanauskiene, J., Startiene, G., 2009. Theoretical reasoning of the use of indicators and indices for sustainable development assessment. Eng. Econ. 63 (3).

Cox, D.T., Gaston, K.J., 2018. Human–nature interactions and the consequences and drivers of provisioning wildlife. Phil. Trans. Biol. Sci. 373 (1745), 20170092.

De Nys, H.M., Kingebeni, P.M., Keita, A.K., Butel, C., Thaurignac, G., Villabona-Arenas, C.-J., Lemarcis, T., Geraerts, M., Vidal, N., Esteban, A., 2018. Survey of Ebola viruses in frugivorous and insectivorous bats in Guinea, Cameroon, and the Democratic Republic of the Congo, 2015–2017. Emerg. Infect. Dis. 24 (12), 2228.

Dong, E., Du, H., Gardner, L., 2020. An interactive web-based dashboard to track COVID-19 in real time. Lancet Infect. Dis. 20 (5), 533–534.

El-Kafrawy, S.A., Corman, V.M., Tolah, A.M., Al Masaudi, S.B., Hassan, A.M., Müller, M.A., Bleicker, T., Harakeh, S.M., Alzahrani, A.A., Alsaaidi, G.A., 2019. Enzootic patterns of Middle East respiratory syndrome coronavirus in imported African and local Arabian dromedary camels: a prospective genomic study. Lancet Planet. Health 3 (12), e521–e528.

Folke, C., Biggs, R., Norström, A.V., Reyers, B., Rockström, J., 2016. Social-ecological resilience and biosphere-based sustainability science. Ecol. Soc. 21 (3).

Galvani, A.P., Bauch, C.T., Anand, M., Singer, B.H., Levin, S.A., 2016. Human–environment interactions in population and ecosystem health. Proc. Natl. Acad. Sci. U.S.A. 113 (51), 14502–14506.

Gibbs, A.J., Armstrong, J.S., Downie, J.C., 2009. From where did the 2009 'swine-origin' influenza A virus (H1N1) emerge? Virol. J. 6 (1), 1–11.

Hassell, J.M., Begon, M., Ward, M.J., Fèvre, E.M., 2017. Urbanization and disease emergence: dynamics at the wildlife–livestock–human interface. Trends Ecol. Evol. 32 (1), 55–67.

Hoff, H., 2011. Understanding the Nexus: Background Paper for the Bonn2011 Conference: The Water, Energy and Food Security Nexus. Stockholm Environment Institute (SEI), Stockholm, Sweden.

Hufnagel, L., Brockmann, D., Geisel, T., 2004. Forecast and control of epidemics in a globalized world. Proc. Natl. Acad. Sci. U.S.A. 101 (42), 15124–15129.

Jones, K.E., Patel, N.G., Levy, M.A., Storeygard, A., Balk, D., Gittleman, J.L., Daszak, P., 2008. Global trends in emerging infectious diseases. Nature 451 (7181), 990–993.

Kallings, L.O., 2008. The first postmodern pandemic: 25 years of HIV/AIDS. J. Intern. Med. 263 (3), 218–243.

Khan, Y., O'Sullivan, T., Brown, A., Tracey, S., Gibson, J., Généreux, M., Henry, B., Schwartz, B., 2018. Public health emergency preparedness: a framework to promote resilience. BMC Public Health 18 (1), 1–16.

Kilpatrick, A.M., Randolph, S.E., 2012. Drivers, dynamics, and control of emerging vector-borne zoonotic diseases. Lancet 380 (9857), 1946–1955.

Leeson, G.W., 2018. The growth, ageing and urbanisation of our world. J. Populat. Ageing 11 (2), 107–115.

Lindahl, J.F., Grace, D., 2015. The consequences of human actions on risks for infectious diseases: a review. Infect. Ecol. Epidemiol. 5 (1), 30048.

Lycett, S.J., Duchatel, F., Digard, P., 2019. A brief history of bird flu. Philos. Trans. Royal Soc. B 374 (1775), 20180257.

Mabhaudhi, T., Nhamo, L., Chibarabada, T.P., Mabaya, G., Mpandeli, S., Liphadzi, S., Senzanje, A., Naidoo, D., Modi, A.T., Chivenge, P.P., 2021. Assessing progress towards sustainable development goals through nexus planning. Water 13 (9), 1321.

Mabhaudhi, T., Nhamo, L., Mpandeli, S., Nhemachena, C., Senzanje, A., Sobratee, N., Chivenge, P.P., Slotow, R., Naidoo, D., Liphadzi, S., 2019. The water–energy–food nexus as a tool to transform rural livelihoods and well-being in southern Africa. Int. J. Environ. Res. Publ. Health 16 (16), 2970.

Madhav, N., Oppenheim, B., Gallivan, M., Mulembakani, P., Rubin, E., Wolfe, N., 2017. Pandemics: risks, impacts, and mitigation. In: Jamison, D.T., Gelband, H., Horton, S. (Eds.), Disease Control Priorities: Improving Health and Reducing Poverty, third ed. The World Bank, Wasington DC.

Martinez-Juarez, P., Chiabai, A., Quiroga Gómez, S., Taylor, T., 2015. Ecosystems and Human Health: Towards a Conceptual Framework for Assessing the Co-benefits of Climate Change Adaptation. Basque Centre for Climate Change (BC3), Bilbao, Spain.

Meadows, D.H., Meadows, D.H., Randers, J., Behrens III, W.W., 1972. The Limits to Growth: A Report to the Club of Rome. A Report Fo the Club of Rome on the Predicament of Mankind. Universe Books, Washington DC.

Miller, F., Osbahr, H., Boyd, E., Thomalla, F., Bharwani, S., Ziervogel, G., Walker, B., Birkmann, J., Van der Leeuw, S., Rockström, J., 2010. Resilience and vulnerability: complementary or conflicting concepts? Ecol. Soc. 15 (3).

Milner-Gulland, E., 2012. Interactions between human behaviour and ecological systems. Phil. Trans. Biol. Sci. 367 (1586), 270–278.

Morse, S.S., 2001. Factors in the emergence of infectious diseases. In: Price-Smith, A.T. (Ed.), Plagues and Politics. Palgrave Macmillan, Springer, London, pp. 8–26.

Morse, S.S., Mazet, J.A., Woolhouse, M., Parrish, C.R., Carroll, D., Karesh, W.B., Zambrana-Torrelio, C., Lipkin, W.I., Daszak, P., 2012. Prediction and prevention of the next pandemic zoonosis. Lancet 380 (9857), 1956–1965.

Mpandeli, S., Naidoo, D., Mabhaudhi, T., Nhemachena, C., Nhamo, L., Liphadzi, S., Hlahla, S., Modi, A., 2018. Climate change adaptation through the water-energy-food nexus in southern Africa. Int. J. Environ. Res. Publ. Health 15 (10), 2306.

Naicker, P.R., 2011. The impact of climate change and other factors on zoonotic diseases. Arch. Clin. Microbiol. 2 (2).

Naidoo, D., Nhamo, L., Mpandeli, S., Sobratee, N., Senzanje, A., Liphadzi, S., Slotow, R., Jacobson, M., Modi, A., Mabhaudhi, T., 2021. Operationalising the water-energy-food nexus through the theory of change. Renew. Sustain. Energy Rev. 149, 111416.

Nardo, M., Saisana, M., Saltelli, A., Tarantola, S., 2005. Tools for Composite Indicators Building. European Commission (EU), Ispra.

Nava, A., Shimabukuro, J.S., Chmura, A.A., Luz, S.L.B., 2017. The impact of global environmental changes on infectious disease emergence with a focus on risks for Brazil. ILAR J. 58 (3), 393−400.

Newsome, T.M., Van Eeden, L.M., 2017. The effects of food waste on wildlife and humans. Sustainability 9 (7), 1269.

Nhamo, L., Mabhaudhi, T., Mpandeli, S., Dickens, C., Nhemachena, C., Senzanje, A., Naidoo, D., Liphadzi, S., Modi, A.T., 2020a. An integrative analytical model for the water-energy-food nexus: South Africa case study. Environ. Sci. Pol. 109, 15−24.

Nhamo, L., Ndlela, B., 2020. Nexus planning as a pathway towards sustainable environmental and human health post Covid-19. Environ. Res. 192, 110376.

Nhamo, L., Ndlela, B., Mpandeli, S., Mabhaudhi, T., 2020b. The water-energy-food nexus as an adaptation strategy for achieving sustainable livelihoods at a local level. Sustainability 12 (20), 8582.

Nhamo, L., Ndlela, B., Nhemachena, C., Mabhaudhi, T., Mpandeli, S., Matchaya, G., 2018. The water-energy-food nexus: climate risks and opportunities in Southern Africa. Water 10 (5), 567.

Nhamo, L., Rwizi, L., Mpandeli, S., Botai, J., Magidi, J., Tazvinga, H., Sobratee, N., Liphadzi, S., Naidoo, D., Modi, A., Slotow, R., Mabhaudhi, T., 2021. Urban nexus and transformative pathways towards a resilient Gauteng City-Region, South Africa. Cities 116, 103266.

Niedertscheider, M., Gingrich, S., Erb, K.-H., 2012. Changes in land use in South Africa between 1961 and 2006: an integrated socio-ecological analysis based on the human appropriation of net primary production framework. Reg. Environ. Change 12 (4), 715−727.

Oppenheim, B., Gallivan, M., Madhav, N.K., Brown, N., Serhiyenko, V., Wolfe, N.D., Ayscue, P., 2019. Assessing global preparedness for the next pandemic: development and application of an Epidemic Preparedness Index. BMJ Global Health 4 (1), e001157.

Patel, S.S., Rogers, M.B., Amlôt, R., Rubin, G.J., 2017. What do we mean by 'community resilience'? A systematic literature review of how it is defined in the literature. PLoS Currents 9.

Patz, J.A., Grabow, M.L., Limaye, V.S., 2014. When it rains, it pours: future climate extremes and health. Ann. Global Health 80 (4), 332−344.

Purvis, B., Mao, Y., Robinson, D., 2019. Three pillars of sustainability: in search of conceptual origins. Sustain. Sci. 14 (3), 681−695.

Rao, M., Sastry, S., Yadar, P., Kharod, K., Pathan, S., Dhinwa, P., Majumdar, K., Sampat Kumar, D., Patkar, V., Phatak, V., 1991. A Weighted Index Model for Urban Suitability Assessment—A GIS Approach. Bombay Metropolitan Regional Development Authority, Bombay.

Rowell, D., 2002. State-space Representation of LTI Systems: Analysis and Design of Feedback Control Systems. Cambridge.

Saaty, R.W., 1987. The analytic hierarchy process—what it is and how it is used. Math. Model. 9 (3–5), 161–176.

Saaty, T.L., 1990. Eigenvector and logarithmic least squares. Eur. J. Oper. Res. 48 (1), 156–160.

Saker, L., Lee, K., Cannito, B., Gilmore, A., Campbell-Lendrum, D.H., 2004. Globalization and Infectious Diseases: A Review of the Linkages. World Health Organization (WHO), Geneva.

Sarti, F.M., Adams, C., Morsello, C., Van Vliet, N., Schor, T., Yagüe, B., Tellez, L., Quiceno-Mesa, M.P., Cruz, D., 2015. Beyond protein intake: bushmeat as source of micronutrients in the Amazon. Ecol. Soc. 20 (4).

Satterthwaite, D., Archer, D., Colenbrander, S., Dodman, D., Hardoy, J., Mitlin, D., Patel, S., 2020. Building resilience to climate change in informal settlements. One Earth 2 (2), 143–156.

Sershen, S., Rodda, N., Stenström, T.-A., Schmidt, S., Dent, M., Bux, F., Hanke, N., Buckley, C., Fennemore, C., 2016. Water security in South Africa: perceptions on public expectations and municipal obligations, governance and water re-use. WaterSA 42 (3), 456–465.

Sharp, P.M., Hahn, B.H., 2011. Origins of HIV and the AIDS pandemic. Cold Spring Harbor Perspect. Med. 6, 8–41 a006841.

Shilling, F., Khan, A., Juricich, R., Fong, V., 2013. Using indicators to measure water resources sustainability in California. In: World Environmental and Water Resources Congress 2013: Showcasing the Future, pp. 2708–2715.

Smith, I., Wang, L.-F., 2013. Bats and their virome: an important source of emerging viruses capable of infecting humans. Curr. Opin. Virol. 3 (1), 84–91.

Stringer, L., Quinn, C., Berman, R., Le, H., Msuya, F., Orchard, S., Pezzuti, J., 2014. Combining Nexus and Resilience Thinking in a Novel Framework to Enable More Equitable and Just Outcomes. Sustainability Research Institute. Paper 73.

Stringer, L.C., Quinn, C., Le, H.T., Msuya, F., Pezzuti, J., Dallimer, M., Afionis, S., Berman, R., Orchard, S., Rijal, M., 2018. A new framework to enable equitable outcomes: resilience and nexus approaches combined. Earth's Fut. 6 (6), 902–918.

Tatem, A.J., Rogers, D.J., Hay, S.I., 2006. Global transport networks and infectious disease spread. Adv. Parasitol. 62, 293–343.

Van den Broeck, W., Gioannini, C., Gonçalves, B., Quaggiotto, M., Colizza, V., Vespignani, A., 2011. The GLEaMviz computational tool, a publicly available software to explore realistic epidemic spreading scenarios at the global scale. BMC Infect. Dis. 11 (1), 1–14.

Vanden Broecke, B., Mariën, J., Sabuni, C.A., Mnyone, L., Massawe, A.W., Matthysen, E., Leirs, H., 2019. Relationship between population density and viral infection: a role for personality? Ecol. Evol. 9 (18), 10213–10224.

Walters, C.E., Meslé, M.M., Hall, I.M., 2018. Modelling the global spread of diseases: a review of current practice and capability. Epidemics 25, 1–8.

Wang, B., Liu, F., Zhang, E.C., Wo, C.L., Chen, J., Qian, P.Y., Lu, H.R., Zeng, W.J., Chen, T., Wei, J.P., 2019. The China National GeneBank— owned by all, completed by all and shared by all. Yi chuan - Hereditas 41 (8), 761–772.

Wang, L.-F., Eaton, B.T., 2007. Bats, civets and the emergence of SARS. In: Childs, J.E., Mackenzie, J.S., J.A, R. (Eds.), Wildlife and Emerging Zoonotic Diseases: The Biology, Circumstances and Consequences of Cross-Species Transmission. Springer, Berlin, Heidelberg, pp. 325–344.

WHO, 2005. Ecosystems and human well-being: health synthesis: a Report of the Millennium Ecosystem Assessment. Health Synthesis. In: Corvalan, C., Hales, S., McMichael, A. (Eds.), World Health Organization (WHO), Genva, Switzerland.

WHO, 2018. Managing Epidemics: Key Facts about Major Deadly Diseases. World Health Organization (WHO), Geneva.

WHO, 2020. Coronavirus Disease 2019 (COVID-19): Situation Report. World Health Organization (WHO), Geneva.

Wong, B., Candolin, U., 2015. Behavioral responses to changing environments. Behav. Ecol. 26 (3), 665–673.

Xu, Z., Shi, L., Wang, Y., Zhang, J., Huang, L., Zhang, C., Liu, S., Zhao, P., Liu, H., Zhu, L., 2020. Pathological findings of COVID-19 associated with acute respiratory distress syndrome. Lancet Respir. Med. 8 (4), 420–422.

CHAPTER 12

Financing WEF nexus projects: perspectives from interdisciplinary and multidimensional research challenges

Maysoun A. Mustafa[1] and Christoph Hinske[2]

[1]*School of Biosciences, University of Nottingham Malaysia, Semenyih, Selangor, Malaysia;*
[2]*System Leadership & Entrepreneurial Ecosystems, School of Finance and Accounting, SAXION University of Applied Sciences, Enschede, Netherlands*

1. Introduction

Science is universal, built on robust principles of the scientific methodology and providing objective evidence about objective truths. It attempts to construct a picture of the world that changes in response to other empirical information. While science remains universal, the mechanisms of translation and infrastructural supports in place might differ from one region to another. It is not a monolithic block and comprises varied knowledge and epistemologies (Borie et al., 2019). This condition is further complicated when dealing with interdisciplinary and multidimensional research, such as balancing the interlinkages of water, energy, and food (WEF).

The efficient management of WEF resources is a considerable challenge that stretches across disciplinary, institutional, and geographical boundaries. It is of immense importance for economies that depend on their natural resources, such as the case is for most of the Global South. Furthermore, it is crucial in addressing national, regional, and international goals aspiring toward sustainability and equitable development. Human actions surpass planetary boundaries and push the environment's capacity to support human's needs, leading to the rapid escalation in energy, food, and water prices due to limited accessibility and availability. Furthermore, the impacts of such rapid growth in demand for natural resources undermine the environment's ability to deliver the essential ecosystem services necessary to support the survival of future generations (IPCC, 2014).

Interdisciplinary research approaches are beneficial instruments to incorporate social and political dimensions and engage diverse actors in the management of natural resources, such as water, energy, and food (Albrecht et al., 2018). Beyond the evaluation of the synergies and trade-offs in using the natural resources of water, energy, and food, the assessment of the WEF nexus offers opportunities for a better understanding of its broader impact on social equity socioecological resilience (Albrecht et al., 2018). The WEF nexus approach allows for an evaluation of the nexus impacts on livelihoods and an assessment of the roles of multiple actors and institutions in improving resource management (Biggs et al., 2015).

In this chapter, we reflect on how scientific approaches, here the nexus approach, can be used as a viable tool for regional integration and addressing multidimensional challenges, following the idea of transformative research. We explore a research approach oriented toward specific social problems and characterized by an explicit claim to intervention; the aim is to catalyze specific change processes and actively involve stakeholders in the research process (Schneidewind et al., 2014).

2. The interlinkages within nexus research

Understanding of natural resources and their capacities to support the provision of water, energy, food, and land requires an appreciation and comprehensive assessments of the trade-offs and synergies between the constrained natural resources (Bleischwitz et al., 2018). Strong linkages exist between WEF sectors, thus emerging a natural need to consider the interactions between them (Nhamo et al., 2018). These are under increasing pressure due to demands from population growth, urbanization, climate change, globalization, and trade (Mustafa et al., 2021).

The WEF nexus is a prime example of this type of interdisciplinary research. Researchers working within the WEF nexus are ideally positioned to enable opportunities for connecting actors from very different spaces. Its adoption, methodology, and metrics differ across different environments where it is adopted. A language around the definition of boundaries must be agreed upon by practitioners of water research, energy research, and food research. These are already spaces where multiple disciplines, technologies, and sciences exist and work collaboratively to address complex challenges within their spaces.

Knowledge generation is often skewed toward natural science disciplines, with quantitative approaches often granted more authority than other approaches to scientific discovery, pointedly with social sciences frequently being limited to vulnerability assessments (Borie et al., 2019; Kovacic, 2018; Donovan and Oppenheimer, 2015). Such imbalance unfortunately cripples progress in addressing complex global challenges that necessitate an understanding of social scientific knowledge and an appreciation of the value of local knowledge (Naess, 2013).

Crucially, it also centers on effective mechanisms of knowledge translation. Knowledge translation is a complex system of interactions that depends on a broad spectrum of actors to accelerate the capture of research benefits. Three conceptually distinct streams categorize knowledge translation; diffusion, dissemination, and implementation, all building on direct interactions between researchers—of very different disciplines and backgrounds—and the users. Disciplines, such as anthropology and sociology, offer valuable contributions as their focus on people, institutions, and social issues provide appropriate interventions for effective engagement with the users (Acevedo et al., 2018). They offer the necessary tools to frame the relevant questions for accurately defining the problems at hand, while concomitantly facilitating knowledge generation and translation and appropriate mechanisms for monitoring and evaluation (Acevedo et al., 2018).

3. Transboundary systems and the need for interdisciplinary spaces

Addressing multidimensional and complex challenges necessitates a depth of knowledge and an array of skills, while also spanning the breadth of knowledge and its integration (Huang et al., 2016). It requires the integration of methodologies and concepts from specialized knowledge bases, integrated on a common understanding. These are usually adopted to solve problems where the solutions fall beyond a single discipline or research practice (Huang et al., 2016). This boundary spanning context is the locus of the WEF nexus approach to bridge and integrate disciplines. An interdisciplinary and transformative research approach systemically integrates actors and enables them to see their blind spots. It does not measure outcomes only by the number of publications. It measures success by the degree the system's structure is changed, and new connections are created and parts are being aligned and integrated, realizing the emergence of a new system's overall state (Schneidewind et al., 2014).

Interdisciplinary research approaches are increasing in prominence, as is the adoption of nexus approaches. Interdisciplinary research brings experts from multiple disciplines combining methods and ideas, thereby providing innovative solutions and suitable approaches for addressing complex transboundary challenges such as climate change and managing competing demands of the WEF nexus. Integrating social and natural sciences is pivotal for addressing such multisectoral challenges. Reflections on these pathways can inform the practitioners of WEF nexus approaches (in research, policy, and management), accelerating scientific discoveries (Huang et al., 2016). Our current 19th-century Prussian education and research systems perform extraordinary well when solving complicated problems, though they fall dramatically short when put into practice to create innovative solutions for complex, ambiguous challenges (Webster, 2017).

The task of meeting the water, energy, and food needs of a growing population over a long-term calls for a careful examination of the demands on natural resources, as well as the available and potential capabilities to adapt and develop technological solutions (Acevedo et al., 2018) A careful assessment of transboundary systems offers a catalyst to explore synergies for mutually beneficial and efficient use of human, financial, technological, and other resources, providing opportunities for integration and negotiation of trade-offs.

Crucially, transboundary systems do not translate to a borderless intellectual space, ignoring social, cultural, legal, and political realms (Shome, 2006). To advocate for such systems is to advocate for the erosion of the nation's power and respect of sovereignty. Transboundary systems are about thinking across, over, and against the lines that may connect or disconnect continuous systems (Shome, 2006). Consequently, interdisciplinary approaches need to expand their efforts in horizontal, boundary-spanning situations. This boundaryless collaboration requires decision-makers in interdisciplinary research approaches to build relationships of corporation and trust with other individuals and organizations. Doing so requires basic skills and systems thinking, as it is the most appropriate problem-solving framework for sustainability. However, Beehner (2019) found that this promising approach is not yet widely applied in research and practice.

Taking the water system as an example, water has the capacity to link users across geopolitical borders. Transboundary water systems extend across national borders as a single system yet they are impacted by various issues stemming from the social, cultural, and legal realms. An estimated 60% of the freshwater supply is supplied by rivers that flow through two or more countries (Saguier et al., 2021; UN, 2014), highlighting the delicate balances needed to govern such a valuable natural resource. Furthermore, these interconnected water systems are essential contributors to livelihood and biodiversity protection (Saguier et al., 2021).

4. Role of funding in fostering interdisciplinary dialogue

Despite increasing interest in research that breaks free of traditional discipline boundaries, there is concern that existing funding structures do not adequately support it. There are few incentives for addressing such challenges through market-driven solutions. Market-driven solutions drive for excellence, delivering solutions efficiently for a demonstrated audience. Research to address challenges for the common good is often driven by national and international policies and agenda, and is more fittingly addressed by public funding or philanthropy. Dubbed as the "Paradox of interdisciplinarity," this phenomenon reflects the precarious positioning of interdisciplinary research. Even though it is encouraged at the policy level, interdisciplinary research is associated with consistently lower funding success (Bromham et al., 2016).

There is a lack of comparable data on the "fundability" of interdisciplinary research, which can be attributed to missing quantitative approaches to effectively measure the degree of interdisciplinarity (Bromham et al., 2016). Questions remain unanswered with regard to how the degree of interdisciplinarity may be defined, and whether some projects may be more interdisciplinary than others. Bromham et al. (2016) conducted a case study of approximately 20,000 proposals submitted to the Australian Research Council and found that the degree of interdisciplinarity was negatively correlated with funding scores while controlling for the primary research field and other factors. Thus, concluding that funding success decreased the more the projects increase in interdisciplinarity. This causal relation might correlate with the perception of higher risk and cost in these proposals or the reviewer's limited grasp of interdisciplinary fields (Bromham et al., 2016). On the other side of the spectrum, Nichols (2014) assessed awards by the National Science Foundation and reported significant success rates for proposal with relatively high degrees of interdisciplinarity.

If anything, these limited and contrasting findings have effectively demonstrated that it is possible to assess the success of interdisciplinary research funding, thereby offering funding agencies opportunities the tools to assess and reflect on their efforts (Acevedo et al., 2018). Funding interdisciplinary research programs is built on tacit knowledge held by national funding agencies (Lyall et al., 2013). These play a vital role in shaping interdisciplinary initiatives and pushing forward academic capability to accelerate scientific discovery, inevitably giving funding agencies some definition power over the form or shape that interdisciplinary research takes (Koenig et al., 2016). However, such agencies may often face organizational constraints that restrict their capacity to support novel and innovative approaches toward interdisciplinary research and collaboration (Koenig et al., 2016).

Additional factors that were identified to impede the progress and success of interdisciplinary research include the significant time spent in collaborative relationship building. Moreover, developing shared languages within the researchers hailing from different backgrounds can often pose a challenge too. For example, watershed and resource boundaries are generally adopted by water researchers to define their boundaries of research. In contrast, multiple and varied spatial scales, processes, and levels may be adopted in defining functional spaces in energy research or by food researchers. Additionally, researchers in interdisciplinary spaces are also tasked with agreeing on common perspectives/objectives, all while building on the expertise and experiences of the different disciplines coming together (Bromham et al., 2016).

Thus, we suggest that it is time for a "naming." We believe that helping researchers find "cross-border" vocabulary and frameworks will enable them to understand their different areas more clearly and apply findings more broadly. In the past century alone, "naming" has profoundly impacted how humanity sees the world. In the 1950s, for example, systems theory was named, which unites system thinkers in previously unrelated areas (such as physics, computer science, biology, engineering, geography, sociology, political science, psychotherapy, and economics). Suddenly, experts who were deeply separated by their jargon, practice, methodology, and standards were able to create a broader, more structured, and more common understanding of an area that they had all explored separately (Hinske, 2014).

An innovative example of how researchers can be encouraged to find a shared language and build collaborative relationships is the German Environment Agency's Factor X book series's publication process. The "Factor X" book series promotes good and best practices to enable significant savings in natural resource use and improve resource efficiency. They use a guided process that brings researchers together in webinars, video recordings, communication training, and interactive digital workshops. By doing so, the German Environment Agency intends to not only produce high-level publications but also maximize the number of new connections, integrating various stakeholders' perspectives (Hinske, personal communication, April 02, 2021).

These are all essential building blocks to guarantee successful interdisciplinary research outcomes but come at a steep cost to the practitioners. The complexities increase evermore when transdisciplinary research approaches are needed, whereby actors from various sectors are now involved, e.g., scientists, policy-makers, activists, community members. Therefore, an essential aspect of successful interdisciplinary work is the availability of systemic funding models that take the complex and dynamic nature of disciplinary research approaches into account. In Chapter 8 of the OECD Outlook for Science, Technology and Innovation 2018, the authors present a promising financing instrument that guarantees selected centers of excellence long-term resource allocations (OECD, 2018). These centers often include researchers and infrastructures from various institutions that promote the interdisciplinary and collaborative context necessary for effective, high-risk "breakthrough research" (OECD, 2018).

5. Shared value within multidimensional challenges

The idea of shared value is explored by Kramer (2016), who posits that multidimensional challenges such as poverty, food insecurity, and climate change cannot be solved without engaging stakeholders in value-creating dialogues. As societal challenges are getting increasingly complex (Beehner, 2019), we require much more significant investment in transformative and interdisciplinary research approaches that can engage more of stakeholders collaboratively. Solving such complex challenges is also in corporations' interest, which cannot continue to prosper and experience continued growth without successfully tackling such systemic issues. There are opportunities for applying the WEF nexus approach as a framework to support intersectoral collaboration and prospects for enhanced regional dialogue and integration (Saidmamatov et al., 2020).

Within spaces as diverse as the WEF nexus, there is a need for an integrated understanding of the interrelationships between the people and natural resource base, involving an interwoven sphere of sociocultural, economic, ecological, and political threads all functioning at different scales (Acevedo et al., 2018). Linear research approaches do not address the complexity of the feedback happening in complex stakeholder settings. They tend to focus on how data generation and subsequent interventions lead to a change in specific means that drive change in a particular aspect.

The nature of intersectoral collaboration and regional integration is sometimes referred to as third-order change; a term derived from theories and observations of learning in single, double, and triple loops. To achieve regional dialogue and integration, the system participants need to improve and adapt their context to their changing realities by examining their underlying assumptions and roles (Flood and Romm, 1996). A nexus approach to research examines the system of causes and effects, feedback and stakeholders, and leads research interventions to generate a much more resilient system with a much more significant and more lasting impact.

Working across disciplinary lines offers opportunities for integrated thinking, allowing the testing and implementation of transformative technologies and solutions, and borrowing and adapting successful concepts from one discipline to another (Acevedo et al., 2018; Horton et al., 2017). Nevertheless, tensions may exist between the multiple actors involved in the management of such complex and critical resources. As Saguier et al. (2021) identify in the management and use of transboundary water systems, involving a multitude of actors and institutions can foster collaboration and maneuver potential or existing conflicts. Crucial in addressing transboundary governance issues are skills related to emotional intelligence as they allow for the alignment and positioning of multiple sectoral stakeholders to advance actions.

This research approach is uniquely positioned to link diverse knowledge sources and institutions, engaging cross-sectoral actors and decision-makers, providing an ideal platform for research that informs policy and policy that informs research. A complete cyclical research framework does not end with guiding policy alone. It also traverses beyond the egos of scientists and acknowledges that interdisciplinary science works well when in an environment of continuous reciprocity. A genuinely egalitarian system where each actor is as important and relevant to the mission. Such framing offers an opportunity for actors within interdisciplinary domains where the presence of multitudes of values may impede or restrict decision-making progress (Borie et al., 2019).

6. The challenge of goal setting

A key challenge within such transboundary research is access to data, particularly metrics or nationally defined or aggregated goals. Ensuring access requires better interactions and communication at national and institutional levels. Such complex and cross-sectoral managements require special attention to sociopolitical systems and governance, to leverage positive impacts on policy-making and sustainability driven strategies (Albrecht et al., 2018).

The nature of the SDGs requires urgent action and partnerships to implement them (Mustafa et al., 2021). The WEF nexus outlines a broad spectrum of goals that are within the sustainable development agenda. This fact emphasizes the need to truly understand the systemic interlinkages between the goals and targets for successful implementation and monitoring. In an assessment by Nilsson et al. (2017), the relationship between food and water resources was found to be quite nuanced, whereby SDG 2 is dependent on SDG 6 (Clean Water and Sanitation for All) and with the potential to negatively impact progress toward SDG 6. Traditional intensive agricultural practices rely heavily on exploiting land and water resources, hindering progress toward SDG 6 (Mustafa et al., 2021). Thus, advancing food security may be achieved at the expense of healthy water systems.

An assessment of progress toward the adoption of the SDGs in the Global South could provide an opportunity for the unveiling of stress points between these goals and balancing competing demands. Moreover, as Acevedo et al. (2018) note, the very notion of "sustainability," which centers on the three aspects of the environment, economy, and society and overlooks the critical dimension of politics as well as multifaceted processes and activities that are central to any discourse on sustainability and resource management. Thereby, the focus shifts away from a siloed approach toward an approach that is respectful of the synergies and trade-offs through coordinated policy interventions, allowing for progress across all pillars supporting each other (Mustafa et al., 2020).

This outcome asks public policy to generate practical benefits across the spectrum of multidimensional resource challenges (Lyall et al., 2013). The interlinkages—both competing and synergistic—between the social and natural spheres exist at multiple scales. There is a need for interdisciplinary narratives to elucidate such examples of interlinkages that necessitate the move beyond disciplinary silos, organisational frameworks, and even geopolitical boundaries.

7. Advancing nexus research

Integration of different science disciplines requires physical, social, and organizational integration structures (Heberlein, 2008), principally when dealing with resource-related issues such as WEF nexus. Policy and public institutions that support such missions within the transboundary systems are far too often lacking (Saguier et al., 2021). There is a strong need for policy and funding instruments to encourage such research further. Practical steps can be taken to promote collaborative working environments, benefiting funders, practitioners, and interdisciplinary research users.

It is valuable to have approaches that incentivize the scientists behind such approaches through efforts and rewards, e.g., funding for interdisciplinary research. However, traditional systems of reward in scientific research may not be adapted for interdisciplinary research. Outputs of interdisciplinary research may be fewer but often come with a promise of broader felt impact. These outputs and impact speak to an entirely different audience of funders; thus, financing WEF nexus approaches would follow a different approach to traditional scientific research.

Research evaluation systems are designed for measuring a narrow range of success, such as publication in peer-reviewed journals (Bromham et al., 2016). This confined focus disadvantages interdisciplinary research proposals, which do not center on promises of such outputs but encompass other aspects that may be less appreciated, such as collaborative networks and data sharing (Bromham et al., 2016). While there has been a growth in interdisciplinary journals, most metrics cater to traditional scientific ends. Some narratives state that interdisciplinary journals may have a lower reputation than older and more established discipline-based journals (Shome, 2006) or that while the number of interdisciplinary journals is increasing, they may not hold the same level of prestige or impact factor as single-discipline journals (Acevedo et al., 2018). Simultaneously, multidisciplinary journals such as *Science* and *Nature* do not necessarily include more interdisciplinary research than other single-discipline journals (Acevedo et al., 2018).

As science moves toward a new era shaped by the Declaration on Research Assessment (DORA) and Plan S, there is increasing hope for interdisciplinary research. The focus moves toward evaluating the social impact and wider contribution of research to the knowledge of the societies they support. The various outputs of scientific research are increasingly recognized and acknowledged, including datasets, new knowledge, training of young scientists, and crucially open science (DORA, 2013; cOAlition, 2018). The desire to move toward a more accurate evaluation of research outputs that do not rely merely on publications as the key research outputs and journal impact factors as the measure of scientific quality is pivotal for advancing interdisciplinary research.

8. Concluding remarks

WEF nexus approaches are in a perilous position, whereby they are driven by public policy and sustainable development agenda but see little rewards and incentives and support from traditional funding agencies. How do we move forward then? The goal is not to position siloed science and interdisciplinary sciences to compete for recognition or funding. It is not about disrupting siloed science or carving out space within the traditional scientific streams. It is, however, about coexisting; integrating knowledge that is generated directly from disciplines and through their interactions with varied actors and domains.

It is about creating new spaces for transboundary scientific mechanisms and translating reward mechanisms, allowing practitioners to build depth of knowledge (siloed science) and work toward knowledge translation or interdisciplinary science. Recognizing that newly generated knowledge may often be too highly specialized and fragmented that on its own may not offer solutions to complex problems. Recognizing the strengths and weaknesses of each domain, encouraging its adoption in the relevant spaces will encourage practitioners to engage in both or either.

This evolution calls for understanding and appreciation of each space to allow reward systems to be transferable or recognizable across the different domains. Domain-specific expertise is essential to drive interdisciplinary research, and interdisciplinary research is valuable to deliver outcomes that have far-reaching societal and economic outcomes.

References

Acevedo, M.F., Harvey, D.R., Palis, F.G., 2018. Food security and the environment: interdisciplinary research to increase productivity while exercising environmental conservation. Glob. Food Security 16, 127–132. https://doi.org/10.1016/j.gfs.2018.01.001.

Albrecht, T.R., Crootof, A., Scott, C.A., 2018. The Water-Energy-Food Nexus: a systematic review of methods for nexus assessment. Environ. Res. Lett. 13, 043002.

Beehner, C.G., 2019. System Leadership for Sustainability. Routledge, London.

Biggs, E.M., Bruce, E., Boruff, B., Duncan, J.M.A., Horsley, J., Pauli, N., McNeill, K., Neef, A., Van Ogtrop, F., Curnow, J., Haworth, B., 2015. Sustainable development and the water–energy–food nexus: a perspective on livelihoods. Environ. Sci. Pol. 54, 389–397.

Bleischwitz, R., Spataru, C., VanDeveer, S.D., Obersteiner, M., van der Voet, E., Johnson, C., Andrews-Speed, P., Boersma, T., Hoff, H., van Vuuren, D.P., 2018. Resource nexus perspectives towards the United Nations sustainable development goals. Nat. Sustain. 1, 737–743. https://doi.org/10.1038/s41893-018-0173-2.

Borie, M., Pelling, M., Ziervogel, G., Hyams, K., 2019. Mapping narratives of urban resilience in the global south. Global Environ. Change 54, 203–213. https://doi.org/10.1016/j.gloenvcha.2019.01.001.

Bromham, L., Dinnage, R., Hua, X., 2016. Interdisciplinary research has consistently lower funding success. Nature 534, 684–687.

cOAlition, S., 2018. Accelerating the Transition to Full and Immediate Open Access to Scientific Publications. www.coalition-s.org/wp-content/uploads/PlanS_Principles_and_Implementation_310519.pdf.

Donovan, A., Oppenheimer, C., 2015. Resilient science: the civic epistemology of disaster risk reduction. Sci. Publ. Pol. 43 (3), 363–374.

DORA, 2013. San Francisco Declaration on Research Assessment. https://sfdora.org/read/.

Flood, R.L., Romm, N.R., 1996. Diversity management. In: Flood, R.L., Romm, N.R.A. (Eds.), Critical Systems Thinking. Springer, Boston, MA, pp. 81–92. https://doi.org/10.1007/978-0-585-34651-9_5.

Heberlein, T.A., 2008. Improving interdisciplinary research: integrating the social and natural sciences. Soc. Nat. Resour. 1, 5–16. https://doi.org/10.1080/08941928809380634.

Hinske, C., 2014. Ecosynomics - Evidence of Post-economic High Performance. Friedrich Ebert Stiftung. http://library.fes.de/pdf-files/bueros/ghana/11296.pdf.

Horton, P., Banwart, S.A., Brockington, D., Brown, G.W., Bruce, R., Cameron, D., Holdsworth, M., Lenny Koh, S.C., Ton, J., Jackson, P., 2017. An agenda for integrated system-wide interdisciplinary agri-food research. Food Secur. 9 (2), 195–210.

Huang, Y., Zhang, Y., Youtie, J., Porter, A.L., Wang, X., 2016. How does national scientific funding support emerging interdisciplinary research: a comparison study of big data research in the US and China. PLoS One 11 (5), e0154509. https://doi.org/10.1371/journal.pone.0154509.

IPCC, 2014. Climate Change 2014: Synthesis Report. Contribution of Working Groups I, II and III to the Fifth Assessment Report of the Intergovernmental Panel on Climate Change [Core Writing Team, R.K. Pachauri and L.A. Meyer (eds.)]. IPCC, Geneva, Switzerland, p. 151.

Koenig, T., Gorman, M.E., 2016. The challenge of funding interdisciplinary research: a look inside public research funding agencies. In: Frodeman, R., Thompson Klein, J., Santos Pancheco, D., Carlos, R. (Eds.), The Oxford Handbook of Interdisciplinarity, second ed. Oxford University Press, Oxford Handbooks, Oxford, pp. 513–524.

Kovacic, Z., 2018. Conceptualizing numbers at the science–Policy interface. Sci. Technol. Hum. Val. https://doi.org/10.1177/0162243918770734.

Kramer, M., 2016. Creating Shared Value for Smallholder Farmers. www.sharedvalue.org/groups/creating-shared-value-smallholder-farmers.

Lyall, C., Bruce, A., Marsden, W., Meagher, L., 2013. The role of funding agencies in creating interdisciplinary knowledge. Sci. Publ. Pol. 40, 62–71. https://doi.org/10.1093/scipol/scs121.

Mustafa, M.A., Mabhaudhi, T., Avvari, M.V., Massawe, F., 2020. Transition towards sustainable food systems: a holistic pathway towards sustainable development. In: Galanakis, C. (Ed.), Food Security and Nutrition. Academic Press, pp. 33–56.

Mustafa, M.A., Mabhaudhi, T., Massawe, F., 2021. Building a resilient and sustainable food system in a changing world – a case for climate-smart and nutrient dense crops. Glob. Food Security 28 (1), 100477.

Naess, L.O., 2013. The role of local knowledge in adaptation to climate change. Wiley Interdiscipl. Rev. Clim. Change 4 (2), 99–106.

Nhamo, L., Ndlela, B., Nhemachena, C., Mabhaudhi, T., Mpandeli, S., Matchay, G., 2018. The water-energy-food nexus: climate risks and opportunities in southern Africa. Water 10 (5), 56.

Nichols, L.G., 2014. A topic model approach to measuring interdisciplinarity at the National Science Foundation. Scientometrics 100 (3), 741–754.

Nilsson, M., Griggs, D., Visbeck, M., Ringler, C., McCollum, D., 2017. A framework for understanding sustainable development goal interactions. In: A Guide to SDG Interactions: From Science to Implementation. International Council for Science, Paris.

OECD, 2018. OECD Science, Technology and Innovation Outlook: Adapting to Technological and Societal Disruption. OECD Publishing, Paris.

Saguier, M., Gerlak, A.K., Villar, P.C., Baigun, C., Venturini, V., Lara, A., dos Santos, M.A., 2021. Interdisciplinary research networks and science-policy-society interactions in the Uruguay River Basin. Environ. Dev. 38, 100601. https://doi.org/10.1016/j.envdev.2020.100601.

Saidmamatov, L., Rudenko, I., Pfister, S., Koziel, J., 2020. Water−energy−food nexus framework for promoting regional integration in central Asia. Water 12, 1896. https://doi.org/10.3390/w12071896.

Schneidewind, U., Singer-Brodowski, M., 2014. Transformative Science: Climate Change in the German Science and University System, second ed.

Shome, R., 2006. Interdisciplinary research and globalization. Commun. Rev. 9, 1−36.

UN, 2014. Factsheet on Transboundary Water. www.un.org/waterforlifedecade/transboundary_waters.shtml.

Webster, K., 2017. The Circular Economy: A Wealth of Flows, second ed. Ellen MacArthur Foundation Publishing, Cowes.

Further reading

Metropolis Schneidewind, U., Singer-Brodowski, M., Augenstein, K., 2016. Transformative science for sustainability transitions. In: Brauch, H., Spring, O.Ú., Grin, J., Scheffran, J. (Eds.), Handbook on Sustainability Transition and Sustainable Peace, Hexagon Series on Human and Environmental Security and Peace, vol. 10. Springer, Cham. https://doi.org/10.1007/978-3-319-43884-9_5.

CHAPTER 13

The Water—Energy—Food nexus as a rallying point for sustainable development: emerging lessons from South and Southeast Asia

Andrew Huey Ping Tan[1], Eng Hwa Yap[2], Yousif Abdalla Abakr[3], Alex M. Lechner[4,5,8], Maysoun A. Mustafa[6] and Festo Massawe[6,7]

[1]*School of Intelligent Manufacturing Ecosystem, XJTLU Entrepreneur College (Taicang), Xi'an Jiaotong-Liverpool University, Suzhou, Jiangsu, People's Republic of China;* [2]*School of Robotics, XJTLU Entrepreneur College (Taicang), Xi'an Jiaotong-Liverpool University, Suzhou, Jiangsu, People's Republic of China;* [3]*Department of Mechanical, Materials and Manufacturing Engineering, University of Nottingham Malaysia, Semenyih, Selangor, Malaysia;* [4]*Lincoln Centre for Water and Planetary Health, University of Lincoln, Lincoln, United Kingdom;* [5]*School of Environmental and Geographical Sciences, University of Nottingham Malaysia, Semenyih, Selangor, Malaysia;* [6]*School of Biosciences, University of Nottingham Malaysia, Semenyih, Selangor, Malaysia;* [7]*Centre for Transformative Agricultural and Food Systems (CTAFS), School of Agricultural, Earth and Environmental Sciences, University of KwaZulu-Natal, Pietermaritzburg, South Africa;* [8]*Monash University Indonesia, Tangerang Banten, Indonesia*

1. Introduction

The world recently entered the 21st century and witnessed explosive economic growth in the past 40 years unlike anything experienced previously. Global concerns have shifted from the industrial revolution, to world wars, to the cold war, and eventually to sustainable development (SD), all within the short span of a century. With global population expected to grow to 9.7 billion by 2050 (United Nations, 2019), global acceptance and application of an action-oriented mindset toward to support SD has become even more important to address resource and economic scarcity, especially in developing countries and regions, such as South and Southeast Asia (SEA). Consequently, the basic resources of water, energy, and food, coined as the "WEF," alongside their interactions and trade-offs, commonly known as the "nexus," has become an extremely important consideration, as the well-being of these resources are few of the many indicators that mark the difference between a developing and developed country (Investopedia, 2020; Surbhi, 2015).

SD is originally defined by Brundtland as "development that meets the needs of the present without compromising the ability of future generations to meet their own needs" (World Commission on Environment and Development, 1987) and will be the operating definition for this chapter. The three pillars of sustainability are the economy, environment, and society (Emas, 2015; Mensah, 2019), which can be considered from the "development" perspective as "sustaining" nature, life support, and community at the same time (United Nations, 2014). Others have urged for five pillars (5Ps) of SD (Mustafa et al., 2021), which are people, planet, prosperity, peace, and partnership. These five pillars were recently discussed by Mustafa et al., in the context of food systems and SD (Mustafa et al., 2021).

The multiple dimensions of sustainability can also be considered from a resource security perspective through the "water—energy—food nexus" paradigm. The use of the term has risen dramatically ever since it was first introduced at the Bonn Nexus Conference in 2011 (Hoff, 2011). In a time when globalization and interconnectedness are experiencing unprecedented growth, the nexus approach has been introduced to aid and support in the transition toward a Green Economy (Hoff, 2011). The nexus approach is a modern attempt to move away from traditional methods of silo-thinking (Brears, 2017; Pullin and AME Study Group on Functional Organization, 1989), by slowly dissolving interdisciplinary boundaries and looking for solutions at the interfaces between disciplines. Initially, the WEF nexus was not clear to many, simply because of its large and all-encompassing approach, which considers many factors across all levels of the society. However, Hoff (2011) provided three guiding principles to apply when employing the nexus approach, namely (1) investing to sustain ecosystem services, (2) creating more with less, and (3) integrating the poorest by accelerating access to resources.

The Sustainable Development Goals (SDGs) are also closely linked to the WEF nexus via Goals 2, 6, and 7 (United Nations, 2021) which strive for zero hunger, clean water and sanitation, and affordable and clean energy. These three interlinked goals were already challenging, even before the COVID-19 pandemic, but have become an even greater challenge as the pandemic continues to cause havoc globally. For the food sector, food insecurity rose from 22.4% in 2014 to 25.9% in 2019 (United Nations, 2021), and climate shocks and locust crises have negatively impacted food production and supply systems (United Nations, 2021). While 2.2 billion people lacked safe drinking water and 4.2 billion people lacked proper sanitation in 2017, the pandemic has further exacerbated this problem as basic handwashing is known to be the most effective method for COVID-19 prevention (United Nations, 2021). The pandemic has also magnified problems in the energy sector, for example, 1 in 4 health facilities are not electrified (United Nations, 2021).

Without a doubt, the WEF nexus approach encompasses many concepts, methods, and techniques for addressing and alleviating water, energy, and food securities. The definitions of these three resource securities may vary widely, owing to the varied means for securing them or the unique goals associated with them, but they are largely bound by the requirements of "four A's"—availability, accessibility, affordability, and acceptability (Ang et al., 2015; Yao and Chang, 2014). For example, energy security includes having sufficient electricity, where water resources are one of the main ingredients in their generation, to power devices, machines, and systems for commercial and industrial applications. Water security is achieved when a population has clean water and acceptable sanitation standard, where it is a result of developed water treatment and distribution consuming large amounts of electricity. At the same time, food security, which includes not experiencing hunger and receiving sufficient nutrients to undertake daily activities, is having enough nutritious food, which requires water and energy to produce. As such, the security of these basic resources eventually comes down to being able to attain them without much difficulty, at reasonable prices, and without severely impacting the environment. Ultimately, the many elements of WEF are linked to the concept of SD and the many challenges which the SDGs address such as the water crises, which is ranked globally as one of the top risks (World Economic Forum, 2015), ever-increasing energy demands (IEA: Directorate of Global Energy Economics, 2015), and proliferation of suffering due to hunger (FAO, 2009); thus the WEF nexus indeed is one of the best models for holistically addressing SD challenges.

Around the world, researchers have employed the nexus approach to investigate the WEF linkages in their respective regions, across a wide range of scales and contexts. For example, Hardy et al. (2012) investigated the WEF relationships by assessing the water needs in power plants, measured in the form of annual water withdrawal. Yang et al. (2009) traced land and water footprint for biofuel, using different types of crops. Karatayev et al. (2017) showed that water stress would occur should current practices of energy systems in Kazakhstan remain the same.

The aim of this chapter is to understand the status of the WEF nexus in South and Southeast Asia. This chapter first reviews past works on the WEF nexus that has been conducted in countries in these regions providing an overview of the diverse perspectives and scales of the approaches taken and findings of researches on the WEF nexus. In the second part, we take a deeper dive, examining a case study for Malaysia, outlining a WEF nexus conceptual framework, how complex systems methods can be used to model WEF systems and the key factors and approaches, which should be considered when using a WEF nexus approach.

238 CHAPTER 13: The Water−Energy−Food nexus as a rallying point

1.1 Brief description of South and Southeast Asia

South Asia, sometimes referred to as the "Indian subcontinent," is a subregion of Asia, consisting of Afghanistan, Bangladesh, Bhutan, India, Maldives, Nepal, Pakistan, and Sri Lanka, as illustrated in Fig. 13.1 (Ryabchikov, 2020). In the northern regions, South Asia is bounded by mountain ranges of Hindu Kush, Karakoram, and Himalayas, and to the south lies the fertile Indo-Gangetic Plains, boasting an area of 2.5 million km^2 (Ryabchikov, 2020).

Southeast Asia (SEA) consists of 11 countries, namely Brunei, Cambodia, Timor-Leste, Indonesia, Laos, Malaysia, the Philippines, Singapore, Thailand, Myanmar, and Vietnam, as depicted in Fig. 13.1. Among them, Indonesia is the largest country, and also it has the largest archipelago in the world (CIA, n.d.). The largest sea within this region is the South China Sea; the final destination for many SEA rivers.

2. A critical review into the WEF of South and Southeast Asia

A number of studies have been undertaken to address the WEF nexus in the regions of South and Southeast Asia (Gathala et al., 2020; Islam et al., 2019; Keskinen et al., 2015; Putra et al., 2020; Rasul, 2014; Saklani et al., 2020; Spiegelberg et al., 2017). These studies have provided a better understanding of WEF in the South and Southeast Asian regions, highlighting important factors surrounding the WEF status and direction in the region.

Putra et al. conducted a systematic analysis of WEF in South Asian countries, namely Bangladesh, Sri Lanka, Nepal, Pakistan, and India, whereby the

FIGURE 13.1

Map of South and Southeast Asia.

interactions between 36 WEF indicators represented by open data (e.g., access to electricity, food supply per-capita; surface water irrigation) were analyzed at the country scale (Putra et al., 2020). This study demonstrated that there are likely to be a number of trade-offs (e.g., where progress in one sector reduces progress in another sector) and synergies (e.g., where both sectors progress in tandem), which can contribute to insecurity and increasing security respectively. A key finding from this study was that trade-offs within energy and water sectors were clear, as opposed to synergies found among the WEF sectors. At the national level, interactions between WEF indicators varied among countries. It was found that trade-offs were prevalent in India, Nepal, and Pakistan, where there were more negative correlations between identified key indicators, but the opposite is true for Bangladesh and Sri Lanka, where these synergies were evidenced by more positive correlations between indicators (Putra et al., 2020). However, the relationship among same indicators could vary among countries. For example, relationship between energy type generation and access to energy were synergistic in a few countries and a trade-off in others. Additionally, Putra et al. (2020) identified some of the most important key indicators contributing either positively or negatively toward holistic WEF well-being. Despite identifying more trade-offs than synergies within the water and energy sector, overall, the share of synergies was larger than trade-offs. The synergies can be leveraged to enhance WEF security, and the trade-offs need to be tackled to achieve WEF security in this region.

In another study, Rasul (2014) leveraged extensive secondary data to perform a regional-scale qualitative analysis on the role of Hindu Kush Himalayan (HKH) ecosystem services in sustaining WEF security nexus downstream driven by flows of benefits within the hydrological system. Ecosystem services naturally provided by the HKH are vital in supporting the production of WEF nexus in HKH as well as its downstream dependencies. Key challenges related to the WEF in this region were highlighted, which include increasing population, reduction of agricultural land, and declining food production (which are also water- and energy-intensive), scarcity of water and energy, and negative impacts of biomass energy (Rasul, 2014). Importantly, this study demonstrated the spatial dependencies of WEF systems whereby the impacts on HKH will be far reaching, into neighboring countries as well, with water and energy benefits delivered via large rivers such as the Brahmaputra and Mekong rivers. However, the sustainability of WEF security in the HKH is a multifaceted challenge, such as land and forests degradation, changes in headwater regions, ever-increasing demand of resources, and profit-motivated enterprise actions. Thus, it is impossible to overstate the importance of effective and proper management of WEF nexus, which requires transdisciplinary, multicontext, and transboundary cooperation and coordination.

Linkages between components of the WEF system are typically driven by spatial dependencies associated with hydrological system within Southeast Asia as well. Keskinen et al. (2015) investigated the WEF nexus in a transboundary river basin, namely the Tonle Sap Lake, situated in the heart of Cambodia, which also demonstrated the complex nature of WEF interactions at the regional scale. Using scenario-driven quantitative research, components of hydrological and water resources were characterized. In addition, livelihoods, food security and important WEF-related impacts in the area were successfully derived. Firstly, they showed that a planned hydropower dam development in the Mekong River Basin would cause significant impact on the water levels, e.g., abnormally higher levels during dry season. Consequently, the total area of floodplain was expected to change over time, thereby demonstrating a critical energy–food link for the region, which could be captured using a WEF approach. Secondly, the socioeconomic assessment of key databases showed obvious links between livelihood and food security, whereby the majority of the population are involved in agriculture and fishing. Subsequently, alternative scenarios completed the findings by showing that construction of the Mekong Dam would eventually impact the food security and livelihood of populations in the area, through changes to the flood pulse, ecosystem productivity, floodplain habitats, dry season water levels, and Tonle Sap's fish production. Constructions of hydropower in one country along the Mekong River have also been proven to affect livelihoods downstream, and calls have been made to halt further development (Save the Mekong Coalition, 2009).

In contrast, Spiegelberg et al. (2017) conducted WEF nexus research in the region of Dampalit Watershed, in the Philippines, by exploring the connectivity between upland farmers and downstream fishers. They found that the behavior of the people and their links to social and natural components meant that the social interlinkages within the Dampalit subwatershed were limited. Consequently, to improve water resources and food security, the people should be brought closer together using a WEF nexus–centralized network.

WEF security needs to be addressed at multiple scales, from the national-scale, accounting for geopolitics to site-scales taking on technical approaches. For example, Saklani et al. (2020) conducted a scholarly review of hydroenergy cooperation in the subregion of BBIN (Bangladesh, Bhutan, India, and Nepal) and examined closely key bilateral relationships of countries with India within the respective sectors of water and energy. They found that India plays a hegemonic role in its water management with neighboring countries, while it is seen as a regional powerhouse when it comes to bilateral energy engagements. The factors for future progress within the context of international cooperation of water and energy sectors include (1) improved information sharing for managing water resources, (2) improved India's regional leadership, (3) held simultaneous diplomatic discussions on W—E nexus, and (4) enhanced regional cooperation between all BBIN countries (Saklani et al., 2020). Gathala et al. (2020) took a technical approach in examining the WEF nexus of cropping practices, including conservation agriculture-based sustainable intensification (CASI), in a selected area in the Eastern Gangetic Plains (EGP), namely India, Bangladesh, and Nepal, by conducting on-farm trials (Islam et al., 2019) and experimental treatments across eight districts in the EGP. They found that indeed CASI would improve the WEF efficiency in the region and uptake of CASI was strongest when it is implemented as part of a larger multistakeholder framework involving potentially researchers, financial institutions, agroindustries, policymakers, and governments (Gathala et al., 2020).

An urban nexus approach is also especially important for South and Southeast Asia given the rapidly growing and urbanizing populations in this region (Lechner et al., 2020). The urban nexus considers the "waste" element, making it an energy, water, food, and waste (EWFW) nexus, as the waste factor contained in the other three domains are substantial and establishing a circular economy, as well as decoupling growth from use of resources is critical (Lehmann, 2018). To establish the urban nexus approach, key steps that must be taken include creating a group of coworking organizations, establishing measurable indicators, providing technical expertise in various areas, continuing targeted subsidies for renewable energies, and upscaling.

Achieving SD will require a WEF nexus approach, which understands the interlinkages (synergies and trade-offs) between water, energy, and food resources at multiple scales through a variety of disciplinary lenses. The WEF studies conducted in South and Southeast Asia have provided a good foundation, which must be considered when designing country-specific and regional policies around water, energy, and food security. Regionally, there is a need for a WEF nexus approach to be incorporated in natural resource management to avoid negative environmental and social impacts commonly associated with lack of intersectoral cooperation. Clearly, the countries covered in this chapter are diverse and are at different levels of development. Comparing with WEF nexus studies performed in other regions, such as in Central Asia (Adnan, 2013), the United States (Finley and Seiber, 2014), Africa (Mukuve and Fenner, 2015), and United Kingdom (Howarth and Monasterolo, 2016), these studies are similar, as they also emphasized and highlight the importance of synergies, trade-offs, and intersectoral indicators, on top of addressing socioeconomic factors to a very large extent. Availability of resources and quality of governance also differ among these countries; however, for WEF to work, there must be interregional and cross-border cooperation based on shared resources and prosperity.

3. Case study: WEF in Malaysia
3.1 Introduction to Malaysia and WEF conceptual framework

Malaysia is a Southeast Asian developing nation, with a population of 33 million (CIA, 2021), situated between the mainland Southeast Asia and insular Southeast Asia. A big part of Malaysia, namely the Malay Peninsula, functions as a bridge between the two parts of Southeast Asia and also shares culture with the surrounding islands (Leinbach, 1999). Having a total land area of 328,657 km^2 and water area of 1190 km^2, Malaysia is blessed with a variety of natural resources, such as petroleum, natural gas, and timber, as well as a total renewable water resource of 580 billion m^3 (CIA, 2021). Malaysia is bordered with Thailand to the north of Peninsular Malaysia, Singapore to the south, and Brunei and Indonesia on the Borneo Island.

The energy sector in Malaysia is made interesting by the dynamic composition of energy types, namely coal, gas, oil, hydropower, and renewable energies, alongside their distribution systems and its policies toward deregulation. The National Grid in Malaysia consists mainly of three electricity distributors, namely Tenaga Nasional Berhad (TNB) in Peninsular Malaysia, Sabah Electricity Sdn. Bhd. (SESB) in Sabah, and Sarawak Energy Berhad (SEB) in Sarawak (Tan et al., 2018). In 2017, Malaysia's electricity consumption was at 116,273 GWh (Energy Commission (Malaysia), 2019). Being an active oil-producing nation, Malaysia has reserves of crude oil and condensates of 4.7 billion barrels (Energy Commission (Malaysia), 2019). Throughout the nation, the total power generation capacity installed, distributed across hydro, natural gas, coal, diesel, renewable energies, and others, is 34,182 MW, of which 29,218 MW are available (Energy Commission (Malaysia), 2019). Malaysia moved from four-fuel energy policy into five-fuel diversification by introducing renewable energies into the mix.

Malaysia's water industry is highly decentralized and is also divided between water treatment and sewage treatment. Moreover, the governance of water in Malaysia is separated between state and federal, where a fair amount of control is granted to the state governments. On the whole, water withdrawn in Malaysia is almost equally distributed between industries, agriculture, and municipalities (FAO, 2005).

The food sector of Malaysia revolves largely around its staple food, rice, where 12% of the GDP is made up of the agricultural sector, which employs 16% of the population (Tan et al., 2018). In similar way to many other regions around the world, the food sector is a key demand driver of the energy and water sector, as growing crops require substantial amount of energy and water. On the other hand, Malaysia does contribute to the food-for-energy relationship, where some portion of energy generation is from biomass and biofuel.

3. Case study: WEF in Malaysia

Tan et al. (2018) proposed a conceptual framework for the Malaysian WEF nexus as depicted in Fig. 13.2. The conceptual framework presented here varies slightly from the original one as proposed by Tan et al. (2018), where the objectives and goals have been moved to the center, being encapsulated and affected by sociological, technological, economic, environmental, and political (STEEP) factors. At the center of the WEF nexus conceptual framework lies the heart of the system, where key objectives and goals would affect decisions made within the WEF sectors, as well as their corresponding consequences. These key objectives, as depicted here, have been kept as a general representation, rather than a specific one. This allows the conceptual framework to be flexibly applied at various levels of government, such as regional, national, or international. For example, the same framework would apply regardless of whether the carbon emission policy is enacted to affect industries in a particular state, across the whole country, or a subcontinental region. The bilateral relationships between resource sectors are represented at three corners of the framework, where their intra- and interrelation activities would revolve around the STEEP factors to either contribute or stray from the total system's objectives. As a whole, the conceptual framework proposes to investigate any WEF issue, by addressing the

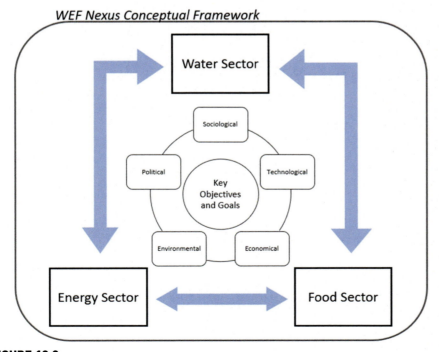

FIGURE 13.2
WEF nexus conceptual framework (Tan et al., 2018).

intersectoral activities that stem from one or more of the STEEP factors, while keeping a close check and alignment to any particular key objectives and goals of a region.

3.2 Complex systems approach and causality

To take a detailed dive at addressing the WEF status in Malaysia, a complex systems approach has been taken to discuss, with aid of systemic mental models and several past works, the intricacies of the Malaysian WEF network. The main drivers of resource demand and stress are the ever-decreasing regional natural resource base as well as the exponential growth of country's population. As depicted by both the research into the energy and water security of Malaysian WEF (Tan et al., 2020; Tan and Yap, 2019), the systems causality around population growth and demand for resources can be represented by a causal loop diagram (CLD) in Fig. 13.3. Essentially, the increment in the country's population as a result of birth and immigration is driving the demand for water, energy, and food, and these pulling forces will be felt across each resources' supply chain (Fig. 13.3).

The total demand for these resources due to population growth can be analyzed by considering the per capita demand of each resource. For electricity, the per capita consumption was at 4595 kWh, in 2019 (Energy Commission (Malaysia), 2019). For water, the per capita consumption in 2018 was 226 L/capita/day (SPAN, 2019). Measuring per capita food demand on the other hand could be conducted in many ways. One way of measuring is to consider the main staple food of the country and measure the amount needed by a person in weight, in a particular year. For Malaysia, the staple food is rice, and it is found that Malaysians on average consume 80 kg/capita/year of rice (Department of Statistics Malaysia, 2013). It is thus easily shown that if the population of the country increases, efforts would be needed to secure the demands of the people, either by expanding generation capacities, improving water systems, or grow more food crops. However, using the main staple food and in particular

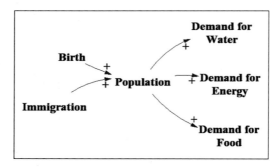

FIGURE 13.3

Population growth and demand for resources (Tan et al., 2020; Tan and Yap, 2019).

rice (commonly used in Malaysia) as a measure of food security is somewhat flawed because the approach does not take into account the nutritional content of the food produced and consumed. Taking a nutritional lens when addressing key WEF issues is important given that the production and the outcomes of consuming different food types (e.g., protein, carbohydrates, oils, micronutrients) will require varied resource considerations.

3.2.1 Energy security

As Tan and Yap (2019) have shown, the initiation of new energy capacity is mainly driven by the gap in requirement. Also, two decisions around the choice of new capacity need to be considered, namely nonrenewables or renewables. Together, a host of variables would affect the decision of whether to expand on the capacities, namely the national renewable energy target, the available energy resource, the demand pull, and so forth. On top of that, the type of conventional energy to expand on is also up for debate, considering the different natural resource available, such as coal, oil, and gas. Additionally, the cost of each energy would be different, which would result in different levelized cost of electricity (Cheok et al., 2021; Tan and Yap, 2019). Coal remains the cheapest option among all the conventional energy but contributes the most negatively in terms of environmental impact (Cheok et al., 2021), and the cost of renewable energy has dramatically fallen across the world (McPhee, 2020). Due to international and national agenda, such as the Paris Agreement under the United Nations Framework Convention on Climate Change as well as the National Policy on the Environment, renewable energies must be considered when planning energy capacity expansion for the country. Taking a balance between cost of energy as well as energy emissions reduction, it is found that Malaysia could afford 20% renewable energy penetration, while keeping cost increase to a minimal (Tan and Yap, 2019).

Malaysia's current energy market is also adopting a single buyer model (SBM), where a single entity purchases power from independent power producers (IPP) and sells them directly to the customer (Lee et al., 2019). This is a transition period for Malaysia toward a fully deregulated model (Teljeur et al., 2016), as privatization of Malaysia's electricity market started in the 1990s, and has improved the competitiveness of unit energy price (Malaysia Energy Commission, 2013). Ideally, the SBM should have more than one distributor. However, there is only one distributor in Peninsula Malaysia, namely TNB, and therefore, end customers do not have options to choose from where they can buy their electricity. Fig. 13.4 shows a completely deregulated energy market model, where competition exists between IPPs as well as between distributors (Teljeur et al., 2016). In between the IPPs and distributors lies a power pool and network where electricity is traded and eventually transacted based on the best price bid. End customers will also get to choose their own distributor.

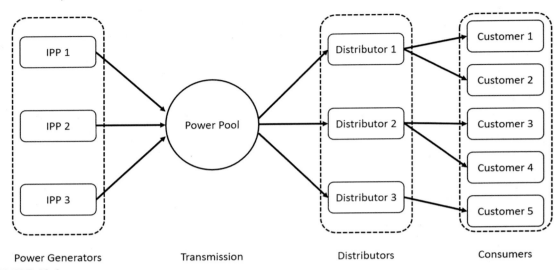

FIGURE 13.4
Deregulated retail competition model (Lovei, 2000).

3.2.2 Water security

Fig. 13.5 shows the urban water cycle, which represents the typical flow of water in an urban setting (Tan et al., 2020). This is a closed system, where if the total internal renewable water resources of the country do not change too drastically, the total water flowing for a particular water system would be the same. Two types of treatment systems exist, namely the water supply and services treatment and the wastewater treatment service. Mainly, surface water and groundwater would form the natural water resources, but an estimation can only be given on their quantity, because in nature, these two types of water can and will naturally seep and flow to and from each other.

Fig. 13.6 shows the water system and tariff loops of Malaysia's water sector, as adapted from Tan et al. (2020). The complexity of the water system is portrayed through two types of end customer, namely industrial and commercial, and two types of tariffs, namely water and sewage tariffs. Of course, the tariffs would then depend on the cost to treat and distribute the water. The calculation for the unit cost of water production is thus similar to how one would calculate for levelized cost of energy (LCOE), namely taking cost of all water production and goods as expended by the water production entity, and divide by total water supply (Tan et al., 2020).

The water governance in Malaysia has observed positive change before, as in 2006, the waste sector was restructured where responsibility was divested into the Ministry of Energy, Green Technology and Water (KeTTHA), the

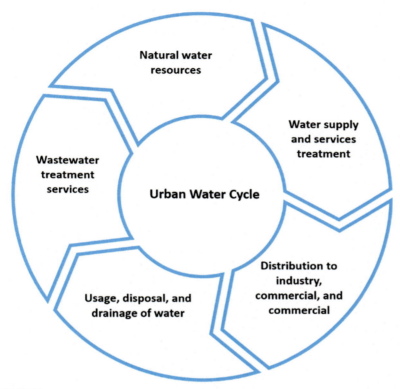

FIGURE 13.5
The urban water cycle.

National Water Resource Council (NWRC), and the National Water Services Commission (Kim, 2012). Historically, privatization has had mixed results throughout Malaysia. In Penang, the privatization of water services to Perbadanan Bekalan Air Pulau Pinang Sdn Bhd (PBAPP) has shown improvement in key indicators such as supply coverage in urban and rural areas, while keeping water tariffs low (Weng, 2002). However, privatization in other state jurisdictions did not improve water services and quality because the process lacked openness and competitiveness (Bonnardeaux et al., 2017). One way forward for Malaysia's water industry would be to centralize governance and streamline water management into a single body (Kim, 2012) and move from water supply management (WSM) mode to water demand management (WDM) mode.

Malaysia has a history of water dealings with its neighboring country, Singapore, as pronounced by four water agreements signed in years 1927, 1961, 1962, and 1990 (Chew, 2019). Singapore depends on Malaysia for its water supply, in exchange for rental and supply of treated water. To support

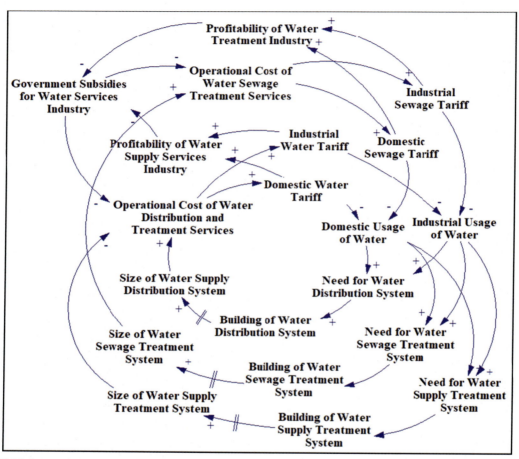

FIGURE 13.6
Water system and tariff loops (Tan et al., 2020).

these agreements, Malaysia provided large areas of land and river (Chew, 2019), for a fee, to Singapore, where the water–land nexus can be observed on an international level.

3.2.3 The energy–water link

From Fig. 13.7, the energy–water nexus is observed to exist between energy generating capacities and water treatment, sewage, and distribution capacities. The greater the operations in the water sector, the greater the usage of energy, whether it is from the usage of conventional energy or renewable energy. Operations of all the generation capacity, as well as the water system, would ultimately affect the nation's total internal renewable water resources. This shows a

3. Case study: WEF in Malaysia   249

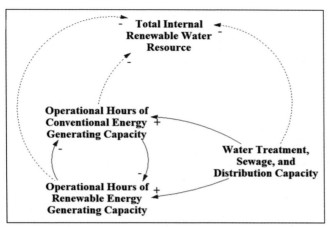

FIGURE 13.7
Energy generation operational hours and the water system (Tan et al., 2020).

bidirectional relationship between the water and energy sectors, where water is used in the production of electricity, and energy is also expended to treat and distribute the water. Taking into account the second laws of thermodynamics, a minimum net loss of energy can only be hoped for as best case for this natural resource—depleting closed loop. Consequently, the efficiency of both the energy and water technologies is important in optimizing and minimizing the net loss of energy during the operation.

3.2.4 Food security
While the food security of Malaysia revolves largely around rice (Tan et al., 2018), it is also important to ensure the nation is properly fed with sufficient nutrients, as described in the food security definition by FAO in 2001 (FAO, 2001). It is thus important for Malaysia to achieve self-sufficiency in rice, currently at 71% (Department of Statistics Malaysia, 2019), and also to ensure that the population are substantially nourished while doing so. Other major foods for Malaysia also include wheat flour, cooking oil, and sugar (Halim, 2015). Malaysia has long history of importing rice from neighboring countries, such as Thailand and Vietnam (Daño and Samonte, 2005). However, recent restrictions imposed on the export by these countries, which may have been caused by the complexity of the COVID-19 pandemic, have prompted Malaysia to import a record amount from India (Jadhav, 2020).

For the past decade, the National Agrofood Policy 2011—2020 (NAP4) was implemented in Malaysia, and served as a guideline for agricultural programs and projects (Ministry of Agriculture Malaysia, 2011). The NAP4 hoped to improve the performance of the agricultural industries, boost national food security, and increase revenue for the food sector. Three objectives were laid out, namely (1) to ensure adequacy of food supply and safety, (2) to develop agrofood into competitive and sustainable industry, and (3) to increase revenue for agricultural entrepreneurs (Abu Dardak, 2019).

A few studies have also been conducted from various perspective on Malaysia, such as improving self-sufficiency level (Bala et al., 2014), urban agriculture (Rezai et al., 2016), and climate change (Al-Amin and Ahmed, 2016). Bala et al. (2014), through simulation, showed that self-sufficiency level of rice would improve through steps such as (1) gradual transition to biofertilizers to maintain the input subsidies on agricultural input, (2) fund research and development of hybrid varieties which provide higher yields, (3) exploring possibilities of increasing cropping intensity as it directly increases rice production, (4) adjusting food policies to facilitate the transfer of agricultural technologies, and (5) making use of available land, which is abundant in Malaysia. Rezai et al. (2016) showed that the dynamics and perception of food security vary according to income group, where higher income earners are more likely to adopt urban agriculture, a modern approach known to increase food security as well. Al-Amin and Ahmed (2016) suggested that climate change impacts on Malaysian food security could be mitigated by taking proper strategies from the perspectives of management, infrastructure, and community engagement.

As highlighted by Hoff (2011), the links between energy and water to food are mainly one directional from the former two to the latter, especially taking consideration the agricultural landscape of Malaysia. In trying to achieve the desired self-sufficiency level of rice in Malaysia, consumptive water by paddy must be considered in the food supply chain, on top of energy expended in the form of electricity and machinery. Fig. 13.8 depicts the possible water and energy use for food production. There are clear links between water and energy resources contributing to the production of food outputs, which establishes the W–F and E–F linkages. However, without analyzing the WEF nexus on a more specific and intricate level, there are only vague links from food resources back for water and energy sectors.

3.3 Water–Energy–Food nexus in Malaysia—challenges and opportunities

Similar to other regions, the WEF nexus of Malaysia offers a unique perspective, considering the dynamics of energy type, electricity deregulation, water withdrawals, the urban water cycle that draws energy throughout the flow, and the contribution of water and energy resources for the production of food. Indeed, deeper intricate relationships could be drawn, and more specific indicators could be investigated at the interface or resource sectors, but intersectoral impacts are already obvious despite only taking a surface assessment. For

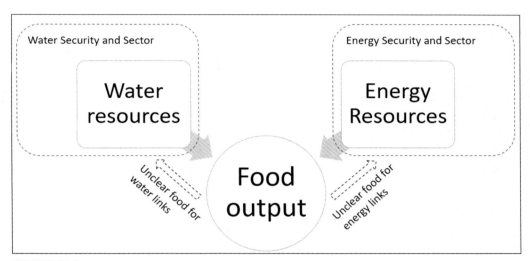

FIGURE 13.8
Water and energy links to food sector.

example, accurate tracking of energy used per water delivered, and water withdrawals per unit energy produced are yet to be available for the country. Thus, this presents an opportunity to open the doors of interdepartmental discussion of the resource sectors, where gains will be realized in the form of improved efficiency, increased population's resource security, and mitigation of climate change impacts.

Malaysia took a step in the WEF direction, by the creation of a Ministry of Energy, Green Technology, and Water, which provided a national master plan for the period of 2017–2030 (KeTTHA, 2017). Albeit interresource discussion between WEF were minimal in the brief, the master plan at the very least addresses the improvement for each sector, which would indirectly have benefits for sectoral trade-offs. The challenge, in this case, is presented to Malaysia for further increasing the integration and transdisciplinary development between the resource sectors, to achieve national holistic resource objectives, which should ultimately improve WEF resource securities.

4. Critical findings and key take-home messages

The WEF nexus approach is a useful and emerging method for holistically addressing the three basic resource securities for any region. From the various WEF studies in various regions, it is undeniable that intrinsic links exist between the three sectors, manifested through the dimensions of STEEP factors.

The downstream impacts are profoundly pronounced and complex to manage especially in regions with transboundary river basins, where river flow across international boundaries. An application of hydropower for energy on the river upstream in one country is enough to negatively impact the livelihoods of downstream food security in a neighboring country (see Section 2). The sociopolitical impacts must be taken into consideration when sectors or nations alike decide on their next technological agenda and implementations. Various WEF techniques exist and have been tried to a certain extent, at least from a research perspective, but a full-fledged practical implementation has yet to be seen, or at least not practiced widely.

The scenario in Malaysia is also unique in its own sense, with the energy and water sectors presenting both opportunities and challenges. Being rich in natural resource may provide stability in terms of power generation, but regulation of the energy market is less than ideal, which causes uncompetitive electricity pricing. Indirectly, this might have caused the financial dependence of the water sector, where external financing is necessitated to sustain the water sector. Taking a long-term approach might be wise, such as investing in future or alternative energy systems, and could potentially relieve the water sector of financial burdens in the long run and eventually reduce food prices. Consequently, energy, water, and food security could be improved together as a whole. However, more concrete data would need to be obtained, especially in a systemic manner, using a nexus approach, albeit using various methodologies, before WEF drives the Malaysian resource security landscape.

Nomenclature

BBIN Bangladesh, Bhutan, India, and Nepal
CASI Conservation agriculture-based sustainable intensification
CLD Causal loop diagram
EGP Eastern Gangetic Plains
EWFW Energy, water, food, and waste
FAO Food and Agriculture Organization
GDP Gross domestic product
HKH Hindu Kush Himalayan
IPP Independent power producer
KeTTHA Ministry of Energy, Green Technology and Water
LCOE Levelized cost of energy
NAP4 National Agrofood Policy 2011–2020
NWRC National Water Resource Council
PBAPP Perbadanan Bekalan Air Pulau Pinang Sdn Bhd
SBM Single buyer model
SD Sustainable development
SDG Sustainable Development Goals
SEA Southeast Asia

SEB Sarawak Energy Berhad
SESB Sabah Electricity Sdn. Bhd.
STEEP Sociological, technological, economic, environmental, and political
TNB Tenaga Nasional Berhad
WDM Water Demand Management
WEF Water–energy–food
WSM Water supply management

References

Abu Dardak, R., 2019. Malaysia's Agrofood Policy (2011–2020) - Performance and New Direction. FFTC Agric. Policy Platf. (Food Fertil. Technol. Center). https://ap.fftc.org.tw/article/1368. (Accessed 16 May 2021).

Adnan, H., 2013. Water, Food and Energy Nexus in Asia and the Pacific. United Nations, Bangkok.

Al-Amin, A.Q., Ahmed, F., 2016. Food security challenge of climate change: an analysis for policy selection. Futures 83, 50–63.

Ang, B.W., Choong, W.L., Ng, T.S., 2015. Energy security: definitions, dimensions and indexes. Renew. Sustain. Energy Rev. 42, 1077–1093.

Bala, B.K., Alias, E.F., Arshad, F.M., Noh, K.M., Hadi, A.H.A.A., 2014. Modelling of food security in Malaysia. Simulat. Model. Pract. Theor. 47, 152–164.

Bonnardeaux, D., Rodriguez, M. de la L.D., Nash, A., Jan, W.S.W., 2017. Water Provision in Malaysia Privatise or Nationalise? Institute for Democracy and Economic Affairs (IDEAS), Kuala Lumpur.

Brears, R.C., 2017. The green economy and the water-energy-food nexus. In: Brears, R.C. (Ed.), The Green Economy and the Water-Energy-Food Nexus Palgrave Macmillan, London, pp. 23–50.

Cheok, J.J., Tan, A.H.P., Yap, E.H., Wong, K.-C., Goh, B.H., 2021. Least cost option for optimum energy penetration in Malaysia. In: Osman Zahid, M.N., Abdul Sani, A.S., Mohamad Yasin, M.R., Ismail, Z., Che Lah, N.A., Mohd Turan, F. (Eds.), Recent Trends in Manufacturing and Materials towards Industry 4.0. Springer Singapore, Singapore, pp. 131–142.

Chew, V., 2019. Singapore-Malaysia Water Agreements. Singapore Infopedia. https://eresources.nlb.gov.sg/infopedia/articles/SIP_1533_2009-06-23.html (Accessed 13 June 2021).

CIA, 2021. The World Factbook - Countries - Malaysia. www.cia.gov/the-world-factbook/countries/malaysia/ (Accessed 23 April 2021).

CIA, n.d. World Factbook - Indonesia [WWW Document]. www.cia.gov/the-world-factbook/countries/indonesia/. (Accessed 23 April 2021).

Daño, E.C., Samonte, E.D., 2005. Public sector intervention in the rice industry in Malaysia. State Interv. Rice Sect. Sel. Ctries. Implic. Philipp. 187–216.

Department of Statistics Malaysia, 2019. Value of Gross Output of Agriculture Sector Registered an Annual Growth Rate of 11.1 Per cent to RM91.2 Billion. www.dosm.gov.my/v1/index.php?r=column/ctheme&menu_id=Z0VTZGU1UHBUT1VJMFlpaXRRR0xpdz09&bul_id=SzR6dm52MUJxV3hmODJGb3dsaHRwZz09 (Accessed 5 October 2021).

Department of Statistics Malaysia - Agriculture, 2013. www.dosm.gov.my/v1/index.php?r=column/ctwo&menu_id=Z0VTZGU1UHBUT1VJMFlpaXRRR0xpdz09 (Accessed 24 April 2021).

Emas, R., 2015. Brief for GSDR 2015 the Concept of Sustainable Development: Definition and Defining Principles. United Nations, New York.

Energy Commission (Malaysia) (Suruhanjaya Tenaga), 2019. Malaysia Energy Statistics Handbook 2019. Suruhanjaya Tenaga (Energy commission), Putrajaya.

FAO, 2009. Declaration of the World Summit on Food Security. World Food Summit 16–18 November 2009, International Institute for Sustainable Development (IISD), Winnipeg.

FAO, 2005. AQUASTAT Malaysia Water Withdrawal 2005. FAO. http://www.fao.org/nr/water/aquastat/countries_regions/Profile_segments/MYS-WU_eng.stm. (Accessed 16 May 2021).

FAO, 2001. The State of Food Insecurity in the World 2001. Food and Agriculture Organization of the United Nations, Rome.

Finley, J.W., Seiber, J.N., 2014. The nexus of food, energy, and water. J. Agric. Food Chem. 62, 6255–6262.

Gathala, M.K., Laing, A.M., Tiwari, T.P., Timsina, J., Islam, M.S., Chowdhury, A.K., Chattopadhyay, C., Singh, A.K., Bhatt, B.P., Shrestha, R., Barma, N.C.D., Rana, D.S., Jackson, T.M., Gerard, B., 2020. Enabling smallholder farmers to sustainably improve their food, energy and water nexus while achieving environmental and economic benefits. Renew. Sustain. Energy Rev. 120, 109645.

Halim, N., 2015. Assessment of Food Security Challenges in Malaysia. FFTC Agric. Policy Platf. (Food Fertil. Technol. Center), pp. 1–4.

Hardy, L., Garrido, A., Juana, L., 2012. Evaluation of Spain's water-energy nexus. Int. J. Water Resour. Dev. 28, 151–170.

Hoff, H., 2011. Understanding the Nexus. Background Paper for the Bonn 2011 Nexus Conference. Stockholm Environment Institute (SEI), Stockholm.

Howarth, C., Monasterolo, I., 2016. Understanding barriers to decision making in the UK energy-food-water nexus: the added value of interdisciplinary approaches. Environ. Sci. Pol. 61, 53–60.

IEA: Directorate of Global Energy Economics, 2015. World Energy Outlook. International Energy Agency (IEA), Paris.

Investopedia, 2020. Developed Economy. http://www.investopedia.com/terms/d/developed-economy.asp. (Accessed 18 May 2021).

Islam, S., Gathala, M.K., Tiwari, T.P., Timsina, J., Laing, A.M., Maharjan, S., Chowdhury, A.K., Bhattacharya, P.M., Dhar, T., Mitra, B., Kumar, S., Srivastwa, P.K., Dutta, S.K., Shrestha, R., Manandhar, S., Sherestha, S.R., Paneru, P., Siddquie, N.E.A., Hossain, A., Islam, R., Ghosh, A.K., Rahman, M.A., Kumar, U., Rao, K.K., Gérard, B., 2019. Conservation agriculture based sustainable intensification: increasing yields and water productivity for smallholders of the Eastern Gangetic Plains. Field Crop. Res. 238, 1–17.

Jadhav, R., 2020. Malaysia Signs Record Rice Import Deal with India: Exporters. Reuters. www.reuters.com/article/india-malaysia-rice-idINKBN22R1BI. (Accessed 20 May 2021).

Karatayev, M., Rivotti, P., Sobral Mourão, Z., Konadu, D.D., Shah, N., Clarke, M., 2017. The water-energy-food nexus in Kazakhstan: challenges and opportunities. Energy Proc. 125, 63–70.

Keskinen, M., Someth, P., Salmivaara, A., Kummu, M., 2015. Water-energy-food nexus in a transboundary river basin: the case of Tonle Sap Lake, Mekong river basin. Water (Switzerland) 7, 5416–5436.

KeTTHA, 2017. Green Technology Master Plan Malaysia 2017-2030. Ministry of Energy, Green Technology and Water (KeTTHA), Putrajaya.

Kim, C.T., 2012. Malaysian Water Sector Reform: Policy and Performance. Wageningen Academic Publishers, Netherlands.

Lechner, A.M., Gomes, R.L., Rodrigues, L., Ashfold, M.J., Selvam, S.B., Wong, E.P., Raymond, C.M., Zieritz, A., Sing, K.W., Moug, P., Billa, L., Sagala, S., Cheshmehzangi, A., Lourdes, K., Azhar, B., Sanusi, R., Ives, C.D., Tang, Y.-T., Tan, D.T., Chan, F.K.S., Nath, T.K., Sabarudin, N.A.B.,

References

Metcalfe, S.E., Gulsrud, N.M., Schuerch, M., Campos-Arceiz, A., Macklin, M.G., Gibbins, C., 2020. Challenges and considerations of applying nature-based solutions in low- and middle-income countries in Southeast and East Asia. Blue-Green Syst. 2, 331–351.

Lee, D.T.L., Tan, A.H.P., Yap, E.H., Tshai, K.Y., 2019. Deregulating the Electricity Market for the Peninsula Malaysia Grid Using a System Dynamics Approach, Communications in Computer and Information Science. Springer, Singapore.

Lehmann, S., 2018. Conceptualizing the urban nexus framework for a circular economy: linking Energy, Water, Food, And Waste (EWFW) in Southeast-Asian cities. In: Peter, D. (Ed.), Urban Energy Transition, Renewable Strategies for Cities and Regions, second ed. Elsevier Ltd, Amsterdam, pp. 371–398.

Leinbach, T.R., 1999. Britannica - Southeast Asia. www.britannica.com/place/Southeast-Asia (Accessed 24 April 2021).

Lovei, L., 2000. The Single-Buyer Model: a dangerous path toward competitive electricity markets. Public Policy Priv. Sect. 225, 4.

Malaysia Energy Commission, 2013. Peninsular Malaysia Electricity Supply Industry Outlook. Suruhanjaya Tenaga, Putrajaya.

McPhee, D., 2020. Renewable Energy Outdoes Fossil Fuels on Price as costs Plummet, New Report Finds. www.energyvoice.com/renewables-energy-transition/243209/renewable-energy-outdoes-fossil-fuels-on-price-as-costs-plummet-new-report-finds/ (Accessed 15 May 2021).

Mensah, J., 2019. Sustainable development: meaning, history, principles, pillars, and implications for human action: literature review. Cogent Soc. Sci. 5 (1).

Ministry of Agriculture Malaysia, 2011. Dasar Agromakanan Negara 2011-2020.

Mukuve, F.M., Fenner, R.A., 2015. Scale variability of water, land, and energy resource interactions and their influence on the food system in Uganda. Sustain. Prod. Consum. 2, 79–95.

Mustafa, M.A., Mabhaudhi, T., Avvari, M.V., Massawe, F., 2021. Transition toward sustainable food systems: a holistic pathway toward sustainable development. In: Galanakis, C.M. (Ed.), Food Security and Nutrition. Academic Press, Massachusetts, pp. 33–56.

Pullin, J., AME Study Group on Functional Organization, 1989. Organizational Renewal-Tearing Down the Functional Silos, pp. 4–14.

Putra, M.P.I.F., Pradhan, P., Kropp, J.P., 2020. A systematic analysis of water-energy-food security nexus: a South Asian case study. Sci. Total Environ. 728, 138451.

Rasul, G., 2014. Food, water, and energy security in South Asia: a nexus perspective from the Hindu Kush Himalayan region{star, open}. Environ. Sci. Pol. 39, 35–48.

Rezai, G., Shamsudin, M.N., Mohamed, Z., 2016. Urban agriculture: a way forward to food and nutrition security in Malaysia. Proc. Soc. Behav. Sci. 216, 39–45.

Ryabchikov, A.M., 2020. Britannica - South Asia. www.britannica.com/place/South-Asia (Accessed 24 April 2021).

Saklani, U., Shrestha, P.P., Mukherji, A., Scott, C.A., 2020. Hydro-energy cooperation in South Asia: prospects for transboundary energy and water security. Environ. Sci. Pol. 114, 22–34.

Save the Mekong Coalition, 2009. Thousands Call for Regional Governments to Save the Mekong. https://archive.internationalrivers.org/es/node/3790.

SPAN, 2019. Water & Sewerage Statistics 2019.

Spiegelberg, M., Baltazar, D.E., Sarigumba, M.P.E., Orencio, P.M., Hoshino, S., Hashimoto, S., Taniguchi, M., Endo, A., 2017. Unfolding livelihood aspects of the water–energy–food nexus in the Dampalit watershed, Philippines. J. Hydrol. Reg. Stud. 11, 53–68.

Surbhi, S., 2015. Difference between Developed Countries and Developing Countries. Key Differences. http://keydifferences.com/difference-between-developed-countries-and-developing-countries.html. (Accessed 17 May 2021).

Tan, A.H.P., Tshai, K.Y., Ho, J.-H., Yap, E.H., 2018. A conceptual framework for assessing Malaysia's Water, Energy and Food (WEF) security nexus. Adv. Sci. Lett. 24, 8822–8825.

Tan, A.H.P., Yap, E.H., 2019. Energy security within Malaysia's water-energy-food nexus — a systems approach. Systems 7, 1–24.

Tan, A.H.P., Yap, E.H., Abakr, Y.A., 2020. A complex systems analysis of the water-energy nexus in Malaysia. Systems 8 (2), 19.

Teljeur, E., Hoven, Z. van der, Kagee, S., 2016. Why we need alternative ESI structures. In: The Electricity Retail Competition Unicorn. 2nd Annual Competition and Economic Regulation (Acer) Conference, Southern Africa, Livingstone, Zambia, March 11-12 2016.

United Nations, 2021. The Sustainable Development Goals Report 2020. Des. Glob. Challenges Goals 1–68. United Nations, New York.

United Nations, 2019. World Population Prospects 2019, World Population Prospects 2019. United Nations, New York.

United Nations, 2014. Prototype Global Sustainable Development Report. United Nations Department of Economic and Social Affairs (UN-DESA), New York.

Weng, C.N., 2002. The Perbadanan Bekalan Air Pulau Pinang Sdn Bhd (PBAPP): A Good Example of Corporate Social Responsibility of a Private Water Company in Malaysia Chan Ngai Weng. Seminar.

World Commission on Environment and Development, 1987. Our Common Future. World Commission (Accessed 17 May 2021).

World Economic Forum, 2015. Global Risks Report 2015, tenth ed. World Economic Forum, Geneva.

Yang, H., Zhou, Y., Liu, J., 2009. Land and water requirements of biofuel and implications for food supply and the environment in China. Energy Pol. 37, 1876–1885.

Yao, L., Chang, Y., 2014. Energy security in China: a quantitative analysis and policy implications. Energy Pol. 67, 595–604.

CHAPTER 14

The water—energy—food nexus: an ecosystems and anthropocentric perspective

Sally Williams[1], Annette Huber-Lee[1], Laura Forni[2], Youssef Almulla[3], Camilo Ramirez Gomez[4], Brian Joyce[1] and Francesco Fuso-Nerini[1]

[1]*U.S. Water Program, Stockholm Environment Institute, Somerville, MA, United States;* [2]*U.S. Water Program, Stockholm Environment Institute, Davis, CA, United States;* [3]*Department of Energy Technology, KTH Royal Institute of Technology, Sweden;* [4]*Division of Nergy Systems KTH-dEH, KTH Royal Institute of Technology, Stockholm, Switzerland*

1. Introduction

It is essential to utilize an intersectoral strategy to accomplish the 2030 sustainable development agenda and its 17 Sustainable Development Goals (SDGs). To accomplish this, a water—energy—food (WEF) nexus approach can be used to assess the positive and negative consequences an action affecting one system can have on the others. The WEF nexus approach begins from the premise that food, water, and energy resources are inextricably interlinked and all face pressures from climate change, population change, and environmental change. Various analytical frameworks have added evidence to the effectiveness of the WEF nexus approach and its ability to provide a holistic portrait of potential future environmental resource realities and help decision-makers implement strategies to tackle these interrelated issues.

As climate change continues to affect the world, regions around the globe are affected in different ways. Areas characterized by various climates, topographies, and ecosystems all exhibit climate change—related challenges, but the presentation of these challenges varies (IPCC, 2021). Latin America and the Middle East and North Africa (MENA) regions are examples of two distinct climatic realities, but case studies from both demonstrate the applicability of the WEF nexus approach and its ability to aid in adaptation strategy development. In the case studies discussed in the following, although the context of the challenges faced differ, the complexity associated with future planning remains the same. In Brazil, Paraguay, and Argentina, a years-long drought is driving up energy prices in Brazil due to the region's reliance on hydropower, while simultaneously decreasing exports from Paraguay because of record low water levels in the Paraná River—the main riverboat export route (Costa,

2021). Concurrently, in Jordan, the country faces an immediate need for renewable energy technologies to meet energy demands for agriculture and domestic use, and to provide for the energy needed for groundwater pumping—a main contributor to the country's overexploitation of aquifers. These two regional case studies illustrate the complex dynamics decision-makers face as they create future adaptation strategies. Each dynamic is made more complex by the uncertainties that accompany it. To attempt to define and address these uncertainties, the WEF nexus approach can be paired with a stakeholder-driven participatory process.

The stakeholder-driven participatory process utilized in the case studies is the robust decision support (RDS) practice. The RDS process, outlined in Fig. 14.1, can be defined as a sustained effort to bring together stakeholders from various interest groups, who contribute their own desired outcomes and experiences, to work collaboratively to formulate the problem faced, define uncertainties, and operationalize objectives (Purkey et al., 2018). Via the RDS framework, viewpoints and expertise from across the water, energy, food, and other relevant sectors are brought together to share knowledge and formulate a future resource management strategy in an interactive, inclusive, and nexus-centered framework.

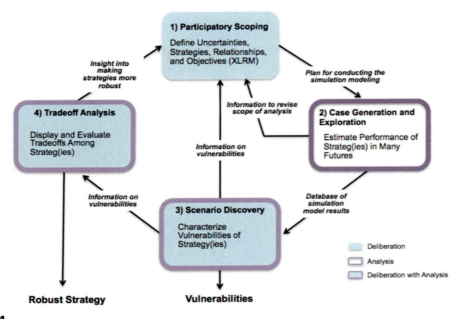

FIGURE 14.1

The robust decision-making framework, an example of the analysis with deliberation approach to decision-making under uncertainty (Purkey et al., 2018).

Applying the RDS process to the WEF nexus approach allows for the integration of water, energy, and food sector considerations in a decision-making framework that aims to be representative of the challenges faced by regional stakeholders, including the broader population. As can be expected, complexity and trade-offs are inherent in this approach; however, they are necessary to tackle climate change issues holistically, with adequate attention paid to both short-term and long-term implications and benefits. As evidence for the interdependencies between the water, energy, and food sectors of the world continues to grow, it is imperative to understand the ecosystems and anthropocentric implications of these interlinkages. The WEF nexus approach is a powerful tool to do so. The case studies outlined in this chapter further corroborate this perspective.

2. Approach

A key component of the WEF nexus methodology is a participatory stakeholder approach to identify key sectoral challenges and uncertainties faced in a given region of interest, as well as to assess modeling scenario findings. The RDS process aims to bring together voices across all relevant sectors and stakeholder groups to build a collaborative and iterative process through which people, not algorithms, drive the ultimate resource management solution that is implemented (Purkey et al., 2018). The RDS process comprises nine steps, carried out across a number of stakeholder workshops:

1. Define decision space: This is a thorough review of past decisions made for the management of WEF resources in the region. This includes examining plans and reports and interviews with decision-makers to define the decision space (i.e., decision context).
2. Map key actors: Once the decision-making context is understood, RDS requires identifying key stakeholders to include in the participatory approach. These actors should represent all relevant parties and offer a holistic representation of resource realities faced in a given region.
3. Problem formulation: Stakeholders are brought together in workshops, the first utilizing the XLRM matrix to frame the decision-making challenge and the uncertainties surrounding the resource management issue identified. The XLRM process involves articulating uncertainties (X), assembling analytical tools (R) to represent management options (L) to produce desired metrics of performance (M).
4. Tool construction: This refers to the build-out of models to represent the findings of the XLRM process. This tool must respond fully to the stakeholders' problem formulation to ensure it is not a "black box"

outcome, which diverts from the focus on balancing the values held by each distinct stakeholder.
5. Scenario definition: Once analytical tools are developed, a set of scenarios must also be formulated to drive model runs with possible future management strategies. These scenarios are based on the articulated planning uncertainties from workshops.
6. System vulnerability: A business-as-usual (BAU) scenario is run to demonstrate the vulnerabilities of the current system to stakeholders. BAU refers to the scenario in which all existing resource management systems and policies are unchanged. After this, stakeholders are asked to identify "could live with" and "would love to have" thresholds.
7. Options analysis: The model ensemble is run again to include representations of the current formulation of the stakeholder-proposed management options.
8. Results exploration: Output from the model is produced to represent each stakeholder-defined performance metric. The use of innovative, interactive data visualization tools to explore the outcome space defined by the desired metrics of performance for each combination of articulated uncertainties and identified management options is critical to the success of the RDS practice. These model outputs are discussed with stakeholders to develop shared knowledge and feedback space.
9. Decision support: Based on the shared insights developed through the participatory exploration of the ensemble model output database, the performance of specific management options can be evaluated relative to the BAU base case and each other. If successful, a set of acceptable resource management options is agreed upon by stakeholders (Fig. 14.2).

A visual of the RDS process is provided in Fig. 14.1. For the case studies of Latin America and MENA, the RDS process was utilized, although the context of these projects forced certain alterations to the traditional RDS model. An outline of

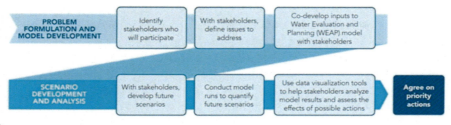

FIGURE 14.2
Process of stakeholder engagement and model building used in the WEF nexus methodology. *WEF*, water—energy—food.

FIGURE 14.3
Timeline for nexus approach in Jordan (The original timeline was extended due to the COVID-19 pandemic).

the RDS timeline for the Stockholm Environment Institute's (SEI) Jordan project is illustrated in Fig. 14.3 and points to the timeline impact COVID-19 had on the stakeholder engagement process. Certain workshops were moved online, although the essence of the RDS process was preserved to the best of the SEI and partner teams' abilities.

3. WEF case studies: MENA and Latin America

3.1 Case study 1: Jordan and Morocco

From 2018 to 2021, SEI-US has worked on a study to support a transformational change into WEF systems planning and management in Jordan and Morocco in the MENA region. In the past, resource management studies and strategies have taken a siloed approach to address one sector's challenges, rather than assessing environmental resources holistically to account for interlinkages. As the region faces increasing water scarcity, MENA has been a focus area for emerging WEF nexus work. This adds difficulty to food production, puts pressure on groundwater aquifers, and increases energy needs for groundwater pumping. In Jordan, nearly three-quarters of the country's land is covered by desert, making it a natural resource-scarce country. It has one of the lowest renewable water resources available per person worldwide (MWI, 2018). Jordan imports roughly 87% of its food, and among the country's agricultural land, only 52% is irrigated, yet this 52% accounts for 90% of agricultural production. Groundwater constitutes the main water source with around 60% share of total withdrawals, followed by surface water at around 27% and treated wastewater at 14% (MWI, 2018). Jordan relies heavily on fossil fuel imports (NEPCO, 2020).

Morocco faces a similar reality. The region is already water-stressed, with projections that point to increased temperatures, decreased rainfall, an extension of drought, and increased extreme events in the future due to climate change (Niang et al., 2014). The country heavily relies on the Souss-Massa River Basin for WEF activities. The basin is the leading area for export-oriented citrus and vegetables in Morocco. It also provides water to more than 2 million people, the majority of whom live in the greater metropolitan area around Agadir. A desalination project is underway to reduce water insecurity in the country, but that requires additional energy for treatment and for transport, and raises legitimate questions of how the Souss-Massa region can achieve this objective while reducing the negative impact on agricultural resources and GHG emissions. To address the climate challenges described for both Jordan and Morocco, as well as to present a more representative set of possible future scenarios to aid in MENA's resource management strategy, a participatory framework was utilized by SEI and partners to identify key challenges, uncertainties, and next steps in the region.

During SEI's work with both Jordan and Morocco, multiple nexus dialogue workshops were delivered to identify pressing challenges and potential solutions and better comprehend the nexus dynamics in the countries. All workshops included key decision-makers and stakeholders for the respective regions. In Jordan, the following priority challenges were identified:

1. Water scarcity (highest priority)
2. Agricultural productivity and water quality
3. Shift to energy independence

Concurrently, in Morroco's Souss-Masa River Basin, stakeholders in workshops outlined a similar set of primary challenges:

1. The Souss Massa is a major center for irrigated agriculture but has faced water scarcity for many years. As a result, there has been a history of overpumping of aquifers that has led to problematic saltwater intrusion along the coast.
2. Grid electricity and butane are the main sources of groundwater pumping. Both are subsidized, which results in both economic and environmental burdens.
3. There is an urgent need to increase agricultural water productivity to ensure the most value is generated per cubic meter of water. However, water still relies largely on pumping and transport, which requires energy.

A representation of the mapping of these key challenges in Jordan is illustrated in Fig. 14.4. As is visible on the map, an action in one sector has effects on other sectors. For example, increased water scarcity (lower left corner) reduces food production (upper right corner).

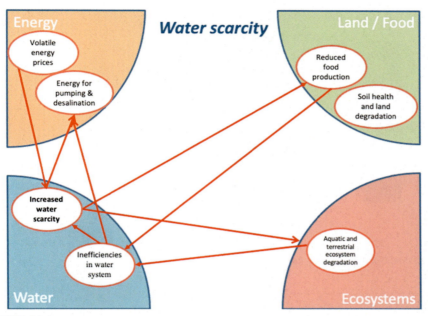

FIGURE 14.4
Mapping of the Jordan nexus challenge—water scarcity.

Researchers developed a WEF nexus model for both countries to allow stakeholders to assess the impact of selected nexus interactions. Modeling tools included WEAP for water management mapping, MABIA to estimate crop production based on water availability, and a GIS-based energy modeling tool. Fig. 14.5 offers a schematic representation of the key drivers of the WEF nexus models.

Results from modeling activities (Table 14.1) allowed the research team to present a set of scenarios to stakeholders based on which decision-makers could collaborate and outline future strategies for resource management. In Jordan, results showed how desalination is needed to address water scarcity, but it must be coupled with low-carbon electricity generation to avoid exacerbating climate change. Reducing nonrevenue water can have positive effects on municipal unmet demand and reduction of energy for water pumping, but it does not improve agricultural water productivity and may have negative feedback effects on the Jordan Valleys aquifer levels. Energy efficiency in the form of the transport, treatment, and pumping of water can offset energy intensive projects such as desalination by substantially reducing the load on the energy system, preventing increased emissions and achieving a more resilient water system.

264 | CHAPTER 14: The water—energy—food nexus

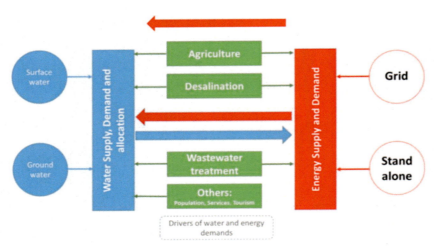

FIGURE 14.5
Schematic representation of the key drivers of the water—energy—food nexus model.

Table 14.1 Scenario analysis summary of results. The No intervention scenario is highlighted with a gray background (the reference case). Colored arrows are used to denote positive or negative differences between tested scenarios and the No intervention scenario for selected indicators.

Scenario	Municipal unmet demand (%) *	Agricultural unmet demand (%) *	Aquifer water table depth (m) **	Energy demand (GWh) *	Agricultural productivity (kg/m3) *
No intervention	17.1%	38.3%	AZ -139 m JV -63 m DS -174 m	3694	3.27
Reduced NRW	12.8% ↓	38.6% —	AZ -139 m — JV -70 m ↑ DS -171 m ↓	3314 ↓↓↓	3.26 —
New Resources	11.8% ↓↓	38.0% —	AZ -139 m — JV -63 m — DS -161 m ↓↓	3912 ↑	3.27 —
Increased water productivity	17.1% —	34.6% ↓	AZ -133 m ↓ JV -51 m ↓↓ DS -167 m ↓↓	3582 ↓	3.78 ↑
Integrated strategies	7.8% ↓↓↓	34.7% ↓	AZ -133 m ↓ JV -58 m ↓ DS -152 m ↓↓↓	3470 ↓↓	3.76 ↑

↓: positive decrease   ↑: positive increase   ↑: negative increase   —: no significant change
AZ: Amman Zarqa aquifer   JV: Jordan Valleys aquifer   DS: Dead Sea aquifer
* Last decade average (2040-2050)   ** Last year value (2050)

The greatest takeaway is that when all interventions are considered together, all of the major drawbacks are reduced and the benefits augmented, producing a more holistic solution to the WEF nexus challenges in Jordan.

In Morocco, nexus modeling efforts helped researchers and stakeholders understand how water, energy, and food systems are largely intertwined and how sectorial solutions will not achieve holistic outcomes (Fig. 14.6). Due to the complex nature of the nexus challenges, none of the tested sectorial solutions could address all challenges (Table 14.2). Key findings included the following:

1. Desalination is an effective measure to address water scarcity, but it has to be coupled with low-carbon electricity generation to not exacerbate climate change (Fig. 14.8);
2. The increased water productivity due to the large-scale adoption of drip irrigation and good agricultural production will improve the valorization of agricultural commodities, but with limited benefits to groundwater drawdown.
3. The phase-out of butane for groundwater pumping and replacing it with solar PV has shown to be an economically and environmentally competitive strategy. However, this shift should go hand in hand with effective measures (e.g., monitoring systems) to avoid ground water overexploitation with free energy source (Fig. 14.7).

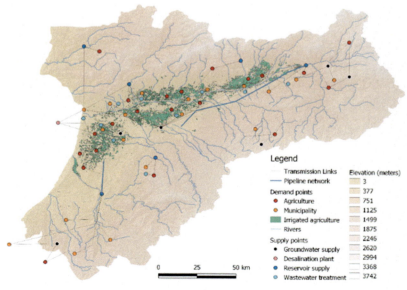

FIGURE 14.6
Illustration of the water and energy models integration for the Souss-Massa Basin.

Table 14.2 Summary results of the scenarios.

Scenario	Impact on groundwater level (m)*	Impact on unmet water demand (%)**	Impact on agricultural productivity (kg/m3)	Impact on energy requirement (GWh)***
Reference Scenario (REF)	Souss: -180m Chtouka: -101m	Agriculture: 46.6%	1.6	516
Desalination Scenario (DES)	Souss: -179.78m — Chtouka: -94m (-7.5m) ⬇	29% ⬇	1.94 ⬆	1016 ⬆⬆
Wastewater reuse scenario (WWR)	Souss: -179.88m — Chtouka: -89.7m (-9.0m) ⬇	28.7% ⬇	1.942 ⬆	1029 ⬆⬆⬆
Increased Water Productivity scenario (IWP)	Souss: -179.55m — Chtouka: -92.3m (-11.5m) ⬇	23.4% ⬇⬇⬇	2.19 ⬆⬆	1001 ⬆⬆
Integrated Strategies scenario (INS)	Souss: -179.56m — Chtouka: -87.3m (-13.7m) ⬇⬇⬇	23.8% ⬇⬇	2.24 ⬆⬆⬆	1007 ⬆⬆

*Last year (2050) value. **Last decade (2040-2050) avg. value ***Entire modelling period (2020-2050) avg. value

(⬆ red) Negative increase. (⬆ green) Positive increase. (⬇) Positive decrease. (—) No significant change.

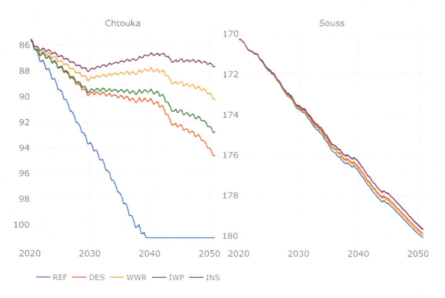

FIGURE 14.7
Change in groundwater table depth (m) for Chtouka and Souss aquifers across all scenarios: reference (REF), desalination (DES), desalination with wastewater reuse (WWR), increased water productivity (IWP), and integrated strategies (INS).

3. WEF case studies: MENA and Latin America

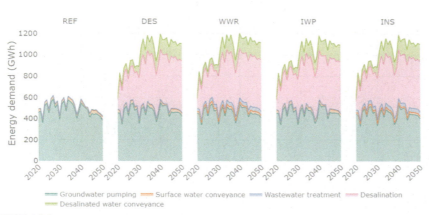

FIGURE 14.8
Energy requirements for different activities across all scenarios: reference (REF), desalination (DES), desalination with wastewater reuse (WWR), increased water productivity (IWP), and integrated strategies (INS).

As is made evident in the aforementioned discussion, utilizing a participatory approach was valuable in assessing challenges and solutions to sustainability challenges in both Morocco and Jordan. Through nexus WEF dialogues and cross-sectorial integration, it was possible to identify pressing issues between these countries' water, energy, and food sectors. Nexus thinking helped to understand how all these systems are largely intertwined and how sectorial solutions will not achieve holistic outcomes. None of the tested sectorial solutions targeted all combined challenges; instead, some solutions negatively affected other sectors. Therefore, integrated strategies are needed to holistically target the challenges among all sectors.

3.2 Case study 2: Argentina and Brazil

The Latin America region has 33% of global water resources, so it is a relatively rich water region that in the last century has experienced institutional changes around integrated water governance structures that can support the coordination of WEF management actions (Flachsbarth et al., 2015).

Historically, the region has had humid conditions from abundant freshwater from rivers like the Amazon (BR), Orinoco (CO), Negro (AR), Parana (AR), and Magdalena (CO), and dryer conditions from arid and semiarid regions in northern Mexico, central pacific coast in Peru and Bolivia, and the southern part of Argentina and Chile (Fig. 14.8). The main economic activities are export-driven and expanding agricultural production and mining (Fig. 14.9).

In the past 50 years, Latin America has promoted a move from centralized (national government) to a more decentralized (interstate) water governance

FIGURE 14.9
Rainfall distribution (left) and vegetation in Latin America (right). *Adapted from Mahlknecht, J., González-Bravo, R., Loge, F.J., 2020. Water-energy-food security: a Nexus perspective of the current situation in Latin America and the Caribbean. Energy 194, 116824. https://doi.org/10.1016/j.energy.2019.116824.*

driven by participatory processes (Baird and Plummer, 2020). For example, Argentina is a water-rich country that had a water governance reform in 1994 with the incorporation of interjurisdictional river basin organizations (Baird and Plummer, 2020).

Argentina is considered a water-rich country; only 4.2% of its available water is used, and most of this use goes to the food sector. However, water availability is unevenly distributed with arid and semiarid territories (75%), large rain-fed areas in the fertile plains, and irrigated agriculture at the foot of the Andes (Barros et al., 2015; Mahlknecht et al., 2020). In the energy sector, Argentina has a high energy per capita consumption compared with other countries in the region, resulting in importing energy from neighboring countries (Sheinbaum et al., 2011). Only 11.9% of energy production is from renewables, and it depends mainly on fossil fuels (53.6% from natural gas and 33.9 from oil derivatives) (Mahlknecht et al., 2020). Natural gas production is expanding partly from market-driven forces but also because of government subsidies (Forni et al., 2021; Mathier et al., 2018; Poder Ejecutivo Nacional, 2020).

Each provincial state in Argentina is responsible for planning and managing its own water resources. The system is not fully decentralized; some enabling legislation gives levels of water authority to the provinces that must coordinate management with other jurisdictions for shared water resources. This coordination between provinces within a shared watershed occurs on an as-needed basis. For example, coordination and organization between provinces in the Andes, where hydropower development and irrigation systems are implemented, happens earlier than in rain-fed flat plains (Baird and Plummer, 2020). The Interjurisdictional Water Basin Authority (AIC) in the Comahue region of Patagonia is an interprovincial organization that decides on water allocations (agriculture, cities, hydrocarbons) and hydropower production around the Rio Negro River watershed (which includes the Rio Neuquén and Limay watersheds). The AIC is composed of representatives of the provinces of Neuquén, Río Negro, Buenos Aires, and the national state participate.

In a 4-year study carried out between 2012 and 2016, the Stockholm Environment Institute collaborated with various stakeholders in northern Patagonia, Argentina, to examine the water resources for the region's basin. Specifically, the project looked at the integrated water resources options for long-term planning in northern Patagonia's Limay, Neuquén, and Negro River basins. To complete this analysis, researchers utilized the RDS process and the WEAP system to capture the complexity of the basin. The work demonstrates how the RDS process can be used in a noncentralized region such as Patagonia, Argentina, with varying jurisdictions and decision-makers. In this context, RDS is useful due to its ability to account for various possible future scenarios and include the inherent uncertainty in future scenario mapping.

The RDS steps utilized in this study are outlined in Table 14.3. The XLRM framework was used to collaboratively formulate the problem with the region's decision-makers (Tables 14.4–14.6). Once the components of XLRM were identified and agreed upon, the research team created a simulation of the different components of the basin system (hydrology, hydropower, other uses, demand priorities, restrictions, etc.) using the WEAP model monthly. This multicomponent simulation model (WEAP) allowed the team to represent the nonlinearity of key variable changes in the system and the nonseparable spatial and temporal dependencies among them.

Once modeling was complete, SEI constructed a decision space visualization for presenting, communicating, and facilitating a discussion of model ensemble output. The model output indicated significant climate change impacts on the surrounding basin region, augmented with irrigated agriculture expansion and increased hydropower production (Fig. 14.10).

In the final model output results for the project, as expected, the strategy that contemplates the combination of all strategies is the one that provides the greater magnitude of impacts (Fig. 14.11). By combining strategies, this integrated approach minimizes negative consequences across all metrics, although in small proportions for some metrics and minimally for Canal Matriz and Canal Valle Inferior. Overall, a series of strategies improved the outcome of the region when they were integrated rather than evaluated in isolation. The

Table 14.3 Standard robust decision support (RDS) methodology. The shaded area represents the steps implemented in this project.

Phase	Preparation and Formulation				Evaluation and Implementation			
Modules	1. Define decision space	2. Map of key actors	3. Problem formulation	4. Tool construction	5. Scenario definition	6. Options analysis	7. Results exploration	8. decision support
Level of participation	Consultation	Information extraction	Participative research	Co-learning	Cooperation	Cooperation	Co-learning	Co-learning
Capacity building focus	Literature review	Survey	Participative workshop	Training in tools use	Regular meetings	Manual for tools use	Visualization training	Participative workshop
Decision-making products	Decision space definition	Actor interaction	Key system elements identified	Model for discussion	Key scenarios identified	Performance metrics	Meetings with decision-makers	Summary to identify financing
Results	Adaptation and development connections	Participative spaces identified	Intersectoral perspective shared	Climate and adaptation evaluation	Shared vision of the future	Estimated tradeoffs	Decision makers informed	Adaptation actions financed

Table 14.4 List of uncertainties (Xs) identified for the study area and corroborated by the local institution's focus group.

Uncertainties	Representation in WEAP $_m$odel
Future climate	Development of five climate projections: ciclo histórico, GFDL[a] 4.5, GFDL 8.5, MIROC[b] 4.5, MIROC 8.5, ESM2[c] 4.5, ESM2 8.5
Potential expansion of agricultural area	Development of two projections of agricultural land under production: (1) reference scenario with current tendencies and (2) potential land expansion
Changes in cropping patterns to higher-value crops	Evaluation of two potential cropping patterns tendencies: (1) traditional use of agricultural land and (2) higher-value crops

[a]Geophysical Fluid Dynamics Laboratory (GFDL).
[b]Model for Interdisciplinary Research on Climate (MIROC).
[c]Earth System Models Part II (ESM2).

Table 14.5 List of strategies (L) identified for the study area and corroborated by the local institutions.

Strategies	Management goal and description
Base	Current management
Reduction in losses	Slow reduction trend of agricultural water distribution losses up to 50% of the current modeled value
Irrigation efficiency	Slow increase in irrigation efficiency for gravity irrigation (35%–40%) and technical irrigation (65%–80%)
Rational water use in cities	Progressive reduction in urban water demands up to 50% of the current modeled value
Reservoir operation	Change in the operating rules for Piedra del Aguila Reservoir (increase in the maximum level of the normal operation of the reservoir in determined months)
Infrastructure development of hydropower plants (high feasibility)	High feasibility of the development of the following hydropower plants: Chihuido I (2025), Michihuao (2027), and Pantanitos (2035).
Infrastructure development of hydropower plants (low feasibility)	Low feasibility of construction of the following hydropower plants: Collón Cura (2030), La invernada (2027), Pini Mahuida (2027), Cerro Rayoso (2027), Huitrin (2027), Chihuido II (2027), and Integral Río Negro (2035).
Irrigation canal 1	Guardia Mitre-Patagones irrigation canal (operating in 2025 with 50 cfs capacity)
Irrigation canal 2	Chelforo-Rio Colorado (operating in 2025 with 50 cfs capacity)
Combined	Reduction in losses + irrigation efficiency + rational use of urban water + changes in reservoir operation rules + high and low feasibility hydropower plants + canal Guardia Mitre Patagones

Table 14.6 List of performance metrics (M) identified for the study area and corroborated by the local institutions.

Objective	Performance metrics	Desired levels of performance
Ecological (flows)	1. Neuquén River: Ortezuelo Grande 2. Neuquén River: San Patricio del Chañar 3. Neuquén River: before Dique Ballester 4. Negro River: Confluencia 5. Negro River: before Canal Norte 6. Negro River, before Bocatoma Beltran 7. Negro River: Desembocadura	100 m^3/s (min) 7 m^3/s (min) 115 m^3/s (min) 400 m^3/s (min) 450/500 m^3/s (min) 300 m^3/s (min) 250 m^3/s (min)
Coverage of agricultural irrigation requirements	Water demand coverages in all irrigation districts (Associaciones de Riego — AR) 8. Anelo 9. Campo Grande 10. Cinco Saltos 11. Los Barreales	85% (min) for all
Water supply and energy production (reservoir volumes)	12. Mari Menuco 13. Cerros Colorados 14. Chocón 15. Piedra del Aguila	Storage level 411.5 hm^3 (min) 38,000 hm^3 (top of buffer) 13,000 hm^3 (top of buffer) 7739.9 hm^3 (top of buffer)
Water supply (maximum flows in canals)	16. Principal Alto Valle Canal 17. Centenario Canal 18. Arroyito Canal 19. Margen Norte Valle Medio Canal 20. Conesa Canal 21. Valle Inferior Canal	80 m^3/s (max) 7 m^3/s (max) 15 m^3/s (max) 6 m^3/s (max) 28 m^3/s (max) 39 m^3/s (max)
Coverage of urban water requirements	Coverages in all urban centers	100% (min)

Argentina project highlights the intersectionality of climate change impacts, sustainable water management, agricultural development, and renewable energy production.

Like most countries in Latin America, Brazil's economy relies on natural resources. Together with Chile, Brazil has the highest electricity access rate in Latin America (Tolmasquim et al., 2021) since the power sector serves more than 50 million customers, granting 97% of the country's households' reliable electricity. About half (45%) of energy generation comes from renewables, largely from biofuels and hydropower (Caiado Couto et al., 2021).

Biofuels in Brazil, which are produced from sugarcane ethanol and soybean biodiesel, have historically being rain-fed; however, recent expansion into new areas is increasing irrigation needs (Caiado Couto et al., 2021; Hernandes et al., 2014; Siegmund-Schultze et al., 2018). After the United States, Brazil is

3. WEF case studies: MENA and Latin America   273

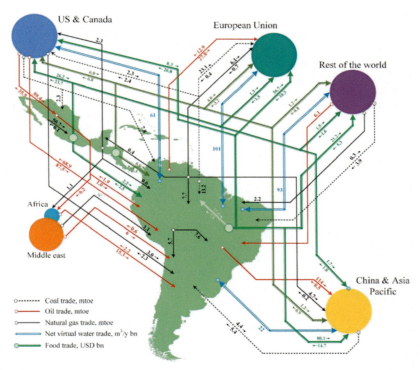

FIGURE 14.10
Major trade movements in Latin America (WEF nexus perspective). *Adapted from Adapted from Mahlknecht, J., González-Bravo, R., Loge, F.J., 2020. Water-energy-food security: a Nexus perspective of the current situation in Latin America and the Caribbean. Energy 194, 116824. https://doi.org/10.1016/j.energy.2019.116824.*

the largest producer of ethanol, and it is the most commonly used biofuel for transportation (Renewable Fuels Association, 2021). Brazil is the third largest hydroelectricity producer in the world after China and Canada, as of 2018, hydropower accounts for 65% of electricity generation (EPE, 2018). However, hydropower generation is expected to fall because it has been vulnerable to shortages in drought years, faced increasing socioenvironmental impacts concerns, and national policies promoting biofuel production (Caiado Couto et al., 2021).

During the summer of 2020—21, the Midwestern and Southeastern parts of the country experienced the worst droughts in 100 years, according to the National Meteorology System (SNM). The droughts resulted in two main problems: less crop productivity and less water input for energy generation in the hydroelectric plants causing increases in prices of food and energy. As of August 2021 (Reuters, 2021), water reserves at hydropower plants have fallen to their lowest level in 91 years causing energy cuts and price increases due to water scarcity. While

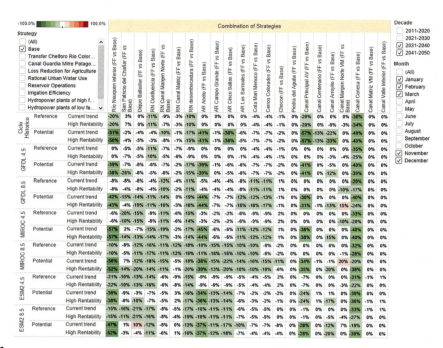

FIGURE 14.11

Impact of strategies on performance metrics (Forni et al., 2018).

the literature indicates that the reliability on water for electricity is expected to fall in terms of expected reductions in hydropower production due to drought, the push for biofuels is expected to put additional pressures on irrigated systems, which are also affected by drought.

Brazil's launch of RenovaBio, a National Biofuels Policy program, in 2019 aimed to increase biofuel production to support Brazil's commitment to the United Nations Climate Change Conference (COP21). Similarly, it also aimed to expand Brazil's share of nonhydropower renewables to between 28% and 33%, as well as expand its domestic use of nonfossil energy sources to at least 23% by 2030 (Barros, 2021). Reducing the dependency on hydropower and increasing the role of solar, wind, and other renewables is key for Brazil to achieve low carbon sustainable use of its electric power.

4. Comparisons of the WEF nexus in MENA and Latin America

The contrasts in average available water per capita for the two MENA countries (Jordan: 70 m³/capita and Morocco: 815 m³/capita) and the two Latin America

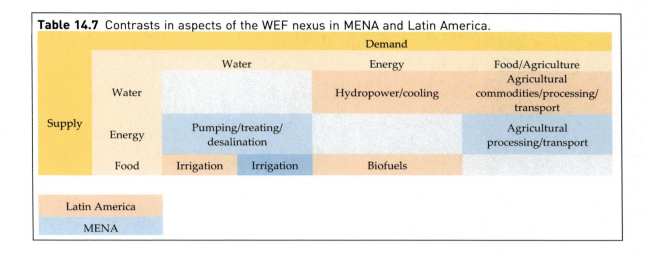

Table 14.7 Contrasts in aspects of the WEF nexus in MENA and Latin America.

countries (Argentina: 6630 m³/capita and Brazil: 27,116 m³/capita) are at two ends of the spectrum with respect to water stress—from the driest to the wettest of countries. Despite the differences in water availability, the use of RDS highlighted the advantages to joint consideration of long-term planning that looks across sectoral silos to find integrated and more resilient strategies.

As shown in Table 14.7, there are contrasts in how the WEF nexus emerges in the two geographies. For example, in Latin America, there are greater water availability issues due to hydropower than in MENA. Similarly, the use of water for biofuels and the use of water as a means of transporting agricultural commodities are also more pronounced in LA compared to MENA. For the water scarce MENA region, energy use for water transport, pumping, and treatment is a dominant characteristic along with the use of energy for agricultural processing and transport.

5. Conclusions

As climate change is taking center stage in terms of the extremes in temperature and extreme events in 2021 and the attention of the world's politicians in COP26 in Glasgow, more holistic approaches to long-term planning must be used. One such approach is RDS, as applied to the MENA and LA contexts using a WEF nexus framing. Despite what appears to be very different contexts geographically, the WEF nexus approach's advantages are of value in both water-rich and water-poor contexts.

References

Baird, J., Plummer, R., 2020. Water Resilience. Springer International Publishing AG.

Barros, S., 2021. Implementation of RenovaBio—Brazil's National Biofuels Policy (Biofuels, Climate Change/Global Warming/Food Security No. BR2021-0008). USDA, Brazil.

Barros, V.R., Boninsegna, J.A., Camilloni, I.A., Chidiak, M., Magrín, G.O., Rusticucci, M., 2015. Climate change in Argentina: trends, projections, impacts and adaptation: climate change in Argentina. Wiley Interdiscip. Rev. Clim. Change 6, 151—169. https://doi.org/10.1002/wcc.316.

Caiado Couto, L., Campos, L.C., da Fonseca-Zang, W., Zang, J., Bleischwitz, R., 2021. Water, waste, energy and food nexus in Brazil: identifying a resource interlinkage research agenda through a systematic review. Renew. Sustain. Energy Rev. 138, 110554. https://doi.org/10.1016/j.rser.2020.110554.

Costa, W., September 27, 2021. Paraguay on the Brink as Historic Drought Depletes River, its Life-Giving Artery. The Guardian from. https://www.theguardian.com/global-development/2021/sep/27/paraguay-severe-drought-depletes-river. (Accessed 25 October 2021).

EPE, 2018. Balanço Energético Nacional 2018 [WWW Document]. EPE. URL https://www.epe.gov.br/pt/publicacoes-dados-abertos/publicacoes/balanco-energetico-nacional-2018 (accessed 10.30.21).

Flachsbarth, I., Willaarts, B., Xie, H., Pitois, G., Mueller, N.D., Ringler, C., Garrido, A., 2015. The role of Latin America's land and water resources for global food security: environmental trade-offs of future food production pathways. PLoS One 10, e0116733. https://doi.org/10.1371/journal.pone.0116733.

Forni, L., Escobar, M., Cello, P., Marizza, M., Nadal, G., Girardin, L., Losano, F., Bucciarelli, L., Young, C., Purkey, D., 2018. Navigating the water-energy governance landscape and climate change adaptation strategies in the northern Patagonia region of Argentina. Water 10 (6), 794. https://doi.org/10.3390/w10060794.

Forni, L.,G., Mautner, M.R., Lavado, A., Fitzpatrick, K., Diaz-Gomez, R., 2021. Watershed Implications of Shale Oil and Gas Production in Vaca Muerta, Argentina.

Hernandes, T.A.D., Bufon, V.B., Seabra, J.E.A., 2014. Water footprint of biofuels in Brazil: assessing regional differences. Biofuel. Bioprod. Biorefin. 8, 241—252. https://doi.org/10.1002/bbb.1454.

IPCC, 2021. Climate Change 2021: The Physical Science Basis. Contribution of Working Group I to the Sixth Assessment Report of the Intergovernmental Panel on Climate Change [Masson-Delmotte. Cambridge University Press. In Press.

Mahlknecht, J., González-Bravo, R., Loge, F.J., 2020. Water-energy-food security: a Nexus perspective of the current situation in Latin America and the Caribbean. Energy 194, 116824. https://doi.org/10.1016/j.energy.2019.116824.

Mathier, D., Méndez, J.M., Bragachini, M., Sosa, N., 2018. La biomasa y la bioenergía distribuida para el agregado de valor en origen".

MWI, 2018. Jordan Water Sector Facts and Figures 2017. Ministry of Water & Irrigation.

NEPCO, 2020. Annual Report of the Year 2019. Tech. rep., National Electricity Power Company, the Hashemite Kingdom of Jordan.

Niang, I., Ruppel, O.C., Abdrabo, M.A., Essel, A., Lennard, C., Padgham, J., Urquhart, P., 2014. In: Barros, V.R., Field, C.B., Dokken, D.J., Mastrandrea, M.D., Mach, K.J., Bilir, T.E., Chatterjee, M., Ebi, K.L., Estrada, Y.O., Genova, R.C., Girma, B., Kissel, E.S., Levy, A.N., MacCracken, S., Mastrandrea, P.R., White, L.L. (Eds.), Climate Change 2014: Impacts, Adaptation, and Vulnerability. Part B: Regional Aspects. Contribution of Working Group II to the Fifth Assessment

Report of the Intergovernmental Panel on Climate Change. Cambridge University Press, Cambridge, United Kingdom and New York, NY, USA, pp. 1199–1265. Africa.

Poder Ejecutivo Nacional (PEN), 2020. Decreto 892/2020: Plan de Promoción de la Producción del Gas Natural Argentino – Esquema de Oferta y Demanda 2020–2024, DECNU-2020-892-APN-PTE.

Purkey, D.R., Escobar Arias, M.I., Mehta, V.K., Forni, L., Depsky, N.J., Yates, D.N., Stevenson, W.N., 2018. A philosophical justification for a novel analysis-supported, stakeholder-driven participatory process for water resources planning and decision making. Water 10 (8), 1009. https://doi.org/10.3390/w10081009.

Renewable Fuels Association, 2021. Alternative Fuels Data Center: Maps and Data–Global Ethanol Production 2019 [WWW Document]. Altern. Fuels Data Cent. URL. https://afdc.energy.gov/data/10331 (accessed 10.30.21).

Reuters, 2021. Brazil Minister Warns of Deeper Energy Crisis amid Worsening Drought. Reuters.

Sheinbaum, C., Ruíz, B.J., Ozawa, L., 2011. Energy consumption and related CO2 emissions in five Latin American countries: changes from 1990 to 2006 and perspectives. Energy 36, 3629–3638. https://doi.org/10.1016/j.energy.2010.07.023.

Siegmund-Schultze, M., Köppel, J., Sobral, M. do C., 2018. Unraveling the water and land nexus through inter- and transdisciplinary research: sustainable land management in a semi-arid watershed in Brazil's Northeast. Reg. Environ. Change 18, 2005–2017. https://doi.org/10.1007/s10113-018-1302-1.

Tolmasquim, M.T., de Barros Correia, T., Addas Porto, N., Kruger, W., 2021. Electricity market design and renewable energy auctions: the case of Brazil. Energy Pol. 158, 112558. https://doi.org/10.1016/j.enpol.2021.112558.

Further reading

Brazil minister warns of deeper energy crisis amid worsening drought | Reuters. (n.d.). Retrieved October 25, 2021, from https://www.reuters.com/world/americas/brazil-minister-warns-deeper-energy-crisis-amid-worsening-drought-2021-08-31/.

Cotosky, E., September 27, 2021. Brazil Suffered from Two Extreme Weather Events in 2021: Floods and Droughts. Climate Scorecard from. https://www.climatescorecard.org/2021/09/brazil-suffered-from-two-extreme-weather-events-in-2021-floods-and-droughts/. (Accessed 25 October 2021).

Karpavicius, L.M., January 21, 2021. Brazil Sources 45% of its Energy from Renewables. Climate Scorecard from. https://www.climatescorecard.org/2021/01/brazil-sources-45-of-its-energy-from-renewables/. (Accessed 25 October 2021).

Masson-Delmotte, V., Zhai, P., Pirani, A., Connors, S.L., Péan, C., Berger, S., Caud, N., Chen, Y., Goldfarb, L., Gomis, M.I., Huang, M., Leitzell, K., Lonnoy, E., Matthews, J.B.R., Maycock, T.K., Waterfield, T., Yelekçi, Ö., Yu, R., Zhou, B. (Eds.), 2021. Climate Change 2021: The Physical Science Basis. Contribution of Working Group I to the Sixth Assessment Report of the Intergovernmental Panel on Climate Change. Cambridge University Press.

Plummer, R., Baird, J., 2021. The emergence of water resilience: an introduction. In: Baird, J., Plummer, R. (Eds.), Water Resilience: Management and Governance in Times of Change. Springer International Publishing, pp. 3–19. https://doi.org/10.1007/978-3-030-48110-0_1.

CHAPTER 15

Water—energy—food nexus approaches to facilitate smallholder agricultural technology adoption in Africa

Michael G. Jacobson
School of Forest Resources, Department of Ecosystem Science and Management, Pennsylvania State University, University Park, PA, United States

1. Introduction

Practitioners of water—energy—food (WEF) nexus have recognized that an integrated approach, which assesses trade-offs and synergies in resource management instead of working in independent silos, is essential to ensure the sustainability of our global, regional, and local WEF systems (Hoff, 2011; Liu et al., 2017; Weitz et al., 2017; Wichelns, 2017; Terrapon-Pfaff et al., 2018; Albrecht et al., 2018; Nhamo et al., 2018). An effective WEF nexus analysis should assess all three sectors simultaneously, providing quantitative and qualitative relationships and linkages between sectors, so that trade-offs and synergies can be analyzed (McCornick et al., 2008). Barriers and critiques of WEF nexus frameworks and approaches include (1) applicable scale, (2) data limitations, (3) entrenched top-down interests not wanting to break down silos, and (4) lack of mainstreaming and operationalization (Allouche et al., 2018; Landauer et al., 2018; McGrane et al., 2018; McCarl et al., 2017; Liu et al., 2017; Termeer et al., 2010; Cash et al., 2006). To address these concerns, creative pathways for operationalizing WEF nexus have been developed using tools such as WEF Tools Index and agent-based modeling multicriteria decision methods (Simpson and Jewitt, 2019; Naidoo et al., 2021). These approaches use stakeholder-driven and local experts' interaction to collect data, which is qualitatively and quantitatively modeled resulting in relatively easy to understand composite WEF Nexus Indices and graphs that show priority areas for intervention.

This chapter discusses how WEF nexus can improve analysis about and solutions for agricultural technology adoption by smallholder farmers, especially in Africa, but in other geographies as well. Babiker et al. (2019) state, "technological and innovative solutions within the WEF nexus, where two, or all three, components of the nexus are integrated as inputs would enhance resource efficiency and expand the available natural resource base, thus bolstering the sustainability and security of the three resources." I start with the premise that technology-based interventions in farming systems can improve agricultural production and ecosystem and livelihoods functioning, but that central to technology interventions is adoption and uptake by farmers. Therefore, regardless of intervention or approach, farmer behavior, risks, and concerns must be addressed. I argue in the following that issues surrounding farmer decision-making about adopting agricultural technologies are well suited to a WEF nexus framework that explicitly examines critical trade-offs and synergies for growing food, namely water and energy. Can the use of a WEF nexus approach by accounting for smallholder farmer trade-offs in their technology adoption decisions lead to a more sustained agricultural transformation? To answer this question, the following sections of the paper discuss (1) the African situation in the context of water, energy, and food crises, (2) an overview of agricultural adoption analysis, and (3) research design elements and considerations for using WEF nexus as a catalyst that leads to more informed smallholder farmer decision-making and sustainable rural transformation in Africa.

2. African context

African society is facing a "perfect storm" of climate change, a growing population, and a shrinking supply of natural resources. The agricultural transition in many African countries has not met expectations despite decades of development programs including the Washington Consensus, the liberation of markets, and structural adjustment (Evenson and Gollin, 2003; World Bank, 2007; Timmer and Akkus, 2008; Alston and Pardey, 2014; Pardey et al., 2016).

Over three-quarters of the farms in most African countries measure less than 2 hectares and an estimated three-quarters of Africa's food is produced by smallholder subsistence farmers (Dixon et al., 2001; HLPE, 2013). Numerous constraints to production proliferate on these small farms including (1) lack of resources and access to markets, (2) lack of secure tenure rights, (3) limited ability of small farmers to participate in the policy process, and (4) urgent need and ability to adapt to rapidly changing conditions such as climate change (Diao et al., 2007; Prasad et al., 2012; Rademacher-Schulz and Mahama, 2012; Benin, 2016; Beegle, 2019; Sahle et al., 2019). Historical and present processes of political and agrarian change, and in particular, land tenure rights or lack thereof—both in the past (colonialism) and in the present (land grabbing)— also influence smallholder farmer conditions. Increasing agricultural productivity is crucial, and this increase will impact and depend upon water, nutrients, and energy.

Africa's population and poverty indicators highlight increased demands on land and water resource supplies. Most African countries are failing to meet both rural and urban improved drinking water access targets set by the UN SDGs, and currently one-fifth of the population faces serious water shortages exacerbated by climate change (Rockström and Falkenmark, 2015; Dos Santos et al., 2017). The projected increase in population by 2030 is expected to lead to, at least, a tenfold increase in water needs for energy production to support agricultural, industrial, social, and economic growth (AUC-AMCOW, 2016). This implies growing competition for available water resources in the future. Approximately 40% of Africans have no access to electricity, and this percentage rises to 80% in rural areas, a statistic that, if maintained, will result in increased energy poverty throughout the continent (IEA, 2014; Szabó et al., 2016; AfDB, 2021). Compounding these resource demands are climate change and rising temperatures in Africa, which is predicted to result in increased rainfall variability and incidences of extreme weather events (De Sherbinin, 2014). The Intergovernmental Panel on Climate Change (IPCC) estimates that 350 to 600 million people in Africa will be at risk of increased water stress by 2050 due to rainfall variability (Niang et al., 2014). Climate change projections indicate a reduction in the productivity of over 50% of agricultural land in southern Africa by 2050 and a reduction of between 10% and 30% in rainfall, a situation that threatens the livelihoods of over 60% of the population living in rural areas relying on natural systems (Besada and Werner, 2015; Mabhaudhi et al., 2019).

Rural transformation in Africa begins with sound agricultural policies (Barrett et al., 2017). The challenge of meeting food requirements is well recognized by African countries through policies such as the African Union (AU) Agenda 2063, the AU Malabo Summit Declaration of June 2014 on Accelerated Agricultural Growth and Transformation, the Comprehensive Africa Agriculture Development Programme (CAADP), and the Africa Water Vision for 2025 (AU, 2014, 2015; AUC-AMCOW, 2016). African governments continually reaffirm these commitments to eliminate hunger and food insecurity, but practical approaches are lacking, and goals remain unfulfilled (Leakey et al., 2021). The escalating gravity of the food security concerns in Africa and the challenges of meeting global attainment of the UN SDGs suggest a need to reassess ways to meet these goals. WEF nexus has relevance to informing evidence-based decision-making in Africa. Resource security remains a pressing challenge for Africa for which policy guidance on WEF interactions is critical. Transdisciplinary research is required involving multiple disciplines, regions, decision-makers, and civil societies in a format that enables cocuration and codesign. As international agencies and African governments confront climate change and acute food insecurity WEF nexus can provide integrated and transformative approaches to address these complex and cross-cutting problems that affect all sectors.

3. Literature review

To better understand the extent of WEF nexus activities in Africa, a Web of Science literature review in 2020 found 90 papers having keywords (water, energy, food, nexus, and Africa), but only 51 showed a clear connection to all three sectors (Jacobson and Pekarcik, 2021). The criteria for the 51 papers were (1) terms food, water, and energy systems explicitly present—in other words quite a few studies used these five terms but only considered one or two of the WEF sectors, (2) empirical analysis in which there was an evidence-based approach to the study, (3) interpretation of information, and (4) discussed trade-offs. This last criterion was important in distinguishing WEF nexus studies from others. Given this was a nonsystematic review, there are certainly uncertainties regarding the full extent of the available literature, but we felt the search was exhaustive enough to provide a solid overview of available WEF nexus publications with a focus on Africa.

For this chapter, extracted from the 51 WEF nexus-related studies were the primary technologies used. Although numerous technologies may be considered in WEF nexus analyses or projects, there is usually a dominant or focal technology driving the rationale to intervene. As Table 15.1 shows, in WEF nexus studies water-related technologies such as irrigation and hydropower were the most common primary foci, followed by renewable energy-related and biomass-related technologies. A third, smaller group of studies focused on land use, cropping, agroforestry, or conservation agriculture-related technologies. The dominance of irrigation and hydropower can be partly explained by WEF nexus' emergent history in the water sector as a successor to the Integrated Water Resources Management (IWRM) approach (Giordano and Shah, 2014). Renewable energy-focused WEF nexus studies emerged in part due to the urgency of its potential synergistic impacts on climate change compared with "traditional" fossil fuel energy sources. Another reason for the dominance of the water sector and, more recently, the energy sector, is perhaps a result of the data modeling tools and scale of most studies. To date, most WEF nexus studies have been carried out at a larger scale and are more top-down in approach, only indirectly reaching local-level, smallholder farmers (Allouche et al., 2018). Hydropower and irrigation, for example, are usually modeled at transboundary-, regional-, or catchment-level scale, lending themselves to national or even transnational analysis. Agriculture production, on the other hand, as noted is mainly carried out by smallholder farmers at a local level where data are limited. The next paragraphs highlight a few examples from papers showing technologies and their associated agricultural production trade-offs.

Irrigation, the most examined water-related technology in WEF nexus papers and one touted by the AU CAADP, is recognized as a necessary transformative

Table 15.1 Common Technology Interventions and Trade-offs (related to agriculture) found in WEF nexus studies in Africa.

Primary technology intervention	# of papers	Example trade-offs identified in studies
Water-related		
Irrigation	25	• Energy use—pumping and access • Efficiency—overextraction • Types—e.g., drip, solar, PV pumps, gravity, diesel • Subsidized energy policies
Hydropower	14	• Optimal water allocation • Electricity demand (population) • Ecosystem impact • Drought/climate—water availability
Wastewater recycling	1	• Energy use for agriculture • Carbon emissions
Virtual water management	1	• Water footprint
Energy-related		
Renewables, e.g., solar photovoltaic technology	12	• Land constraints • Water withdrawal
Biofuels/biomass	7	• Deforestation/environmental conservation • Large-scale versus domestic production • Land availability
Rural electrification, grid, and off-grid solutions	4	• Land availability
Waste-to-energy pathways	4	• GHG emissions • Recycling • Land availability
Food-related		
Agriculture mechanization	7	• Land use and watershed planning • Water availability • Energy use
Sustainable agriculture	7	• Water use • Land tenure
Land-use planning	3	• Optimal allocation

technology to boost agricultural yields of African smallholder farmers. African governments have embraced irrigation since most African farmers depend on unreliable rain-fed systems, where the duration of the rainy seasons and the distribution of rains over time vary, even more so in recent times from climate change (AUC-AMCOW, 2016). One solution is the use of renewable energy sources such as solar pumps; however, this has been shown to result in overextraction and expanded water withdrawals, causing subsequent drops in aquifer levels (Flammini et al., 2014). Even introducing water and energy-efficient technologies such as drip irrigation have not necessarily reduced overall water consumption (Doukkali and Lejars, 2015; Jobbins et al., 2015; Xie et al., 2020). Significant energy use to pump and distribute water can be risky in regions already facing energy insecurity (Mabhaudhi et al., 2018). For example, in a WEF nexus case study of Ethiopia's sugarcane production, Hailemariam et al. (2019) found that modern irrigation technologies, while able to prevent overextraction of water, require increased total energy consumption. The sites that used modern irrigation technologies (sprinkler, pivot, drip) had an 18%—21% greater total energy consumption (MJ/ha-year) than those that used traditional systems. Many irrigation-related WEF nexus studies focus on larger agriculture scale projects, e.g., foreign investment and export cash crops. Smallholder farmer adoption of irrigation technologies is severely conditional and constrained by access to capital and maintenance costs and other resources (money, land, subsidies, etc.).

Hydropower, the second most common primary technology noted in WEF nexus papers, is an important source of electricity for many African countries but competes for water with agriculture, municipal, and industrial uses and ecological purposes. These trade-offs have been well documented in WEF nexus case studies but usually at a larger scale than addressing local smallholder farmer trade-offs (Johnson et al., 2018; Allam and Eltahir, 2019; Ferrini and Benavides, 2020). Related to irrigation and hydropower is the importance of watershed/landscape planning for intensifying cropping, without which can lead to mismanaged irrigation (e.g., excessive groundwater extraction) and rainfall water resources, severe deforestation, and overgrazing of rangelands and pastures (Tian et al., 2018). One such WEF nexus study using a panel of selected sub-Saharan Africa countries from 1980 through 2013 found that increased cereal yields and its agricultural value-added led to significantly increased water poverty indicators (Ozturk, 2017). In other studies, it was found that accounting only for blue water in conventional water resource planning and interventions, ignoring green (rain-fed agriculture) water and virtual (equivalent of water embedded in imported food products) results in exclusion or underestimation of important resources for food security and accurate food trade balance (Chahed et al., 2015; Allan et al., 2015). WEF nexus has also been used to assess trade-offs between expanding irrigation versus improved fertilizer application. While both interventions improved agricultural productivity in the Mékrou river basin of West Africa, the benefits of fertilization were almost twice as strong as irrigation, mainly due to the cost-heavy upfront irrigation investment (Udias et al., 2018).

Many of the WEF nexus studies that had an energy-related intervention focused on trade-offs between alternative production sources, especially fossil fuel versus renewable energy for irrigation. For example, solar irrigation in Kenya, while improving agricultural yields, was found to also increase the overextraction of groundwater, causing subsequent drops in aquifer levels (Flammini et al., 2014). A study comparing stand-alone solar photovoltaic (PV) to diesel for irrigation water pumping throughout sub-Saharan Africa found that, although solar PV pumping is more environmentally sustainable, "the cost-effectiveness varies by land suitability and crop type" (Xie et al., 2020). Likewise, in Tunisia, while solar PV is economically feasible to use for off-grid pumping due to a fade-out of diesel subsidies, highly subsidized electricity tariffs for pumping still prevented grid-connected solar pumping irrigation systems from being economically feasible (Keskes et al., 2019).

Another set of energy-focused WEF nexus studies examined trade-offs in biomass production, mainly for household cooking (and some power), such as from biogas, fuelwood, and charcoal. When there is heavy dependence on traditional biomass for energy sources in rural households, very common in Africa, there are trade-offs between environmental conservation (deforestation, biodiversity, soil productivity, water quality, and quantity losses), energy access, and agricultural productivity—which has implications for food security (Hoffman et al., 2017; Kougias et al., 2018; Stein et al., 2018; Martins, 2018). Another trade-off is the usage of livestock dung as a fuel source for domestic energy but can deplete soil fertility, which negatively impacts the value of harvested crops (Mekonnen et al., 2017). A solution to counteract the negative trade-offs of dung and crop residue for domestic biofuel use is the introduction of agroforestry that can provide synergies if grown on-farm and the trees did not compete with cropland (Imasiku and Ntagwirumugara, 2020). Alternatively, dung can be used in a circular economy approach for biogas (microdigester) and then use the bioslurry as a fertilizer. Biogas produced from organic waste is lauded as an ideal sustainable source of household energy. Biogas is viewed as having synergies with recycling food waste and mitigating deforestation, but upfront capital costs hinder adoption (Phimister et al., 2014).

Food-related technology interventions, albeit less common in WEF nexus studies, focused on issues such as land degradation, large-scale land investments, sustainable agriculture, climate change, agroforestry, and land tenure. Impacts of large-scale land investments were found to exacerbate water availability, dominate on-farm energy use, and affect local food security (Boote et al., 2014). Converting large forest and savanna areas to cropland, combined with intensification practices, has affected water quality by enhancing nutrient exports to riverine systems and increasing water shortages (Tian et al., 2018). Land tenure systems are a common barrier to adopting irrigation technologies among smallholder farmers (Jobbins et al., 2015). Land-use conflicts, for example, between pastoralist and agricultural communities in the Sahelian band, have resulted in an increased scarcity of resources such as fertile land and water (Ferrini and Benavides, 2020). Climate change will challenge sustainable agriculture directly through changes in weather, seasonality, and extreme events. African climate change policies have also recognized land use and land change (LULC) as one of the greatest contributors to GHG emissions and a threat to agricultural and environmental sustainability (Conway et al., 2015; Gogoi et al., 2019). For example, Rademacher-Schulz and Mahama (2012) noted in their WEF nexus study that "the decline in crop production for own consumption; shifts in the rainy season; unemployment; longer drought periods followed by unreliable harvest; and increase in drought frequency" result in households using migration as a risk management strategy in Ghana.

In summary, the literature review of WEF nexus studies in Africa illustrates comparatively less "nexus thinking" has been carried on food (agricultural) specific production technologies per se than both water and energy technologies. In addition, the literature shows most studies focused on large-scale interventions. Another observation is that most papers show trade-offs and less so synergies. An example, among many, is where the introduction of higher-yielding water extraction pumps in Ethiopia focused on water implications but did not address energy use trade-offs between those pumps and more typically used conventional centrifugal pumps (Gebregziabher et al., 2014; Mottaleb, 2018). A few studies did provide examples of synergies but were few and far between: for example, using biodigesters for energy use and then using their waste for fertilizer (Smith et al., 2015) or how the use of alien invasive plants for biomass to make charcoal support biodiversity, incentivize communities to remove these plants, and support a greater circular economy (Hoffman et al., 2017). But even in these papers, the question lacking is how to ensure smallholder adoption of these synergetic activities.

4. Farmer technology adoption

A common and driving theory in agricultural research is that new practices and innovative technologies will result in technology adoption and subsequently intensive farming practices for improving land and labor productivity, leading to labor-saving technologies and excess rural labor, and thus enabling the promotion of rural off-farm jobs, rural income, urbanization, and industrialization (Boserup, 1965; Binswanger and Ruttan, 1978; Ruthenberg, 1980; Diao et al., 2008; Barrett et al., 2017). A manifestation of the theory (aka Boserup hypothesis) was the Green Revolution in the 1960s where advances in crop breeding led to a successful push for high yield variety adoption in Asia and

Latin America, saving millions of lives (Pingali, 2012). However, recent studies show there was an uneven distribution of benefits from the Green Revolution across Asia and that the focus on staple "starch" crops has had negative impacts on health and nutrition (Moritzer, 2008; DeFries et al., 2018). During the Green Revolution, adoption efforts were implemented through the introduction of "technological fixes," such as "packages," for farmers (Lipton, 1989). Packages could include parts of or all of the following: improved seed and cultivars, fertilizers, irrigation, pest and disease control, access to outside support, credit, and finance. Assessing reasons for lack of progress in African agriculture development compared with other regions are numerous (including colonization, corruption, etc.), and the intensification technologies and "packages" of the Green Revolution are less suited to the environmental and social conditions of Africa (McIntyre et al., 2009).

It is well known that adopting an agricultural technology or multiple technologies is not treated by the farmer as a unique decision but is bundled with other technological as well as socioeconomic considerations. Farmers make choices based on numerous observed and unobserved preferences. There is a vast literature on factors influencing farmer adoption of various production-enhancing technologies. These include agroecological conditions, tenure, farm size, labor availability, credit and supply constraints, information exchange links, value chains, and institutions, policies, and knowledge exposure (Feder et al., 1985; Shively, 1997; Khanna, 2001; Pattanayak et al., 2003; Mercer, 2004; Jones et al., 2019; Koundouri et al., 2006; Moser and Barrett, 2006; Doss, 2006; Kassie et al., 2013; Maertens and Barrett, 2013; Meijer et al., 2014; Sheahan and Barrett, 2017; De Janvry et al., 2017; Mukasa, 2018; Zimmer et al., 2018; Kebebe, 2019). Risk aversion by farmers is very pertinent, especially upfront investment (capital, and labor) costs, with benefits much later, which are common in environmental-related technology adoption. Traditional metrics such as yields, productivity, farm incomes, employment, or trade may not be seen for years, and on-farm financial decisions are certainly influenced by off-farm considerations (Giller et al., 2011). Typically, these adoption studies are done ex post through experimental and nonexperimental design. Experimental design constructs the counterfactual through random assignment of individuals to treatment and control groups. Without the ability to create treatment groups, nonexperimental design uses tools such as "difference and differences," propensity score matching, regression discontinuity, instrumental variables, and panel data. Regardless of the method, it is recognized that there is much unobserved behavior, even at the farmer's plot level (Bulte et al., 2014).

Although farm technology impacts were traditionally assessed using simple and measurable criteria such as production, productivity, farm incomes, employment, or trade, more recent methods integrate farmer preferences and behavior through more holistic approaches. Various schools of thought promoting technologies for sustainable farming systems include sustainable intensification (SI), conservation agriculture (CA), climate-smart agriculture (CSA), and agroecology. These approaches are commonly touted as ways to meet growing food demand while simultaneously minimizing environmental impacts (Campbell et al., 2014; Loos et al., 2014; Pretty and Bharucha, 2014; Baudron et al., 2012; Lipper, 2014; Vanlauwe et al., 2014; Altieri et al., 2015; van Ittersum et al., 2016; Aggarwal et al., 2018; Pretty, 2018; Koch et al., 2019; Mutenje et al., 2019 Loconto et al., 2020; Jat et al., 2020). Although claiming to provide pathways and processes for food security, these approaches are not without critiques. CA technologies such as no-till agriculture, cover crops, and maize pits have had relatively wide promotion but very little uptake in Africa (Stevenson et al., 2019). Likewise, agroecology also focuses on "principles" at the field level, namely, crop diversification, intercropping, agroforestry, integrating crop and livestock, and soil management measures and has shown positive outcomes and is comprehensive in its approach but also has sporadic uptake by farmers (Kerr et al., 2021). Typical adoption issues with these CA and agroecology practices include crop residue competition, labor demand, and lack of external inputs (Giller et al. 2009, 2015; Lahmar et al., 2012, Knowler, 2015). SI is criticized for being too narrowly focused on food production and not wider considerations such as food accessibility (Loos et al., 2014). Others note the lack of research prioritization and urgency in SI meeting global food systems goals by 2050 (Cassman and Grassini, 2020). CSA aims to meet the food security goals of sustainable agricultural intensification in the context of climate change, but positive impacts are often short-lived, and one study shows that any gains revert to their previous condition or worse (Jagustović et al., 2021).

Notwithstanding the conceptual relevance of and the nuances between these schools of thought, in practice, many of these seemingly holistic approaches still focus on specific or singular agricultural objectives without considering whole-farm systems and the diverse socioecological issues for constructive decision-making. As an example, Amadu et al. (2020) argue that "CSA still uses broad categories like soil and water conservation and erosion control without considering whole farm and external considerations. In so doing CSA reflects specificity rather than understanding the full diversity of CSA practices across contexts in turn limits adoption impacts on resilience beyond the local context." Missing from many of these analyses, however, is assessing and recognizing specific trade-offs that farmers face, especially for water and energy use decisions. Promotion of SI, CA, CSA, and agroecology technologies says little about farmer choices and trade-offs in the wider context, especially off-farm considerations, and seldom discusses factors that influence adoption in one sector and its impact on another sector. Understanding underlying causes and interactions of vulnerability, risk aversion, and power dynamics is seen as a prerequisite to introducing new technologies.

The decision by a farmer to adopt a technology is an adaptive and dynamic process, constantly moving and changing over time, but it is also highly context-specific (Feola et al., 2015; Bernard et al., 2014). WEF nexus thinking begets the need for understanding their decision-making across biophysical (e.g., soil organic matter, rainfall, physical yields) and social/institutional (e.g., household income, consumption, markets, tenure, education, labor use, power relations, gender) conditions, and circumstances faced by farmers (Feola and Binder, 2010; Maertens and Barrett, 2013; Jain et al., 2015). Intertemporal decision-making (i.e., trade-offs) can be captured explicitly through comprehensive WEF nexus frameworks (Mabhaudhi et al., 2019). A holistic approach to promoting technological interventions—in which comparative economic advantage and environmental repercussion on society, local communities, and farmers are considered—is necessary.

5. Research designs for incorporating a priori assessment

This section discusses some broad design approaches for incorporating a priori assessments of farmers' agricultural technology adoption decision-making process within a WEF nexus framework. Three key threads led to this design thinking, namely, (1) WEF nexus studies to date focus on larger-scale issues than at the local household/farmer level, (2) the lack of clear trade-off analysis in most current agricultural interventions approaches and methods such as CSA and SI, and (3) most assessments of farmer adoption are ex post monitoring and/or impact evaluations.

Data requirements will vary according to the WEF nexus issues being tackled, but its availability is essential for evaluating trade-offs and synergies across sectors (Termeer et al., 2010; Landauer et al., 2018). In part due to more data available at larger scales, very few WEF nexus studies have directly looked at farmer/household trade-offs. Africa is known as a data-poor continent, and furthermore, data generated for nexus modeling are often scattered across networks or institutions and rarely consolidated. For an agricultural technology that would require farmer adoption, data must be collected and modeled in the farmer's context-specific social landscape bounded by other actors, which include extension agents, rural development agents, local authorities, or agribusinesses. Numerous WEF nexus methodologies discussed in the following can be used to engage farmers and their key networks. Recent efforts using theories of change approaches synthesize available biophysical and socioeconomic data at relevant scales (Nhamo et al., 2020; Naidoo et al., 2021). A first step is collecting data using well-recognized mixed methods such as key informants, focus groups and surveys, and/or agent-based modeling that includes Delphi process expert panels. The next step is integrating the data in system models to develop indices that can be benchmarked and used to assess trade-offs. Examples of these approaches include WEF tools Composite Index (Simpson and Jewitt, 2019) and trade-offs and analytic hierarchy process in a multicriteria decision-making (MCDM) process (Mabhaudhi et al., 2019). These models and resulting indices and outputs link relationships among WEF resources and provide a vivid synopsis of the state of WEF resources and their trade-offs to help decision-makers, but as a next step, they should be made accessible to all stakeholders, especially farmers, in their process of considering adoption of agricultural technologies.

While farmer decision-making is influenced by larger watershed, regional, and international factors, it is imperative that WEF nexus studies contextualize to local-level farmers and their households' constraints and circumstances. Adapting and simplifying WEF nexus outputs, sharing them with individual farmers at their household level, and receiving feedback for them are often missing steps in the process of designing and implementing agricultural projects. Once WEF nexus outputs are adapted to the farmer context, before project implementation, they can be shared through meetings, house, and farm visits (through extension agencies or NGOs for example) that explain how the trade-offs can impact their land uses and livelihoods. Not only should farmers be able to see and discuss how the trade-offs and expected outcomes will impact their water, energy, and food resources, but in this process, project implementers and policymakers can receive important feedback that could revise policies, interventions, and solutions.

6. Conclusion

This chapter was intended to start to discussion that expresses the potential of WEF nexus to support decision-making about technology adoption among smallholder farmers. This argument can be made in any region, and Africa was used given its acute food insecurity and land-use issues. The next step beyond broad design considerations discussed earlier is the "how" and that requires further analysis and case studies. Using a WEF nexus approach in the context of agriculture technology adoption I argue could lead to more successful agricultural transformation in Africa as envisioned in the Green Revolution. A better understanding of African farming systems (including the use of water and energy) and farmer behavior patterns in their social and biophysical environment, at multiple scales, could improve decision support systems and lead to more sustainable land-use management. Policies or projects that introduce new agricultural technologies to promote change require understanding farmer behavior actions beyond the household but within their socio-ecological special and temporal context. Recognizing that Africa is not a homogeneous region, it is critical that local-level solutions and synergies produce opportunities to add to the current body of WEF nexus research. More emphasis should be placed on site-specific studies instead of the plethora of top-down large-scale WEF nexus ones.

Agricultural systems worldwide are already operating beyond the environmental/planetary limits for sustainability (Pretty et al., 2018). An immense challenge is addressing the seemingly conflicting goals of increasing agricultural yields on existing land while, at the same time, using fewer harmful inputs to lessen impacts on the environment (Pretty et al., 2011; Cassman and Grassini, 2020). Given additional challenges posed to farmers, such as climate change and the multitude of different agricultural systems, there is a more urgent need for integrated transdisciplinary thinking that links behavior patterns to the social and biophysical environment and multiple scales. Beyond the farmer-level analysis, WEF nexus offers sustainable pathways to support the transition to a resilient and sustainable rural and urban environment. Africa's demographic crisis, especially its burgeoning youth population, can be both an opportunity and a challenge for agricultural transformation. Rural outmigration impacts agricultural production at regional scales, while youth entrepreneurship represents great opportunities for their engagement in agriculture, food systems, and agribusiness (Mueller and Thurlow, 2019; Thieme and Kovacs, 2015; Millios, 2018). Therefore, greater priority should be placed on WEF nexus research and development for plans and policies related to the adoption and introduction of appropriate technologies, and the smallholder farmer should be the center of that analysis.

References

AfDB, 2021. African Development Bank, Light up and Power Africa — A New Deal on Energy for Africa. Abidjan.

Aggarwal, P.K., Jarvis, A., Campbell, B.M., Zougmoré, R.B., Khatri-chhetri, A., Vermeulen, S.J., 2018. The climate-smart village approach: framework of an integrative strategy. Ecol. Soc. 23 (1).

Albrecht, T., Crootof, A., Scott, C., 2018. The Water-Energy-Food Nexus: a systematic review of methods for nexus assessment. Environ. Res. Lett. 13 (4).

Allam, M., Eltahir, E., 2019. Water-energy-food nexus sustainability in the upper blue nile (UBN) basin. Front. Environ. Sci. 7.

Allan, T., Keulertz, M., Woertz, E., 2015. The water-food-energy nexus: an introduction to nexus concepts and some conceptual and operational problems. Int. J. Water Resour. Dev. 31 (3), 301—311.

Allen, E., et al., 2015b. Land use intensification alters ecosystem multifunctionality via loss of biodiversity and changes to functional composition. Ecol. Lett. 18 (8), 834—843.

Allouche, J., Middleton, C., Gyawali, D., 2018. The Water-Food-Energy Nexus: Power, Politics and Justice. Earthscan-Routledge.

Alston, J., Pardey, P., 2014. Agriculture in the global economy. J. Econ. Perspect. 28 (1), 121—146.

Altieri, M., et al., 2015. Agroecology and the design of climate change-resilient farming systems. Agron. Sustain. Dev. 35, 869–890.

Amadu, F., McNamara, P., Miller, C., 2020. Understanding the adoption of climate-smart agriculture: a farm-level typology with empirical evidence from southern Malawi. World Dev. 126 (C).

AU, 2014. Synthesis of the Malabo Declaration on African Agriculture and CAADP. Addis Ababa.

AU, 2015. African union Agenda 2063: the Africa We Want. Addis Ababa.

AUC-AMCOW, 2016. African Water Resources Management Priority Action Programme 2016–2025. Addis Ababa and Abuja.

Babiker, B., et al., 2019. Nexus Assessment for Sudan: Synergies of the Water, energy, and Food Sectors. Nexus Regional Dialogue Programme, Eschborn.

Barrett, C., et al., 2017. On the Structural Transformation of Rural Africa. Policy Research Working Paper, No. WPS 7938. World Bank Group, Washington, D.C.

Baudron, F., Tittonell, P., Corbeels, M., Letourmy, P., Giller, K.E., 2012. Comparative performance of conservation agriculture and current smallholder farming practices in semi-arid Zimbabwe. Field Crop. Res. 132, 117–128.

Beegle, K., Luc, C., 2019. Accelerating Poverty Reduction in Africa. World Bank, Washington, DC.

Benin, S. (Ed.), 2016. Agricultural Productivity in Africa: Trends, Patterns, and Determinants. International Food Policy Research Institute (IFPRI).

Bernard, F., et al., 2014. Social actors and unsustainability of agriculture. Curr. Opin. Environ. Sustain. 6, 155–161.

Besada, H., Werner, K., 2015. An assessment of the effects of Africa's water crisis on food security and management. Int. J. Water Resour. Dev. 31, 120–133.

Binswanger, H., Ruttan, V., 1978. Induced Innovation: Technology, Institutions and Development. The Johns Hopkins University Press, Baltimore, MD.

Boote, D.N., Smither, J.C., Lyne, P.W., 2014. The development and application of an energy calculator for sugarcane production in South Africa. Proc. Annu. Congr. South Afr. Sugar Technologists' Assoc. 87, 459–463.

Boserup, E., 1965. The Conditions of Agricultural Growth. Aldine, Chicago.

Bulte, E., et al., 2014. Behavioral responses and the impact of new agricultural technologies: evidence from a double-blind field experiment in Tanzania. Am. J. Agric. Econ. 96 (3), 813–830.

Campbell, B.M., et al., 2014. Sustainable intensification: what is its role in climate smart agriculture? Curr. Opin. Environ. Sustain. 8, 39–43.

Cash, D., et al., 2006. Scale and cross-scale dynamics: governance and information in a multilevel world. Ecol. Soc. 11 (2), 8.

Cassman, K.G., Grassini, P., 2020. A global perspective on sustainable intensification research. Nat. Sustain. 3, 262–268.

Chahed, J., Besbes, M., Hamdane, A., 2015. Virtual-water content of agricultural production and food trade balance of Tunisia. Int. J. Water Resour. Dev. 31 (3), 407–421.

Conway, D., et al., 2015. Climate and southern Africa's water–energy–food nexus. Nat. Clim. Change 5, 837–846.

De Janvry, A., Sadoulet, E., Suri, T., 2017. Field experiments in developing country agriculture. In: Handbook of Economic Field Experiments, vol. 2. Elsevier, pp. 427–466.

De Sherbinin, A., 2014. Climate change hotspots mapping: what have we learned? Clim. Change 123 (1), 23–37.

DeFries, R., et al., 2018. Impact of historical changes in coarse cereals consumption in India on micronutrient intake and anemia prevalence. Food Nutr. Bull. 39 (3), 377–392.

Diao, X., Hazell, P.B.R., Resnick, D., Thurlow, J., 2007. The Role of Agriculture in Development: Implications for Sub-Saharan Africa. Research Reports 153. International Food Policy Research Institute (IFPRI).

Diao, X., Headey, D., Johnson, M., 2008. Toward a green revolution in Africa: what would it achieve, and what would it require? Agric. Econ. 39, 539–550.

Dixon, J., et al., 2001. A. Farming Systems and Poverty: Improving Farmers' Livelihoods in a Changing World. FAO, Rome, Italy.

Dos Santos, S., et al., 2017. Urban growth and water access in sub-Saharan Africa: progress, challenges, and emerging research directions. Sci. Total Environ. 607, 497–508.

Doss, C., 2006. Analyzing technology adoption using microstudies: limitations, challenges, and opportunities for improvement. Agric. Econ. 34, 207–219.

Doukkali, M.R., Lejars, C., 2015. Energy cost of irrigation policy in Morocco: a social accounting matrix assessment. Int. J. Water Resour. Dev. 31 (3), 422–435.

Evenson, R.E., Gollin, D., 2003. Assessing the impact of the green revolution, 1960 to 2000. Science 300 (5620), 758–762.

Feder, G., Just, R., Zilberman, D., 1985. Adoption of agricultural innovations in developing countries: a survey. Econ. Dev. Cult. Change 33 (2), 255–298.

Feola, G., Binder, C., 2010. Towards an improved understanding of farmers' behavior: the integrative agent-centered (IAC) framework. Ecol. Econ. 69 (12), 2323–2333.

Feola, G., et al., 2015. Researching farmer behaviour in climate change adaptation and sustainable agriculture: lessons learned from five case studies. J. Rural Stud. 39, 74–84.

Ferrini, L., Benavides, L., 2020. Nexus Blog: Opportunities for Mutual Learning? the Grand Ethiopian Renaissance Dam and Fomi Dam: Transboundary Relations and Nexus Thinking across the Sahel. Accessed at: https://www.water-energy-food.org/news/nexus-blog-opportunities-for-mutual-learning-the-grand-ethiopian-renaissance-dam-and-fomi-dams-transboundary-relations-and-nexus-thinking-across-the-sahel.

Flammini, A., et al., 2014. Walking the Nexus Talk: Assessing the Water–Energy–Food Nexus. Food and Agriculture Organization of the United Nations.

Gebregziabher, G., Giordano, M., Langan, S., Regassa, N.,E., 2014. Economic analysis of factors influencing adoption of motor pumps in Ethiopia. J. Dev. Agric. Econ. 6 (12), 490–500.

Giller, K.E., et al., 2009. Conservation agriculture and smallholder farming in Africa: the heretics' view. Field Crop. Res. 114 (1), 23–34.

Giller, K.E., Tittonell, P., Rufino, M., van Wijk, M., 2011. Communicating complexity: integrated assessment of tradeoffs concerning soil fertility management within African farming systems to support innovation and development. Agric. Syst. 104 (2), 191–203.

Giller, K.E., Andersson, J.A., Corbeels, M., Kirkegaard, J., Mortensen, D., Erenstein, O., Vanlauwe, B., 2015. Beyond conservation agriculture. Front. Plant Sci. 6 (October).

Giordano, M., Shah, T., 2014. From IWRM back to integrated water resources management. Int. J. Water Resour. Dev. 30 (3), 364–376.

Gogoi, P.P., Vinoj, V., Swain, D., et al., 2019. Land use and land cover change effect on surface temperature over Eastern India. Sci. Rep. 9, 8859.

Hailemariam, W., et al., 2019. Water–energy–food nexus of sugarcane production in Ethiopia. Environ. Eng. Sci. 36 (7), 798–807.

HLPE, 2013. Investing in Small Holder Agriculture for Food Security. A Report by the High Level Panel of Experts on Food Security and Nutrition of the Committee of World Food Security. Rome.

Hoff, H., 2011. Understanding the Nexus. Background Paper for the Bonn 2011 Conference: The Water, Energy and Food Security Nexus. Stockholm Environment Institute, Stockholm.

Hoffmann, H.K., Sander, K., Brüntrup, M., Sieber, S., 2017. Applying the Water-Energy-Food Nexus to the Charcoal Value Chain. Front. Environ. Sci. 5. https://doi.org/10.3389/fenvs.2017.00084.

IEA, 2014. Africa Energy Outlook: A Focus on energy Prospects in Sub-saharan Africa. International Energy Agency (IEA), World Energy Outlook Special Report, Paris, France.

Imasiku, K., Ntagwirumugara, E., 2020. An impact analysis of population growth on energy- water-food-land nexus for ecological sustainable development in Rwanda. Food Energy Secur. 9 (1).

Jacobson, M., Pekarcik, G., 2021. Water energy Food Nexus in Africa: A Literature Review Forthcoming in B. Floor Handbook on the Water-energy-Food Nexus. Edward Elgar Publishing.

Jagustović, R., Papachristos, G., Zougmoré, R., Kotir, J., Kessler, A., Ouédraogo, M., Ritsema, C., Dittmer, K., 2021. Better before worse trajectories in food systems? An investigation of synergies and tradeoffs through climate-smart agriculture and system dynamics. Agric. Syst. 190.

Jain, L., Kumar, H., Singla, R.K., 2015. Assessing mobile technology usage for knowledge dissemination among farmers in Punjab. Inf. Technol. Dev. 21 (4), 668–676.

Jat, M.L., et al., 2020. Conservation agriculture for sustainable intensification in South Asia. Nat. Sustain. 3, 336–343.

Jobbins, G., et al., 2015. To what end? Drip irrigation and the water–energy–food nexus in Morocco. Int. J. Water Resour. Dev. 31 (3), 393–406.

Johnson, O., et al., 2018. In: Kemsey, J. (Ed.), Exploring the Water-Energy-Food Nexus in Rwanda's Akagara Basin. Stockholm Environment Institute, Stockholm, Sweden.

Jones, M., et al., 2019. Factor Market Failures and the Adoption of Irrigation in Rwanda. Policy Research Working Paper Series No. 9092. The World Bank.

Kassie, M., et al., 2013. Adoption of interrelated sustainable agricultural practices in smallholder systems: evidence from rural Tanzania. Technol. Forecast. Soc. Change 80 (3), 525–540.

Kebebe, E., 2019. Bridging technology adoption gaps in livestock sector in Ethiopia: a innovation system perspective. Technol. Soc. 57 (November 2018), 30–37.

Kerr, R.B., Madsen, S., Stüber, M., Liebert, J., Enloe, S., 2021. Can agroecology improve food security and nutrition? A review. Global Food Sec. 29.

Keskes, T., Zahar, H., Ghezal, A., Bedoui, K., 2019. Impact of Solar Pumping Systems in Tunisia. The Nexus Regional Dialogue Programme.

Khanna, M., 2001. Sequential adoption of site-specific technologies and its implications for nitrogen productivity: a double selectivity model. Am. J. Agric. Econ. 83 (1).

Knowler, D., 2015. Farmer adoption of conservation agriculture: a review and update. In: Farooq, M., Siddique, K.H.M. (Eds.), Conservation Agriculture. Springer, Berlin, pp. 621–642.

Koch, J., Schaldach, R., Göpel, J., 2019. Can agricultural intensification help to conserve biodiversity? A scenario study for the African continent. J. Environ. Manag. 247, 29–37.

Kougias, I., et al., 2018. Water-Energy-Food Nexus Interactions Assessment: Renewable energy Sources to Support Water Access and Quality in West Africa. Luxembourg, European Commission.

Koundouri, P., Nauges, C., Tzouvelekas, V., 2006. Technology adoption under production uncertainty: theory and application to irrigation technology. Am. J. Agric. Econ. 88 (3), 657–670.

Lahmar, R., Bationo, B., Dan Lamso, N., Guéro, Y., Tittonell, P., 2012. Tailoring conservation agriculture technologies to West Africa semi-arid zones: Building on traditional local practices for soil restoration. Field Crop. Res. 132, 158–167.

Landauer, M., Juhola, M., Klein, J., 2018. The role of scale in integrating climate change adaptation and mitigation in cities. J. Environ. Plann. Manag. 62 (5), 741–765.

Leakey, R., Mabhaudhi, T., Gurib-Fakim, A., 2021. African lives matter: wild food plants matter for livelihoods, justice, and the environment—a policy Brief for agricultural reform and new crops. Sustainability 13, 7252.

Lipper, L., et al., 2014. Climate-smart agriculture for food security. Nat. Clim. Change 4 (12), 1068—1072.

Lipton, M., 1989. New Seeds and Poor People. The Johns Hopkins University Press, Baltimore, MD.

Liu, J., et al., 2017. Challenges in operationalizing the water—energy—food nexus. Hydrol. Sci. J. 62 (11), 1714—1720.

Loconto, A., Desquilbet, M., Moreau, T., Couvet, D., Dorin, B., 2020. The land sparing — land sharing controversy: tracing the politics of knowledge. Land Use Pol. 96 (May), 1—13.

Loos, J., et al., 2014. Putting meaning back into "sustainable intensification". Front. Ecol. Environ. 12 (6), 356—361.

Mabhaudhi, T., et al., 2018. Prospects for improving irrigated agriculture in southern Africa: linking water, energy and food. Water 10 (12), 1881.

Mabhaudhi, T., et al., 2019. The water—energy—food nexus as a tool to transform rural livelihoods and well-being in southern Africa. Int. J. Environ. Res. Publ. Health 16 (16), 2970.

Maertens, A., Barrett, C.B., 2013. Measuring social networks' effects on agricultural technology adoption; measuring social networks' effects on agricultural technology adoption. Am. J. Agric. Econ. 95 (2), 353—359.

Martins, R., 2018. Nexusing charcoal in south Mozambique: a proposal to integrate the nexus charcoal-food-water analysis with a participatory analytical and systemic tool. Front. Environ. Sci. 6. https://doi.org/10.3389/fenvs.2018.00031.

McCarl, B., et al., 2017. Data for WEF nexus analysis: a review of issues. In: Current Sustainable/ Renewable Energy Reports, vol. 4, pp. 137—143.

McCornick, P., Awulachew, S., Abebe, M., 2008. Water—food—energy—environment synergies and tradeoffs: major issues and case studies. Water Pol. 10 (S1), 23—36.

McGrane, S., et al., 2018. Scaling the nexus: towards integrated frameworks for analysing water, energy and food. Geograph J. ISSN 14754959.

McIntyre, B.D., Herren, H.R., Wakhungu, J., Watson, R.T., 2009. Agriculture at a Crossroads. International Assessment of Agricultural Knowledge, Science and Technology for Development (IAASTD): Global Report. Synthesis Report. Island Press, Washington DC, p. 590.

Meijer, S., et al., 2014. The role of knowledge, attitudes and perceptions in the uptake of agricultural and agroforestry innovations among smallholder farmers in sub-Saharan Africa. Int. J. Agric. Sustain. 13.

Mekonnen, D., et al., 2017. Food versus fuel: examining tradeoffs in the allocation of biomass energy sources to domestic and productive uses in Ethiopia. Agric. Econ. 48 (4), 425—435.

Mercer, D., 2004. Adoption of agroforestry innovations in the tropics: a review. Agrofor. Syst. 61, 311—328.

Millios, L., 2018. Advancing to a Circular Economy: three essential ingredients for a comprehensive policy mix. Sustain. Sci. 13 (3), 861—878.

Moritzer, E., 2008. Curse or blessing? Kunststoffe Int. 98 (3), 38—39.

Moser, C., Barrett, C., 2006. The complex dynamics of smallholder technology adoption: the case of SRI in Madagascar. Agric. Econ. 35, 373—388.

Mottaleb, K.A., 2018. Perception and adoption of a new agricultural technology: evidence from a developing country. Technol. Soc. 55, 126—135.

Mueller, V., Thurlow, J. (Eds.), 2019. Youth and Jobs in rural Africa: Beyond Stylized Facts. International Food Policy Research Institute (IFPRI) and Oxford University Press, New York, NY.

Mukasa, A.N., 2018. Technology adoption and risk exposure among smallholder farmers: panel data evidence from Tanzania and Uganda. In: World Development, vol. 105(C). Elsevier, pp. 299–309.

Mutenje, M., et al., 2019. A cost-benefit analysis of climate-smart agriculture options in Southern Africa: balancing gender and technology. Ecol. Econ. 163.

Naidoo, D., Nhamo, L., Mpandeli, S., Sobratee, N., Senzanje, A., Liphadzi, S., Slotow, R., Jacobson, M., Modi, A.T., Mabhaudhi, T., 2021. Operationalising the water-energy-food nexus through the theory of change. Renew. Sustain. Energy Rev. 149, 111416.

Nhamo, L., et al., 2018. The water-energy-food nexus: climate risks and opportunities in southern Africa. Water 10 (5), 567.

Niang, I., et al., 2014. Africa. In: Climate Change 2014: Impacts, Adaptation and Vulnerability, Contribution of Working Group II to the Fifth Assessment Report of the Intergovernmental Panel on Climate Change. Cambridge University Press, Cambridge.

Ozturk, I., 2017. The dynamic relationship between agricultural sustainability and food-energy-water poverty in a panel of selected sub-Saharan African countries. Energy Pol. 107, 289–299.

Pardey, P., Chan-Kang, C., Dehmer, S., Beddow, J., 2016. Agricultural R&D is on the move. Nature 537, 301–303. https://doi.org/10.1038/537301a.

Pattanayak, S., et al., 2003. Taking stock of agroforestry adoption studies. Agrofor. Syst. 57, 173–186.

Phimister, E., et al., 2014. How Can We ensure better Use of Organic Waste Materials for Food, Energy Production and Water Use in Sub-Saharan Africa? Nexus Network Thinkpiece Series.

Pingali, P.L., 2012. Green revolution: impacts, limits, and the path ahead. Proc. Natl. Acad. Sci. U.S.A 109 (31), 12302–12308.

Prasad, G., et al., 2012. Towards the development of an energy-water-food security nexus-based modelling framework as policy and planning tool for South Africa. In: Strategies to Overcome Poverty and Inequality Conference. University of Cape Town, Cape Town.

Pretty, J., Bharucha, Z.P., 2014. Sustainable intensification in agricultural systems. Ann. Bot. 114 (8), 1571–1596.

Pretty, J., Toulmin, C., Williams, S., 2011. Sustainable intensification in African agriculture. Int. J. Agric. Sustain. 9 (1), 5–24.

Pretty, J., Benton, T.G., Bharucha, Z.P., Dicks, L.V., Flora, C.B., Godfray, H.C.J., 2018. Global assessment of agricultural system redesign for sustainable intensification. Nat. Sustain. 1 (8), 441–446.

Rademacher-Schulz, C., Mahama, E.S., 2012. Rainfall, Food Security, and Human Mobility: Case Study Ghana, vol. 3. UNU-EHS, Bonn, Germany.

Rockström, J., Falkenmark, M., 2015. Agriculture: increase water harvesting in Africa. Nature 519, 283–285.

Ruthenberg, H., 1980. Farming Systems in the Tropics, third ed. Clarendon Press, Oxford.

Sahle, M., et al., 2019. Quantifying and mapping of water-related ecosystem services for enhancing the security of the food-water-energy nexus in tropical data-sparse catchment. Sci. Total Environ. 646, 573–586.

Sheahan, M., Barrett, C.B., 2017. Ten striking facts about agricultural input use in Sub-Saharan Africa. Food Pol. 67, 12–25.

Shively, G., 1997. Consumption risk, farm characteristics, and soil conservation adoption among low-income farmers in the Philippines. Agric. Econ. 17 (2–3), 165–177.

Simpson, G., Jewitt, G., 2019. The development of the water-energy-food nexus as a framework for achieving resource security: a review. Front. Environ. Sci. 7.

Smith, J.U., Fischer, A., Hallett, P.D., Homans, H.Y., Smith, P., Abdul-Salam, Y., Emmerling, H.H., Phimister, E., 2015. Sustainable use of organic resources for bioenergy, food and water provision in rural sub-Saharan Africa. Renew. Sustain. Energy Rev. 50, 903–917. https://doi.org/10.1016/j.rser.2015.04.071.

Stein, C., Pahl-Wostl, C., Barron, J., 2018. Towards a relational understanding of the water-energy-food nexus: an analysis of embeddedness and governance in the Upper Blue Nile region of Ethiopia. Environ. Sci. Pol. 90, 173–182.

Stevenson, J., et al., 2019. Farmer adoption of plot- and farm-level natural resource management practices: between rhetoric and reality. Global Food Sec. 20, 101–104.

Szabó, S., et al., 2016. Identification of advantageous electricity generation options in sub-Saharan Africa integrating existing resources. Nat. Energy 1, 16140.

Termeer, C., Dewulf, A., Lieshout, M., 2010. Disentangling scale approaches in governance research: comparing monocentric, multilevel, and adaptive governance. Ecol. Soc. 15 (4), pp. 29–29.

Terrapon-Pfaff, J., et al., 2018. Energising the WEF nexus to enhance sustainable development at local level. J. Environ. Manag. 223, 409–416.

Thieme, T., Kovacs, E., 2015. Services and Slums: Rethinking Infrastructures and Provisioning across the Nexus. Nexus Network Thinkpiece Series.

Tian, H., et al., 2018. Optimizing resource use efficiencies in the food–energy–water nexus for sustainable agriculture: from conceptual model to decision support system. Curr. Opin. Environ. Sustain. 33, 104–113.

Timmer, P., Akkus, S., 2008. The Structural Transformation as a Pathway Out of Poverty: Analytics, Empirics and Politics. Centre for Global Development. Working Paper No. 150.

Udias, A., et al., 2018. A decision support tool to enhance agricultural growth in the Mékrou river basin (West Africa). Comput. Electron. Agric. 154, 467–481.

Van Ittersum, M.K., et al., 2016. Can sub-Saharan Africa feed itself? Proc. Nat. Acad. Sci. USA 113 (52), 14964–14969.

Vanlauwe, B., Coyne, D., Gockowski, J., Hauser, S., Huising, J., Masso, C., Nziguheba, G., Schut, M., Van Asten, P., 2014. Sustainable intensification and the African smallholder farmer. Curr. Opin. Environ. Sustain. 8, 15–22.

Weitz, N., et al., 2017. Closing the governance gaps in the water-energy-food nexus: insights from integrative environmental governance. Global Environ. Change 45, 165–173.

Wichelns, D., 2017. The water-energy-food nexus: is the increasing attention warranted, from either a research or policy perspective? Environ. Sci. Pol. 69, 113–123.

World Bank, 2007. World Development Report 2008. Agriculture for development, Washington, D.C.

Xie, H., Ringler, C., You, L., 2020. Last mile energy access for productive energy use in agriculture in Sub-Saharan Africa: what and where is the potential?. In: Presented at the AGU in San Francisco, CA, 09-13 December 2019.

Zimmer, H.C., et al., 2018. Why do farmers still grow corn on steep slopes in northwest Vietnam? Agrofor. Syst. 92, 1721–1735.

CHAPTER 16

Building capacity for upscaling the WEF nexus and guiding transformational change in Africa

Tendai P. Chibarabada[1], Goden Mabaya[1], Luxon Nhamo[2], Sylvester Mpandeli[2], Stanley Liphadzi[2], Krasposy K. Kujinga[1], Jean-Marie Kileshye-Onema[1,3], Hodson Makurira[4], Dhesigen Naidoo[2] and Michael G. Jacobson[5]

[1]Waternet, Mt Pleasant, Harare, Zimbabwe; [2]Water Research Commission of South Africa (WRC), Pretoria, South Africa; [3]School of Industrial Engineers, University of Lubumbashi, Lubumbashi, DR Congo; [4]Department of Construction and Civil Engineering, University of Zimbabwe, Harare Zimbabwe; [5]School of Forest Resources, Department of Ecosystem Science and Management, Pennsylvania State University, University Park, PA, United States

1. Introduction

The water—energy—food (WEF) nexus has been gaining prominence since its presentation at the World Economic Forum in 2011 by the Stockholm International Water Institute (SIWI) (Hoff, 2011). The approach came to the fore as an innovative approach that leads to sustainable development through the ability to integrate distinct but interlinked sectors, thus creating opportunities for the water, energy, and agriculture sectors to harmonize their activities and minimize duplications (Nhamo et al., 2020; Terrapon-Pfaff et al., 2018). In essence, nexus planning has also demonstrated its potential to minimize the negative impacts that individual sectors on their own can have on socioecological systems, especially during this time when consumerism and waste production trends have escalated (Arora and Gagneja, 2020). The uniqueness of the WEF nexus as an integrated approach is embodied in its distinctive essence, promoting long-lasting transformational change and sustainable development (Nhamo et al., 2018). The approach shares its holistic ability by promoting integrated management of resources and their access by all for a fitting human sustainable development (Liu et al., 2018; Nhamo et al., 2019). This is of uttermost importance as access to clean water, modern and clean energy services, and nutritious and sufficient food is at the very center of the fight against global poverty and the efficient implementation and attainment of the Sustainable Development Goals (SDGs) (UNGA, 2015). Given this importance, the WEF nexus is envisaged to play a pivotal role in poverty reduction and achieve the 2030 global agenda for sustainable development and Africa Union's agenda 2063. It is assumed that the WEF nexus will allow the sectors to drive their activities in "integrated" ways.

There is an urgent need to improve existing nexus planning analytical models, provide practical and empirical evidence, enhance uptake at a policy level, and promote capacity development at all levels from curriculum development to practitioner, policymaking, and institutional arrangements (Hoff et al., 2019; Wehn et al., 2021). COVID-19 pandemic experiences have demonstrated that focusing on a single sector during a crisis only exacerbates other sectors' stressors (Nhamo and Ndlela, 2021). Decision-makers have traditionally viewed the world from a linear perspective, thinking that a click of a button would solve existing challenges and get everything back on track (Fogarassy and Finger, 2020; Jørgensen and Pedersen, 2018). However, recent experiences have demonstrated the need for a shift from the norm of pursuing linear models to adopting and implementing circular and transformative approaches that provide long-lasting and integrated pathways toward sustainability (Salvioni and Almici, 2020). The shift from linear to circular modeling involves creating an enabling environment and capacity development to promote uptake and operationalization (Shilomboleni and Plaen, 2019). The shift is necessitated since linearity often forgets the interconnectedness of systems and how their systemic properties shape their interactions, interdependencies, and interrelationships and improve the resource use efficiency while minimizing the environment's degradation. Circularity (circular economy, nexus planning, sustainable food systems, and scenario planning), while a seemingly slow process, integrates and simplifies socioecological systems, indicates priority areas for intervention, and reduces risk and vulnerability (Nhamo and Ndlela, 2021).

The World Economic Forum in its global risk report of January 2021 identified the livelihoods crises as one of the top threats to the world as it speeds up the social cohesion erosion (World Economic Forum, 2011). The adoption of nexus planning in resource management is a value addition in finding solutions to the cross-cutting challenges and the related shocks facing humankind. The fundamental essence of the WEF nexus is its ability to highlight synergies and trade-offs, thus informing the best interventions and promoting the use of other transformative approaches such as scenario planning and circular economy (Mpandeli et al., 2018). As the approach enhances and secures the three intricately connected resources (water, energy, and food) in an integrated way, it is critical to ensure accessibility and affordability of essential and basic resources to all and ensure a balanced socioeconomic development (Hoff, 2011). Thus, the WEF nexus informs transformational change that improves the livelihoods of usually disadvantaged communities in the context of eliminating inequality and reducing poverty (Hoff et al., 2019;

Mabhaudhi et al., 2019). Capacity building through skills training and development of models has been observed across the water, food, and energy sectors, but there is not much evidence that these efforts have adopted the nexus approach and, hence, may not assist to optimize the benefits of such investments.

This chapter discusses pathways to upscaling and developing capacity in WEF nexus perspectives to achieve sustainable resources use and management. Upscaling and developing capacity are envisaged to benefit informed adoption and operationalization of the WEF nexus. Skilled human resources, relevant institutions, and an improved understanding of the intricate interlinkages among resources are critical to innovation and developing appropriate pathways for operationalizing the nexus concept.

2. Status of WEF nexus research in Africa

There is a notable increase in annual research related to the WEF nexus produced in the Global South between 2013 and 2020. 2017 and 2018, where the most productive years, reaching the top-notch of 14 published articles in international journals (Botai et al., 2021). The annual percentage growth of WEF-related publications between 2013 and 2020 in Africa is 6%. Although WEF nexus research has been slow on the African continent as compared with other continents (Botai et al., 2021), the few available research publications indicate a steady understanding of the concept, which has resulted in the development of models that are providing empirical evidence and pathways to operationalize the approach (Nhamo et al., 2018, 2019; Mabhaudhi et al., 2019).

2.1 Understanding drivers of change

Food security, human health, urbanization, energy production, industrial development, economic growth, and ecosystems are all water dependent (UNESCO, 2020). Sub-Saharan Africa is home to more than two-thirds of the world's food-insecure population (FAO et al., 2015). The region grapples with all forms of malnutrition and is not on track to achieving SDG 2 (FAO et al., 2015). Agriculture is the main source of livelihood and is challenged by physical and economic water scarcity, shifting rainfall patterns, and higher incidences of droughts and floods (UNESCO, 2020). The Global South also faces inadequate modern energy services, with less than 20% of the population having access to electricity and approximately 80% still using wood energy. The core sources of electricity in the Global South are hydropower, oil, and coal, which are overdependent on water availability (Power et al., 2016). Sub-Saharan Africa has entered an "Urban Age," characterized by rapid urbanization and uncontrolled migration (Chirisa and Bandauko, 2015; Smit, 2016).

Urbanization creates energy demand for industrial production, cooling, and mobility. Urbanization is not met with infrastructure to deliver water and sanitation services (Chirisa and Bandauko, 2015). Environmental activists and scientists are also concerned that many ecosystems, particularly forests and wetlands, are also at risk. The degradation of ecosystems will not only lead to biodiversity loss but also affect the provision of water-related ecosystem services, such as water purification, carbon capture and storage, and natural flood protection, as well as the provision of water for agriculture, fisheries, and recreation (Burkhard et al., 2014; Power et al., 2016). The WEF approach creates the anticipation to implement appropriate tools, knowledge, and legislation to enable ecological restorative justice, especially in communities devastated by extractive sectors.

The basis of any nexus approach is an attempt to address societal challenges while balancing the different ecosystem resources' uses. This requires a better understanding of the dynamic interrelationships between water, food, climate, environment, and energy. Improved food security has dominated most of the WEF nexus studies in Africa (Botai et al., 2021), with irrigation being a common technology for food security in the global south (Jacobson and Pekarcik, 2021). Agriculture yields can be increased by more than 50% through efficient irrigation technologies, thus providing a clear synergy between water and food (Hamidov and Helming, 2020; Mabhaudhi et al., 2018). A primary WEF nexus trade-off of irrigation is the significant energy required to pump and distribute water, which is a problem for most sub-Saharan Africa as energy is scarce (Hamidov and Helming, 2020; Mabhaudhi et al., 2018). Precision irrigation technologies have been advocated for improved water use and higher water efficiency but sometimes have negative energy consequences as they consume more energy than traditional irrigation systems (Hailemariam et al., 2019). This is exacerbated by farmers' lack of knowledge and extension services on irrigation scheduling and maintenance of these technologies (Hailemariam et al., 2019; Jobbágy et al., 2011). Solar, gravity, and wind pumped irrigation systems have been proposed to counteract the issue of energy demand for irrigation, but they may have the unintended consequence of overextraction of water (Jobbins et al., 2015; Serrano-Tovar et al., 2019). Solar and wind pumped systems are also expensive, raising concerns for scaling these technologies (Beaton et al., 2019).

Energy demand in sub-Saharan Africa is increasing because of rapid urbanization and industrialization (Wolde-Rufael, 2005). However, the energy supplies are dynamic as they reflect the region's energy resource endowment. For example, in North and West Africa, the energy supply is driven by oil and gas reserves, while in southern Africa, the energy supply is dominated by hydroelectric power plants (Monyei and Akpeji, 2020). Hydropower captures attention in WEF nexus studies given its importance in energy availability yet faces competing uses for that water, such as irrigation (agriculture) and ecosystem services and health (Dombrowsky and Hensengerth, 2018; Zhang et al., 2018). Hydropower is also sensitive to climate shocks as evidenced by significant reductions in hydropower generation potential in the Southern Africa region in the past 5 years due to power quality rainfall seasons. The accurate determination of optimal water allocation allows for synergies across the water, energy, and food sectors. Another important energy source in the Global South is biomass used in rural areas for cooking (Heinimö and Junginger, 2009; Muller, 2008). A clear trade-off for biomass is environmental degradation, which results in soil erosion, runoff, and siltation.

Research on the WEF nexus has shown a trajectory from a focus on water and food to water and energy and, more recently, holistic frameworks addressing water, energy, food, climate, and the environment (Botai et al., 2021). While there has been a growing understanding of nexus trade-offs and synergies, land-related issues about meeting food and energy security need remain underrepresented. There is a need for WEF nexus research and frameworks to better address land and climate change threats for a more holistic approach to sustainability and promote enhanced uptake, especially at the policy level.

2.2 WEF nexus planning

The interconnectedness of WEF, and the challenges that arise thereof, requires systemic and transformative approaches such as circular economy, scenario planning, sustainable food systems, and nexus planning to manage trade-offs and synergies and achieve sustainability (Nhamo et al., 2019). Nexus planning is an approach used to understand these complex interactions and inform policy on priority interventions that enhance sustainable socioecological outcomes (Ericksen, 2008; Nhamo et al., 2019). Numerous WEF nexus plans have been proposed for the Global South (Bieber et al., 2018; Meyer, 2019; Nhamo et al., 2019). There is no single "nexus," and nexus is about compromise, trade-offs, and synergies, and nexus should make economic and socioeconomic sense.

Nexus planning has had its share of criticisms, particularly for lack of an operational framework (Albrecht et al., 2018). WEF nexus approaches are still hindered by a lack of clarity on an applicable spatial scale, data availability, and operability. Others have even branded it as a repackaging of the integrated water resources management (IWRM) (Allouche et al., 2014; Benson et al., 2015). Studies reveal that institutional and political constraints are deterring efforts to produce WEF system synergies through technological interventions during nexus planning (Daher et al., 2018; van Gevelt, 2020). Developing institutional frameworks that build on existing structures and integrates all nexus principles will be key. Discourse on the WEF nexus planning, especially in Southern Africa, has not sufficiently addressed institutional capacity needs, despite the significance of an enabling institutional environment and governance frameworks for multisectoral planning and decision-making. Hardly any mutual learning occurs across sectors, inhibiting the cocreation of knowledge from developing multisectoral projects that can generate benefits for all three WEF nexus sectors. There is a lack of collaboration and data sharing, while available data are often incomplete or inaccessible.

Therefore, in the context of sub-Saharan Africa's complex geopolitical setting, there is a need for a structured approach based on practical steps to move nexus dialogues into action. There is a need for a nexus planning framework for the Global South to act as a guide to help define nexus problems and subsequently design relevant investment projects. Additionally, Southern African states will have to demonstrate their readiness to adopt relevant agreements, policies, and strategies, which would enable long-term action plans on the WEF nexus. The use, for example, of a Nexus Capacity Score Card (Meyer, 2019), supplemented with information that could be gathered during training workshops, nexus dialogues, nexus study tours, and other stakeholder engagements in Southern Africa, can allow capacity building needs to be identified. While WEF nexus planning can address global challenges and achieve sustainability at a global level, it should, however, be mostly practical at a local scale (household and community) where adaptation and resilience-building processes should take place (Mpandeli et al., 2018) and after that cascade to natural spatial scales (catchments) and jurisdictional scales where policies and governance structures that affect households and communities are formulated and implemented (Landauer et al., 2019).

The temporal scale is equally critical in nexus planning. It is essential in scenario planning (an important component of nexus planning) and for interpreting future climatic and environmental changes, resource availability, or population projections at different time intervals (Bhave et al., 2018). Awareness and knowledge about data analysis tools, such as nexus models, scenario development, and other tools, will be key to enable multisectoral planning and decision-making. To avoid duplication of data collection efforts in the region, there is a need to ensure efficiency in planning processes, promote the cocreation of knowledge among WEF sectors and across global south states, and close the gap between existing knowledge, tools and their utilization, mutual peer-to-peer learning events, and regular multisectoral dialogues. Improved, well-organized monitoring mechanisms will be needed to identify necessary modifications of policies and priority interventions. Also, enhancing capacity, building consensus, developing relevant legislation, and transferring best practices will be crucial.

Strong institutions will be crucial to ensure the long-term sustainability and mainstreaming of nexus planning perspectives. There is a need for a holistic approach for holding regular multisectoral dialogues, building a knowledge base through nexus assessments and analysis, sharing best practices, mutual learning, and engagement at political and policy levels. Also, demonstrating the applicability and value of the WEF Nexus planning in the context of sub-Saharan Africa (e.g., through pilot projects) will be key. However, nexus planning and realizing such investments are only possible if the relevant capacities and skill set are available.

3. Development of a conceptual framework for WEF nexus upscaling and capacity development

Fig. 16.1 presents a proposed conceptual framework that provides pathways that guide informed and sustainable WEF upscaling, capacity building, and operationalization. As it stands, policy fragmentation remains the main challenge to overcome in terms of WEF nexus adoption, capacity development, and ultimate operationalization (Weitz et al., 2017). This requires a radical shift from linearity to circularity and attain sustainable development. The framework offers detailed, practical, and integrated insights, from multidisciplinary perspectives, on achieving sustainable upscaling and uptake of the WEF nexus approach through informed transformational processes that lead to sustainable development.

3.1 Upscaling and uptake of WEF nexus

Upscaling WEF nexus is the process of replication, spread, or adaptation to increase scale of impact. Upscaling can be described as vertical, horizontal, or deep upscaling. Vertical upscaling is the process of engaging local, regional,

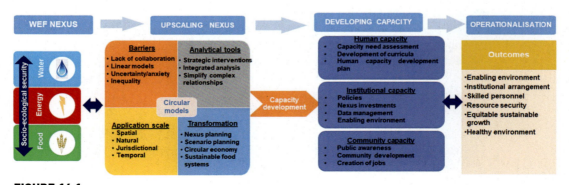

FIGURE 16.1

A transformational conceptual framework to guide informed and sustainable capacity development for rapid uptake and operationalization of the WEF nexus approach.

national, and international policymakers, donors, and development institutions to advocate policy change and to redirect institutional resources (Harvey, 2000; Pachicho and Fujisaka, 2004). Horizontal upscaling, also known as outscaling, is adapting to new contexts and geographically expanding technologies (Pachicho and Fujisaka, 2004), while deep scaling is the process of changing relationships, cultural values, and beliefs (Riddell and Moore, 2015). In this section, we discuss considerations for the successful upscaling of the WEF nexus in sub-Saharan Africa in the vertical and horizontal perspectives. We reflect on the experiences gained so far and how it can be upscaled, and on lessons learned by implementing similar integrated concepts.

Successful upscaling is dependent on how the WEF nexus will be perceived, whether at a policy or community level. New concepts, innovations, or technologies need to respond to the current problems of the targeted beneficiaries. If it is upscaling to the policy level, the approach is more likely to succeed if it aligns with national, regional, and global development goals (Klingner et al., 2003; Passioura, 2010). WEF approaches attempt to address agendas in the Global South (resource security, poverty alleviation). Mabhaudhi et al. (2019) applied a WEF nexus analytical livelihoods model with complex systems understanding to assess rural livelihoods, health, and wellbeing in southern Africa. The model guides decision-making processes by identifying priority areas needing intervention, enhancing synergies, and minimizing trade-offs necessary for resilient rural communities. WEF nexus approaches also provide a solid understanding of synergies and trade-offs associated with SDG targets 2, 3, 6, and 7 (Mabhaudhi et al., 2019; Mitra et al., 2020). More than 100 interlinkages exist between SDGs 2, 3, 6, and 7, implying that a nexus approach is best suited to achieve these targets compared with silo approaches (Mitra et al., 2020). While the role of the WEF nexus approach in addressing water, energy, and food is clear, critics of the approach argue that the scope is narrow and fails to address climate change and the environment, which are associated with sustainable development (Pandey and Shrestha, 2017; Rasul and Sharma, 2016). In the Global South, IWRM was promoted for coordinated development and management of water, land, and related resources to maximize economic and social welfare equitably without compromising ecosystems' sustainability (Al Radif, 1999). This concept was adopted for professionals in the water sector to manage water resources more effectively (OECD, 2014; UNEP, 2014). For WEF to be adopted, there is a need for a clear distinction with IWRM.

Currently, WEF nexus research is mostly embedded in scientific publications, which are only accessible to researchers. To successfully upscale research findings, there is a need to identify dissemination and outreach channels tailor-made for different stakeholders such as practitioners, policymakers, nongovernment organizations (NGOs), etc. Methods of outreach vary depending on the targeted audience and may take many different channels. The basics of any dissemination plan are why—the purpose of dissemination, what—the message to be disseminated, to whom—the audience, how—the method, and when—the timing (Ross-Hellauer et al., 2020). Various academic papers are proposing the WEF nexus framework to guide policy development and governance structures (Benson et al., 2015; Gain et al., 2015; Martinez et al., 2018; Zhang et al., 2018; Zisopoulou et al., 2018; Simpson and Jewitt, 2019), indicating that strategic dissemination plans need to be developed to reach policymakers and other actors of influence such as NGOs and development agencies. There is now a need to determine if the current evidence is enough to influence policy and resource allocation. A point of caution is that policymaking processes are also political not always based on scientific evidence in the Global South. Looking at ways that have been previously successful in speeding the uptake of research findings can provide useful lessons.

Researchers who have developed frameworks, models, and other analytical tools are often looking to outscale their innovations, especially to other researchers. This transfer is often susceptible to differences in scale (field, region, catchment, cities), biophysical and political factors, and development priorities (McGrane et al., 2019). The inability of WEF nexus frameworks, models, and any other analytical tools to capture variability in space and time will hinder researchers' successful adoption. Lack of extensive data sets has been one of the major factors preventing available models from capturing interactions among nexus components and temporal and spatial variability (McGrane et al., 2019). Lack of the necessary skills may also be limited to the adoption of specialized and technical tools and models. To develop tools that address different interactions with the nexus and consider spatial and temporal variability, there is a need for collaborative efforts among researchers with different expertise and from different regions.

Partnerships are not only important at the researcher level. Nexus thinking breaks the sectorial silo approaches and moves away from the narrow vision to a more holistic view. Formal and informal partnerships among all the stakeholders are then crucial to break the silos as it gathers different strengths and divergent views (Hoolohan et al., 2018). Networks are also well suited for knowledge dissemination, shared learning strategies, and access to funding. If partnerships are going to succeed, effective communication and strong feedback loops are required (Naber et al., 2017). This implies sharing and accessing all knowledge and information, thus creating a transparent environment and building trust (Hoolohan et al., 2018). The WEF's ability to integrate distinct but interlinked sectors adds a level of complexity, and unlocking these complexities requires appropriate knowledge and skills. Upscaling needs to be driven by individuals and organizations that have the appropriate capacity.

4. Capacity development for upscaling and uptake of WEF nexus

Despite its potential, the WEF nexus is yet to upgrade from theory to practice fully. The full adoption of the WEF nexus approach is still hindered by the limited number of trained professionals and implementing agencies. However, recent developments in South Africa have shown a steady increase in the number of people engaged in WEF nexus activities (Fig. 16.2). Capacity building should be targeted for technical aspects and the soft skills needed to strengthen institutions, communicate effectively, and drive policy and economic processes. It should also be broad enough to embrace nonacademic settings such as community groups and cultural structures.

Planning and realizing WEF nexus investments is only possible if the relevant capacities and skill sets are available. Capacity building is key to providing organizations and stakeholders with the tools to achieve WEF security goals sustainably. The recommended capacity building efforts should be able to enhance institutional frameworks and empower key stakeholders and decision-makers by providing them with the required knowledge and skills. They should also be able to form an integral part of the structured process that promotes nexus thinking and supports creating an enabling environment for new investments and upscaling of nexus projects. This section points out the capacity needs assessment, WEF nexus curricula issues, and the implementation plan for targeted capacity building for WEF nexus upscaling. However, this recommended capacity building plan should not be viewed as a static capacity building plan but rather as a guideline to develop appropriate activities and strategies that can be further adjusted if necessary.

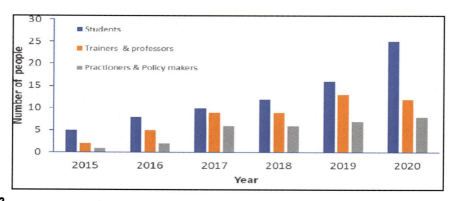

FIGURE 16.2

The number of individuals involved in the WEF activities that promoted capacity building over 6 years period in South Africa.

4.1 Capacity needs assessment

Capacity needs assessments can be conducted by listening to stakeholders' views and experiences. The capacity needs must focus on existing WEF projects' main features and goals; decisional processes supporting WEF nexus—based approaches; challenges and barriers in designing, implementing, and monitoring; lessons learned, upscaling opportunities, knowledge gaps; "resilient qualities" of WEF nexus. As a guideline, capacity needs assessments can be framed in the following six parts:

- Experience of the partners and facilitators in applying WEF nexus solutions and main features of pilot projects (goals, risk addressed, technological characteristics)
- The decision processes fostering the adoption of WEF nexus—based approaches (history, rationale, aims, stakeholders)
- Challenges and barriers in designing and implementing WEF nexus and the performances of the pilot projects
- Monitoring WEF nexus: evaluations of the experiences, benefits, and externalities, an innovation introduced.
- Lessons learned, upscaling opportunities and limits, knowledge gaps.
- Resilience and WEF nexus—which "resilient qualities" characterize WEF nexus? Do WEF nexus seek resilience?

Recommendations drawn from the assessment must inform future nexus activities in the region and facilitate strategic targeting of capacity building efforts. However, the assessment should not be viewed as a static plan but rather serve as an adaptive guide, which can be tailored to specific needs considering new information or the level/scale of a given intervention. More importantly, the key stakeholders should carry out a self-assessment of their existing institutional capacities. The use of the nexus capacity scorecard (Meyer, 2019) supplemented with information gathered during training workshops, nexus dialogues, nexus study tours, and other stakeholder engagements can allow capacity building needs to be identified.

4.2 Building WEF nexus curricula for upscaling

The following aspects of nexus capacity building highlight key curricula topics that have the potential value for upscaling the WEF nexus approach.

4.2.1 WEF nexus tools

The principal aim of much of the WEF nexus capacity building is to develop tools to assess and communicate the connections and interdependencies between the three WEF component systems. The tools are classified in terms of their objectives: sustainability assessment, modeling (including optimization), and visualization. Various techniques for each have been developed and applied in nexus initiatives, each with its different strengths and weaknesses. Sustainability assessments tools provide descriptions of resource use and availability across the WEF sectors (Byrne, 2016; Endo et al., 2015; Flammini et al., 2014). Optimization explores policy and management options, structuring nexus problems as the culmination of multiple, often conflicting criteria. The principal objective is to identify the best solution from a set of alternative management options. Numerous optimization methods have been used in the WEF nexus, including multicriteria decision analysis (Flammini et al., 2014) and multiobjective optimization methods (Hurford et al., 2014). Visualization techniques convey system state and dynamics, explore trade-offs, and communicate multidimensionality and interaction. Various visualization techniques have been applied in WEF nexus initiatives, including Sankey diagrams (Bajželj et al., 2014), interactive maps (Hadka et al., 2015), and multidimensional surfaces for exploring Pareto-optimal fronts (Hurford and Harou, 2014).

The review of these tools has shown that they do not always provide practical solutions, as uncertainties surrounding future conditions sometimes leave optimal solutions vulnerable to failure (Rosenberg, 2015). Future capacity building efforts could partially address uncertainty by coupling optimization methods such as multiobjective optimization methods with deep uncertainty approaches to assess potential interventions' performance under different plausible futures (Herman et al., 2014). Presently, many WEF nexus models are quantitative and engage with stakeholders only as end users of the technical information provided. The technical complexity and computational requirements of optimization models are often given as the justification to reduce the accessibility of these tools to nonacademic parties (Hoolohan et al., 2018). The inclusion of stakeholders in the development of nexus tools has also been limited. Involving stakeholders in the development and training of such tools will enhance their effectiveness through a shared understanding that aids the design of appropriate interventions and encourage trust in decision support tools (Howarth and Monasterolo, 2017).

4.2.2 WEF nexus governance

The social and institutional dimensions of nexus systems are often overlooked in a research space dominated by WEF nexus tools. Yet, understanding governance systems are essential to analyze nexus challenges and possible design solutions. Governance involves the diffuse networks of actors, institutions, and actions within WEF systems. Thus, the capacity building efforts focusing on governance should acknowledge the role of actors, institutions, and actions in the present and future management of WEF systems. The efforts should appreciate the benefits of involving actors across the WEF nexus space throughout the process. Governance capacity building must be articulated within at least three discourses on nexus governance: institutions, framing, and agency. The compartmentalization of WEF systems within departmental silos should be avoided. Future nexus studies should, therefore, employ inter-, multi-, and transdisciplinary methods to examine institutional constraints within the nexus (Howarth and Monasterolo, 2017). Nexus framings should avoid narrowing nexus conceptualization to the only securitization of water, energy, and food resources to avoid the risk of closing discourses around broader objectives such as well-being, equity, and justice. Recommended frameworks should embrace complexity and recognize the context dependency of governance arrangements (Pahl-Wostl et al., 2010). Understanding how underlying governance systems variously constrain, enable, and direct the scale and speed of change is necessary to understand how innovations might be scaled up to enhance their impact. The importance of the inclusion of stakeholders in these efforts should be emphasized to allow the framing of nexus challenges to be better understood and to provide a means of establishing a broader problem frame that better accommodates a diversity of system perspectives, thus counteracting a tendency toward siloed governance systems (Pahl-Wostl, 2009).

4.2.3 WEF multiscaling and interscaling

The notion of scale is important to nexus capacity building. WEF impacts should be considered at a range of scales. The spatial scales that matter to each component of the WEF nexus vary in importance. For example, water supply systems tend to be more localized than food supply systems because of extensive global supply chains for both inputs (e.g., fertilizers, chemicals) and outputs (i.e., food). However, their effects may be experienced on different scales. They are managed, for example, the embedded water in food systems (Kumar and Singh, 2005) or the global impacts of emissions associated with localized production (Bradley et al., 2013). Therefore, more research and capacity building efforts should be directed toward understanding national-scale interactions and impacts of nexus systems either to understand nexus issues within a single country (Conway et al., 2015) or to compare between nations (Mushtaq et al., 2009). Given the importance of water within the WEF nexus, another common scale for capacity building is that of the river basin, where some are located within a single country and others on transboundary. At the largest scale, the global impacts of existing WEF systems must be examined and used to inform the community and high-level strategic visions for addressing overarching challenges such as climate change or sustainable development (Flammini et al., 2014).

Temporal scales are similarly challenging and should be carefully articulated. Capacity building efforts must view long-term outlook as an essential part of nexus thinking but address nexus issues with short-term and medium-term goals (Yang et al., 2016). It is also important that research acknowledges the multiple dimensions of nexus challenges, as it has implications for how nexus challenges and potential solutions are framed. The scale also presents an additional layer of complexity when seeking to engage with stakeholders, as stakeholders' priorities at different scales may be different. Emphasis should be placed on selecting capacity building methods that provide a platform for the negotiation of conflict and collaboration and aid in the identification of scalar issues, trade-offs, or interdependencies, and aid in the resolution of conflicting interests. It is also worth noting that sometimes the nexus' purpose is to understand the potential for upscaling. In these cases, it is important to consider the implications of specific innovations, policies, or case studies to understand the implications for upscaling or processes through which upscaling may occur. Thus, capacity building efforts should strive to bring out a clearer understanding of the dynamics and disconnects in multiscalar systems, including governance structures, and bring out a broader appreciation of challenges faced by actors at different levels.

4.2.4 WEF nexus scenario planning

Capacity building efforts, particularly nexus research, must be intrinsically future focused and must be designed in such a way to avail pathways for mitigating and adapting to future resource management challenges. The future challenges of climate change, population growth, and rising demand for energy, food, and water should form part of the curricula problem. Furthermore, it is necessary to account for the dynamic context of WEF systems when examining impacts. The inherent subjectivity and uncertainty involved in choosing which future scenarios to explore, coupled with the nexus's complex nature, calls for new and innovative methodological approaches. The most common method for understanding the conditions of future systems is forecasting techniques. However, forecasting techniques' predictive power tends to diminish under conditions of substantial uncertainty (Quist and Vergragt, 2006), especially for the WEF nexus, where the future contains multiple sources of uncertainty. Scenario planning methods such as the backcasting approach (Robinson, 2003) can be employed in such circumstances. Scenario approaches offer scope to investigate the possible implications of changes that depart from existing trends without assuming any power of prediction and qualitative and/or quantitative (Swart et al., 2004). Involving stakeholders in scenario development, refining assumptions, and understanding alternative scenarios' implications are also merits. Adequate engagement with stakeholders in developing and interrogating scenarios aids acceptability of the visions' scenarios and enhances the usability and relevance of the outputs.

4.3 Capacity building implementation strategy
In this section, institutional, organizational, and individual capacity building actions are illustrated in the short, medium, and long term.

4.3.1 Institutional capacity building plan
In the short term, there is a need to agree on the definition of water—energy—food security nexus for the Global South and find new innovative ways to engage all three WEF sectors. Comparative policy analysis, including how the national policies influence regional institutional frameworks and cooperation, must identify opportunities for benefit sharing across sectors and countries, considering socioeconomic gains and losses. In the medium term, institutions must broaden stakeholder engagement and, based on policy analyses, take steps to achieve policy coherence. Institutional structures at local scales and community levels should also be considered. The integration of nexus perspectives into national and regional planning processes and the reconciliation of national priorities with regional political priorities to enable joint decision-making must be carried out. In the long term, institutions should have established a clear nexus governance framework at the regional level and enable legal frameworks, strategies, programs, and policies.

4.3.2 Organizational capacity building
In the short term, relevant national and regional organizations to take the nexus dialogue forward must be identified. There is a need to develop guidance on designing and implementing nexus interventions and supporting the preparation of nexus project proposals. Tools and methodologies for informed nexus decision-making must be collected. Training on nexus investments with a particular focus on the links between nexus and financing options and gains would be required. In the midterm, ownership can be generated by involving key stakeholders in pilot studies. Appropriate WEF nexus approaches and methodologies for the Global South must be developed, and knowledge must be cocreated. Implementation of nexus projects can be supported. In the medium-term capacities of river basin organizations, multisectoral planning and decision-making must be strengthened while nexus focal points in national ministries must be nominated. Capacities of regional organizations and national authorities to undertake effective monitoring and evaluation of nexus interventions and capacities of relevant organizations to undertake multisectoral planning and decision-making should be strengthened. There is a need to fully integrate nexus approaches and nexus governance into regional organizations in the long term.

4.3.3 Individual capacity needs
There is a need to develop targeted training workshops/modules on nexus planning, assessment, trade-off analysis, visualization, and quantification of

shared benefits. Implementation challenges of nexus interventions also need to be assessed. Knowledge exchange and mutual learning between different types of stakeholders need to be increased. Strategic study tours and exchange visits to other regions can enhance the global nexus network and enable cross-regional learning. There is a need to integrate nexus perspectives into university degree programs and civil servant training in the long term.

4.4 Enabling environment for capacity building

A guide for creating an enabling environment that fosters the integration of nexus thinking is articulated as follows. Local, national, and regional nexus dialogues that enable knowledge and best practice sharing must be continued in the short term. These should take place regularly and allow for open discussions on nexus challenges, while technical nexus expert dialogues that provide guidance and advice must be established. Access to clear, informative, and tailored data and information on the WEF nexus needs to be provided. Feasibility studies on moving nexus into action will need to be implemented. Guidance on nexus investment opportunities in the Global South, especially through blended financing, must be developed as well. Multisectoral learning workshops are also useful. There should be the identification of priority intervention areas based on national and regional nexus assessments in the medium term. Small pilot applications of the nexus approach to demonstrate its benefits in the Global South context will have to be implemented. At the same time, best practice cases from the region must be collated. Monitoring and evaluation capacities can be strengthened. There is also a need to develop appropriate data and information sharing frameworks. Dialogue with international financial institutions and other investors to discuss common interests can be facilitated. The creation of an online portal for best practice and data information sharing would require to be supported. In the long term, institutionalized, regular multisectoral dialogues funded by states could be availed.

In the Southern Africa region, WaterNet as a regional network of university departments and research and training institutes, given its successful track record in IWRM, is strategically posed to build regional institutional and human capacity in WEF nexus through training, education, research, and outreach by harnessing the complementary strengths of member institutions in the region and elsewhere (Jonker et al., 2012; Kileshye Onema et al., 2020). WaterNet has participated and/or coordinated research involving the WEF nexus through regional projects such as Challenge Programme, AfriAlliance, BRECCIA, and short-term professional training programs for water sector practitioners and middle managers. With the SADC responsibility for capacity building, WaterNet is well poised to coordinate the various capacity building initiatives in the SADC region and serve as a focal point for south–south collaboration activities involving the WEF nexus.

5. Conclusion

This chapter provides pathways to upscaling and developing capacity in WEF nexus perspectives to achieve sustainable resources use and management. Research on the WEF nexus has shown a trajectory from a focus on water and agriculture to water and energy and, more recently, holistic frameworks addressing water, energy, food, climate, and the environment. While there has been a growing understanding of nexus trade-offs and synergies, land-related issues about meeting food and energy security remain misrepresented. The impact of climate change has not been fully explored, yet it poses a challenge to WEF nexus frameworks that have been proposed. There is a need for WEF nexus research and frameworks to be more considerate of land and climate change threats for enhanced uptake, especially at the policy level. There is scope for upscaling of the WEF nexus at both the community and policy levels. It aligns well with development agendas on food, water, and energy security. Upscaling of WEF nexus research is hindered by spatial and temporal scale issues, data availability, trained professionals, and implementing agencies. Capacity building efforts should enhance institutional frameworks and empower key stakeholders and decision-makers by providing them with the required knowledge and skills. Important is to start nurturing nexus thinking and supporting the creation of an enabling environment for new investments and upscaling of nexus projects.

References

Albrecht, T.R., Crootof, A., Scott, C.A., 2018. The water-energy-food nexus: a systematic review of methods for nexus assessment. Environ. Res. Lett. 13 (4), 043002.

Allouche, J., Middleton, C., Gyawal, D., 2014. Nexus Nirvana or Nexus Nullity? A Dynamic Approach to Security and Sustainability in the Water—Energy—Food Nexus (No. 63; STEPS Working Paper).

Al Radif, A., 1999. Integrated water resources management (IWRM): an approach to face the challenges of the next century and to avert future crises. Desalination 124 (1–3), 145–153.

Arora, A.K., Gagneja, A.P.S., 2020. The Association of Hyper-Competitiveness and Consumerism with Ecological and Social Degradation: A Need for a Holistic and Responsible Approach for Environmental and Psychosocial Rehabilitation. 2020 Zooming Innovation in Consumer Technologies Conference (ZINC), pp. 327–332.

Bajželj, B., Richards, K.S., Allwood, J.M., Smith, P., Dennis, J.S., Curmi, E., Gilligan, C.A., 2014. Importance of food-demand management for climate mitigation. Nat. Clim. Change 4 (10), 924–929.

Beaton, C., Jain, P., Govindan, M., Garg, V., Murali, R., Roy, D., Bassi, A., Pallaske, G., 2019. Mapping Policy for Solar Irrigation across the Water-Energy-Food (WEF) Nexus in India. International Institute for Sustainable Development.

Benson, D., Gain, A.K., Rouillard, J.J., 2015. Water governance in a comparative perspective: from IWRM to a 'nexus' approach? Water Altern. (WaA) 8 (1), 756–773.

Bhave, A.G., Conway, D., Dessai, S., Stainforth, D.A., 2018. Water resource planning under future climate and socioeconomic uncertainty in the Cauvery River Basin in Karnataka, India. Water Resour. Res. 54 (2), 708–728.

Bieber, N., Ker, J.H., Wang, X., Triantafyllidis, C., van Dam, K.H., Koppelaar, R.H.E.M., Shah, N., 2018. Sustainable planning of the energy-water-food nexus using decision making tools. Energy Pol. 113, 584–607.

Botai, J.O., Botai, C.M., Ncongwane, K.P., Mpandeli, S., Nhamo, L., Masinde, M., Adeola, A.M., Mengistu, M.G., Tazvinga, H., Murambadoro, M.D., 2021. A review of the water–energy–food nexus research in Africa. Sustainability 13 (4), 1762.

Bradley, P., Druckman, A., Jackson, T., 2013. The development of commercial local area resource and emissions modelling–navigating towards new perspectives and applications. J. Clean. Prod. 42, 241–253.

Burkhard, B., Kandziora, M., Hou, Y., Müller, F., 2014. Ecosystem service potentials, flows and demands-concepts for spatial localisation, indication and quantification. Landsc. Online 34, 1–32.

Byrne, D., 2016. Qualitative Comparative Analysis: A Pragmatic Method for Evaluating Intervention. CECAN Evaluation and Policy Practice Note (EPPN) No. 1 for Policy Analysts and Evaluators.

Chirisa, I., Bandauko, E., 2015. African cities and the water-food-climate-energy nexus: an agenda for sustainability and resilience at a local level. Urban Forum 26 (4), 391–404.

Conway, D., van Garderen, E.A., Deryng, D., Dorling, S., Krueger, T., Landman, W., Lankford, B., Lebek, K., Osborn, T., Ringler, C., 2015. Climate and southern Africa's water–energy–food nexus. Nat. Clim. Change 5 (9), 837–846.

Daher, B., Mohtar, R.H., Pistikopoulos, E.N., Portney, K.E., Kaiser, R., Saad, W., 2018. Developing socio-techno-economic-political (STEP) solutions for addressing resource nexus hotspots. Sustainability 10 (2), 512.

Dombrowsky, I., Hensengerth, O., 2018. Governing the water-energy-food nexus related to hydropower on shared rivers–the role of regional organizations. Front. Environ. Sci. 6, 153.

Endo, A., Burnett, K., Orencio, P.M., Kumazawa, T., Wada, C.A., Ishii, A., Tsurita, I., Taniguchi, M., 2015. Methods of the water-energy-food nexus. Water 7 (10), 5806–5830.

Ericksen, P.J., 2008. Conceptualizing food systems for global environmental change research. Global Environ. Change 18 (1), 234–245.

FAO, IFAD, & WFP, 2015. The State of Food Insecurity in the World: Meeting the 2015 International Hunger Targets: Taking Stock of Uneven Progress. FAO, Rome. https://dx.doi.org/I4646E/1/05.15.

FAO, IFAD, UNICEF, WFP, & WHO, 2015FAO (Ed.), The State of Food Security and Nutrition in the World 2020: Transforming Food Systems for Affordable Healthy Diets. Food & Agriculture Organisation (FAO), Italy, Rome, pp. 1–62. ISBN 978-92-5-108785-5.

Flammini, A., Puri, A., Pluschke, L., Dubois, O., 2014. Walking the Nexus Talk: Assessing the Water-Energy-Food Nexus. Environment and Natural Resources Management Working Paper. FAO, Rome.

Fogarassy, C., Finger, D., 2020. Theoretical and Practical Approaches of Circular Economy for Business Models and Technological Solutions. Multidisciplinary Digital Publishing Institute, Basel.

Gain, A.K., Giupponi, C., Benson, D., 2015. The water–energy–food (WEF) security nexus: the policy perspective of Bangladesh. Water Int. 40 (5–6), 895–910.

Hadka, D., Herman, J., Reed, P., Keller, K., 2015. An open source framework for many-objective robust decision making. Environ. Model. Software 74, 114–129.

Hailemariam, W.G., Silalertruksa, T., Gheewala, S.H., Jakrawatana, N., 2019. Water–energy–food nexus of sugarcane production in Ethiopia. Environ. Eng. Sci. 36 (7), 798–807.

Hamidov, A., Helming, K., 2020. Sustainability considerations in water–energy–food nexus research in irrigated agriculture. Sustainability 12 (15), 6274.

Harvey, L.D.D., 2000. Upscaling in global change research. Climatic Change 44 (3), 225–263. https://doi.org/10.1023/A:1005543907412.

Heinimö, J., Junginger, M., 2009. Production and trading of biomass for energy–an overview of the global status. Biomass Bioenergy 33 (9), 1310–1320.

Herman, J.D., Zeff, H.B., Reed, P.M., Characklis, G.W., 2014. Beyond optimality: multistakeholder robustness tradeoffs for regional water portfolio planning under deep uncertainty. Water Resour. Res. 50 (10), 7692–7713.

Hoff, H., 2011. Understanding the Nexus: Background Paper for the Bonn 2011 Conference: The Water, Energy and Food Security Nexus.

Hoff, H., Alrahaife, S.A., el Hajj, R., Lohr, K., Mengoub, F.E., Farajalla, N., Fritzsche, K., Jobbins, G., Özerol, G., Schultz, R., 2019. A nexus approach for the MENA region—from concept to knowledge to action. Front. Environ. Sci. 7, 48.

Hoolohan, C., Larkin, A., McLachlan, C., Falconer, R., Soutar, I., Suckling, J., Varga, L., Haltas, I., Druckman, A., Lumbroso, D., 2018. Engaging stakeholders in research to address water–energy–food (WEF) nexus challenges. Sustain. Sci. 13 (5), 1415–1426.

Howarth, C., Monasterolo, I., 2017. Opportunities for knowledge co-production across the energy-food-water nexus: Making interdisciplinary approaches work for better climate decision making. Environ. Sci. Pol. 75, 103–110. https://doi.org/10.1016/j.envsci.2017.05.019. In this issue.

Hurford, A., Harou, J., 2014. Visualising Pareto-Optimal Trade-Offs Helps Move beyond Monetary-Only Criteria for Water Management Decisions. EGU General Assembly Conference Abstracts, p. 13494.

Hurford, A.P., Huskova, I., Harou, J.J., 2014. Using many-objective trade-off analysis to help dams promote economic development, protect the poor and enhance ecological health. Environ. Sci. Pol. 38, 72–86.

Jacobson, M., Pekarcik, G., 2021. Water-energy-food nexus approaches and initiatives in Africa. In: Brouwer, F. (Ed.), Handbook on the Water-Energy-Food Nexus. Edward Elgar Publishing, Cheltenham.

Jobbágy, J., Simoník, J., Findura, P., 2011. Evaluation of efficiency of precision irrigation for potatoes. Res. Agric. Eng. 57 (special issue), S14–S23.

Jobbins, G., Kalpakian, J., Chriyaa, A., Legrouri, A., el Mzouri, E.H., 2015. To what end? Drip irrigation and the water–energy–food nexus in Morocco. Int. J. Water Resour. Dev. 31 (3), 393–406.

Jonker, L., van der Zaag, P., Gumbo, B., Rockström, J., Love, D., Savenije, H.H.G., 2012. A regional and multi-faceted approach to postgraduate water education – the WaterNet experience in Southern Africa. Hydrol. Earth Syst. Sci. 16 (11), 4225–4232.

Jørgensen, S., Pedersen, L.J.T., 2018. The circular rather than the linear economy. In: RESTART Sustainable Business Model innovationPalgrave Macmillan, Cham, pp. 103–120.

Kileshye Onema, J.M., Chibarabada, T.P., Kujinga, K., Tariro Saruchera, D., 2020. How capacity development led to the establishment of a tri-basin agreement in the Southern African Development Community. Environ. Sci. Pol. 108, 14–18.

Klingner, J.K., Ahwee, S., Pilonieta, P., Menendez, R., 2003. Barriers and facilitators in scaling up research-based practices. Except. Child. 69 (4), 411–429.

Kumar, M.D., Singh, O.P., 2005. Virtual water in global food and water policy making: is there a need for rethinking? Water Resour. Manag. 19 (6), 759–789.

Landauer, M., Juhola, S., Klein, J., 2019. The role of scale in integrating climate change adaptation and mitigation in cities. J. Environ. Plann. Manag. 62 (5), 741–765.

Liu, J., Hull, V., Godfray, H.C.J., Tilman, D., Gleick, P., Hoff, H., Pahl-Wostl, C., Xu, Z., Chung, M.G., Sun, J., 2018. Nexus approaches to global sustainable development. Nat. Sustain. 1 (9), 466–476.

Mabhaudhi, T., Mpandeli, S., Nhamo, L., Chimonyo, V.G.P., Nhemachena, C., Senzanje, A., Naidoo, D., Modi, A.T., 2018. Prospects for improving irrigated agriculture in southern Africa: linking water, energy and food. Water 10 (12), 1881.

Mabhaudhi, T., Nhamo, L., Mpandeli, S., Nhemachena, C., Senzanje, A., Sobratee, N., Chivenge, P.P., Slotow, R., Naidoo, D., Liphadzi, S., 2019. The water–energy–food nexus as a tool to transform rural livelihoods and well-being in southern Africa. Int. J. Environ. Res. Publ. Health 16 (16), 2970.

Martinez, P., Blanco, M., Castro-Campos, B., 2018. The water–energy–food nexus: a fuzzy-cognitive mapping approach to support nexus-compliant policies in Andalusia (Spain). Water 10 (5), 664.

McGrane, S.J., Acuto, M., Artioli, F., Chen, P., Comber, R., Cottee, J., Farr-Wharton, G., Green, N., Helfgott, A., Larcom, S., 2019. Scaling the nexus: towards integrated frameworks for analysing water, energy and food. Geogr. J. 185 (4), 419–431.

Meyer, K., 2019. Building an Enabling Environment for Water, Energy and Food Security Dialogue in Central Asia. International Union for Conservation of Nature and Natural Resources (IUCN), Belgrade.

Mitra, B.K., Sharma, D., Kuyama, T., Pham, B.N., Islam, G.M.T., Thao, P.T.M., 2020. Water-energy-food nexus perspective: pathway for sustainable development goals (SDGs) to country action in India. APN Sci. Bull. 10 (1), 34–40. https://doi.org/10.30852/sb.2020.1067.

Monyei, C.G., Akpeji, K.O., 2020. Repurposing electricity access research for the global south: a tale of many disconnects. Joule 4 (2), 278–281.

Mpandeli, S., Naidoo, D., Mabhaudhi, T., Nhemachena, C., Nhamo, L., Liphadzi, S., Hlahla, S., Modi, A.T., 2018. Climate change adaptation through the water-energy-food nexus in southern Africa. Int. J. Environ. Res. Publ. Health 15 (10), 2306.

Muller, A., 2008. Sustainable agriculture and the production of biomass for energy use. Climatic Change 94 (3), 319–331.

Mushtaq, S., Maraseni, T.N., Maroulis, J., Hafeez, M., 2009. Energy and water tradeoffs in enhancing food security: a selective international assessment. Energy Pol. 37 (9), 3635–3644.

Naber, R., Raven, R., Kouw, M., Dassen, T., 2017. Scaling up sustainable energy innovations. Energy Pol. 110, 342–354.

Nhamo, L., Mabhaudhi, T., Mpandeli, S., Dickens, C., Nhemachena, C., Senzanje, A., Naidoo, D., Liphadzi, S., Modi, A.T., 2020. An integrative analytical model for the water-energy-food nexus: South Africa case study. Environ. Sci. Pol. 109, 15–24. https://doi.org/10.1016/J.ENVSCI.2020.04.010.

Nhamo, L., Mabhaudhi, T., Mpandeli, S., Nhemachena, C., Senzanje, A., Naidoo, D., Liphadzi, S., Modi, A.T., 2019. Sustainability indicators and indices for the water-energy-food nexus for performance assessment: WEF nexus in practice–South Africa case study. Environ. Sci. Policy 109, 15–24.

Nhamo, L., Ndlela, B., 2021. Nexus planning as a pathway towards sustainable environmental and human health post Covid-19. Environ. Res. 192, 110376.

Nhamo, L., Ndlela, B., Nhemachena, C., Mabhaudhi, T., Mpandeli, S., Matchaya, G., 2018. The water-energy-food nexus: climate risks and opportunities in southern Africa. Water 10 (5), 567.

OECD, 2014. Integrated Water Resources Management in Eastern Europe, the Caucasus and Central Asia European Union Water Initiative National Policy Dialogues Progress Report 2013.

Pachicho, D., Fujisaka, S., 2004. Scaling Up And Out: Achieving Widespread Impact Through Agricultural Research. Centro Internacional de Agricultura Tropical. CIAT Publications, Cali, Colombia.

Pahl-Wostl, C., 2009. A conceptual framework for analysing adaptive capacity and multi-level learning processes in resource governance regimes. Global Environ. Change 19 (3), 354–365.

Pahl-Wostl, C., Holtz, G., Kastens, B., Knieper, C., 2010. Analyzing complex water governance regimes: the Management and Transition Framework. Environ. Sci. Pol. 13 (7), 571–581.

Pandey, V.P., Shrestha, S., 2017. Evolution of the nexus as a policy and development discourse. In: Water-Energy-Food Nexus: Principles and Practices, 1, pp. 11–20.

Passioura, J.B., 2010. Scaling up: the essence of effective agricultural research. Funct. Plant Biol. 37 (7), 585–591.

Power, M., Newell, P., Baker, L., Bulkeley, H., Kirshner, J., Smith, A., 2016. The political economy of energy transitions in Mozambique and South Africa: the role of the Rising Powers. Energy Res. Social Sci. 17, 10–19.

Quist, J., Vergragt, P., 2006. Past and future of backcasting: the shift to stakeholder participation and a proposal for a methodological framework. Futures 38 (9), 1027–1045.

Rasul, G., Sharma, B., 2016. The nexus approach to water–energy–food security: an option for adaptation to climate change. Clim. Pol. 16 (6), 682–702.

Riddell, D., Moore, M.L., 2015. Scaling Out, Scaling up, Scaling Deep: Advancing Systemic Social Innovation and the Learning Processes to Support it. JW McConnell Family Foundation and Tamarack Institute, Toronto and Waterloo, ON.

Robinson, J., 2003. Future subjunctive: backcasting as social learning. Futures 35 (8), 839–856.

Rosenberg, D.E., 2015. Blended Near-Optimal Alternative Generation, Visualization, and Interaction for Water Resources Decision Making.

Ross-Hellauer, T., Tennant, J.P., Banelytė, V., Gorogh, E., Luzi, D., Kraker, P., Pisacane, L., Ruggieri, R., Sifacaki, E., Vignoli, M., 2020. Ten simple rules for innovative dissemination of research. PLoS Comput. Biol. 16 (4), 4.

Salvioni, D.M., Almici, A., 2020. Transitioning toward a circular economy: the impact of stakeholder engagement on sustainability culture. Sustainability 12 (20), 8641.

Serrano-Tovar, T., Suárez, B.P., Musicki, A., Juan, A., Cabello, V., Giampietro, M., 2019. Structuring an integrated water-energy-food nexus assessment of a local wind energy desalination system for irrigation. Sci. Total Environ. 689, 945–957.

Shilomboleni, H., Plaen, R. de, 2019. Scaling up research-for-development innovations in food and agricultural systems. Dev. Pract. 29 (6), 723–734. https://doi.org/10.1080/09614524.2019.1590531.

Simpson, G.B., Jewitt, G.P.W., 2019. The development of the water-energy-food nexus as a framework for achieving resource security: a review. Front. Environ. Sci. 7, 8.

Smit, W., 2016. Urban governance and urban food systems in Africa: examining the linkages. Cities 58, 80–86.

Swart, R.J., Raskin, P., Robinson, J., 2004. The problem of the future: sustainability science and scenario analysis. Global Environ. Change 14 (2), 137–146.

Terrapon-Pfaff, J., Ortiz, W., Dienst, C., Gröne, M.C., 2018. Energising the WEF nexus to enhance sustainable development at local level. J. Environ. Manag. 223, 409–416.

UNEP, 2014. Towards Integrated Water Resources Management. International Experience in Development of River Basin Organisations.

UNESCO, UN-Water, 2020. United Nations World Water Development Report 2020: Water and Climate Change. UNESCO, Paris.

UNGA, 2015. Transforming Our World: The 2030 Agenda for Sustainable Development, Resolution Adopted by the General Assembly (UNGA). United Nations General Assembly (UNGA), New York.

van Gevelt, T., 2020. The water–energy–food nexus: bridging the science–policy divide. Curr. Opin. Environ. Sci. Health 13, 6–10. https://doi.org/10.1016/J.COESH.2019.09.008.

Wehn, U., Vallejo, B., Seijger, C., Tlhagale, M., Amorsi, N., Sossou, S.K., Genthe, B., Kileshye Onema, J.M., 2021. Strengthening the knowledge base to face the impacts of climate change on water resources in Africa: a social innovation perspective. Environ. Sci. Pol. 116, 292–300.

Weitz, N., Strambo, C., Kemp-Benedict, E., Nilsson, M., 2017. Closing the governance gaps in the water-energy-food nexus: insights from integrative governance. Global Environ. Change 45, 165–173.

Wolde-Rufael, Y., 2005. Energy demand and economic growth: the African experience. J. Pol. Model. 27 (8), 891–903.

World Economic Forum, 2011. The Global Risks Report 2021. World Economic Forum, Cologny.

Yang, Y.C.E., Wi, S., Ray, P.A., Brown, C.M., Khalil, A.F., 2016. The future nexus of the Brahmaputra River Basin: climate, water, energy and food trajectories. Global Environ. Change 37, 16–30.

Zhang, X., Li, H.Y., Deng, Z.D., Ringler, C., Gao, Y., Hejazi, M.I., Leung, L.R., 2018. Impacts of climate change, policy and Water-Energy-Food nexus on hydropower development. Renew. Energy 116, 827–834.

Zisopoulou, K., Karalis, S., Koulouri, M.-E., Pouliasis, G., Korres, E., Karousis, A., Triantafilopoulou, E., Panagoulia, D., 2018. Recasting of the WEF Nexus as an actor with a new economic platform and management model. Energy Pol. 119, 123–139.

CHAPTER 17

WEF nexus narratives: toward sustainable resource security

Tafadzwanashe Mabhaudhi[1,2], Aidan Senzanje[3], Albert Modi[1], Graham Jewitt[4] and Festo Massawe[5]

[1]Centre for Transformative Agricultural and Food Systems (CTAFS), School of Agricultural, Earth and Environmental Sciences, University of KwaZulu-Natal, Pietermaritzburg, South Africa; [2]International Water Management Institute (IWMI-GH), West Africa Office, Accra, Ghana; [3]Bioresources Engineering Programme, School of Engineering, University of KwaZulu-Natal (UKZN), Pietermaritzburg, South Africa; [4]IHE Delft Institute for Water Education, Delft, The Netherlands; [5]School of Biosciences, University of Nottingham Malaysia, Semenyih, Selangor, Malaysia

1. The WEF nexus

The water—energy—food (WEF) nexus has developed into a vital transformative and circular approach since 2011 after it was presented at the World Economic Forum by the Stockholm Environment Institute in anticipation of the Sustainable Development Goals (SDGs), which came into effect in 2015 (FAO, 2014; Hoff, 2011; UNGA, 2015). This was the same period when the SDGs were being formulated in response to the continued insecurity of water, energy, and food resources (FAO, 2014; Liphadzi et al., 2021). The three resources, termed WEF resources, are critical for human well-being, poverty reduction, and sustainable development, the reason why all the 17 SDGs are developed around the three resources (Mabhaudhi et al., 2021; UNGA, 2015). The need for the formulation of the SDGs around WEF resources was also motivated by global projections indicating that the demand for the three resources will increase significantly in the coming years due to population growth, economic development, international trade, urbanization, globalization, diversifying diets, depletion of natural resources, technological advances, and climate change (FAO, 2014; Hoff, 2011). This is happening when unsustainable use of resources contributes significantly to planet warming. In return, the water and food resources are impacted negatively by the adverse effects of climate change (Mpandeli et al., 2018). As a result, policymakers needed an approach capable of integrating the management and governance of the three interlinked sectors (Mabhaudhi et al., 2019; Nhamo et al., 2018; Rasul and Sharma, 2016). This is mainly informed by the cross-sectoral and cross-cutting challenges currently bedevilling humankind.

This book dedicates to providing the WEF nexus narratives from 2015 to date, focusing on achieving sustainability, circularity, and transformational change. The WEF nexus has since evolved into a three-dimensional approach used either as an analytical tool, a conceptual framework, or a discourse (Nhamo et al., 2020). As an analytical tool, the nexus systematically applies quantitative and qualitative methods to understand interactions among WEF resources; as a conceptual framework, it simplifies an understanding of WEF linkages to promote coherence in policymaking and enhances sustainable development, and as a discourse, it is a tool for problem framing and promoting cross-sectoral collaboration (Albrecht et al., 2018; Mabhaudhi et al., 2019, 2021; Naidoo et al., 2021; Nhamo et al., 2021a,b). Thus, the approach has become a tool for understanding the complex and dynamic interlinkages between water, energy, and food security issues. In this regard, it can also be used to monitor the performance of the WEF nexus related 2030 Global Agenda of the Sustainable Development Goals (SDGs), particularly SDGs 2 (zero hunger), 6 (clean water and sanitation), and 7 (affordable and clean energy) (Mabhaudhi et al., 2021; Stephan et al., 2018). It has grown into an innovative integrated systems approach through which cross-sectoral sustainability indicators can be derived. Nexus planning encompasses governance and manages trade-offs, synergies, and thresholds through science for human wellbeing and protection of the environment (Mohtar and Daher, 2016; Nhamo and Ndlela, 2020; Nhamo et al., 2021b). This section provides a summary of each chapter.

2. Key messages

Chapter 1 focuses on the evolution of the WEF nexus before and after 2011. The chapter demonstrates that water, energy, and food are vital human well-being resources. Yet, they are under increased pressure to meet the growing demand from an increasing population at a time of worsening insecurity due to depletion of natural resources and degradation of ecosystems and the environment as a whole. These challenges prompted the formulation of the SDGs in 2015, which aim to achieve sustainability in resource management by 2030. As the SDGs recognize that developments in one sector impact other sectors and that any proposed development must balance socioeconomic and environmental sustainability, the WEF nexus was then accepted as an approach to achieve sustainability. Also, as the three resources are the most impacted by climate change, they provide a close link between adaptation, climate system, human society, and the environment. The intricate interlinkages between water, energy, and food resources with the related relationships with socioeconomic development, healthy ecosystems, human development, and sustainable development stimulated the rapid growth of the WEF nexus concept. The authors recognize that the concept existed before 2015. Still, its progression increased after the World Economic Forum of 2011 after the

Stockholm Environment Institute presentation in anticipation of the SDGs. Thus, this chapter sets the tone of the whole book as it focuses on the importance of the approach in establishing the interconnectedness of resources and as a guide for coherent policy decisions that lead to sustainable development.

Chapter 2 focuses on quantitative WEF nexus analysis approaches and their data requirements. The authors list several WEF nexus assessment tools that include Climate, Land, Energy Use, and Water Strategies (CLEWS) (Howells et al., 2013), Water, Energy, Food Nexus Tool 2.0 (Daher and Mohtar, 2015), REMap (Ferroukhi et al., 2015), Multi-Scale Integrated Analysis of Societal and Ecosystem Metabolism (MuSIASEM) (Giampietro et al., 2009), International Institute for Sustainable development (IISD) Water–Energy–Food Security Analysis Framework (Bizikova et al., 2013), Transboundary Basin Nexus Assessment (TBNA) Methodology (Roidt and de Strasser, 2018), among others. Importantly, the authors highlight the importance of the availability and quality of data for each of the available tools.

In Chapter 3, the authors introduce an Earth Observation-WEF (EO-WEF) geotool for WEF resources spatial data visualization and generation. This is informed by the challenges associated with the unavailability and collection of WEF resources—related data and the issues related to proprietary data evolution over time. However, remote sensing has filled this gap by providing most WEF nexus components relatively low-cost and various temporal and spatial resolutions. Furthermore, cloud computing platforms such as the Google Earth Engine (GEE) is facilitating the processing of huge datasets. The authors highlight the role of the EO-WEF geotool in multisectoral data acquisition, processing, visualization analysis, and display, demonstrating its capability at the catchment level.

Chapters 4–9 discuss and demonstrate the application of the WEF nexus at different spatial scales. Chapter 4 discusses issues related to the scale of application of the approach. The authors discuss recent WEF nexus research at varying spatial scales from the household, to national, and up to the global scale. The authors highlight the vast diversity in nexus issues and challenges, along with an accompanying diversity of research and assessment approaches to tackling and better understanding these issues, and then discuss the interactions between these scales, and how policy developed at one scale may impact other scales, potentially in unanticipated and detriment ways. Interestingly, the study establishes the cross-spatial scalar nexus interactions and interrelationships concerning temporal scale, in terms of both policy setting and implementation and impacts to people. Chapter 5 discusses the tools and indices for WEF nexus analysis. The chapter acknowledges that key WEF nexus foci issues and challenges are extremely diverse and change depending on the local situation and setting, the scale at which the nexus is analyzed, and even according to the sector used as the nexus entry point. As a result, various approaches and methods for studying, assessing, and analyzing the nexus can be adapted to suit the specific case under investigation. Each approach has strengths and limitations. Then, Chapter 6 applies the WEF nexus approach at the transboundary scale in the Songwe River Basin between Malawi and Tanzania. The chapter focuses on the qualitative analysis of the WEF nexus system. It explains the process of identifying the major sectors and the main interlinkages between them, and potential synergies and trade-offs, assessing how decisions made in a sector may influence others. Chapters 7—9 demonstrate the usefulness and application of the WEF nexus approach at the local, regional, and dam scales, respectively.

Chapter 10 moves further to apply the WEF nexus as a tool to assess progress toward SDGs, mainly Goals 2 (food), 6 (water), and 7 (energy). The chapter presents an overview of how the WEF nexus is applied to achieve the three SDGs. Then, Chapter 11 adds the health dimension in the nexus thinking. Chapter 11 is informed by the immense and adverse changes occurring due to anthropogenic modifications, posing the risk of novel infectious diseases transmission on humans. The chapter provides pathways that enhance the preparedness and improve the resilience against novel pathogens by assessing vulnerability and the available options to reduce risk through the water—health—ecosystem—nutrition (WHEN) nexus.

Then, Chapter 12 addresses the issues related to financing the WEF nexus projects. The chapter is based on the experiences of interdisciplinary research approaches followed and how previous challenges have carved pathways toward research investment. By doing so, the study identifies critical variables that are either hurdles or leverage points for WEF nexus implementation. As a means to enhance the findings of this chapter, the following section, Chapter 13 provides a detailed WEF nexus success in South and Southeast Asia, and provides an overview on how a multisectoral approach is being adopted as a rallying point for sustainable development. Then, Chapter 14 cements the financing aspect by focusing the WEF nexus application on ecosystems and anthropocentric perspective.

As the WEF nexus is an integrated approach that addresses cross-sectoral and interlinked challenges, Chapter 15 emphasizes analyzing smallholder farmer trade-offs in agricultural production decision-making processes. The chapter, therefore, discusses issues related to irrigation, fertilizer use, and the adoption of solar energy in agricultural production. The focus is on agricultural production as a key factor in food security.

The concluding chapter, Chapter 16, addresses the issues related to capacity building to upscale the WEF nexus and guide transformational change. The chapter highlights the systematic nature of nexus planning as a catalyst for a long-lasting transformation by identifying trade-offs and synergies, which is key to achieving sustainability. WEF nexus research is steadily increasing in the Global South, showing an understanding of the concept. This has resulted in developing models that provide empirical evidence and pathways to operationalize the approach. Chapter 16, therefore, highlights the pathways, opportunities, and challenges toward upscaling and building capacity for the WEF nexus in the Global South.

3. Conclusion

The WEF nexus narratives presented in this book offer readers and practitioners an opportunity to reflect on the intersectoral relations among WEF resources and their stakeholders and call for multisectoral actions (using systems thinking approach), moving away from a single sector and disciplinary silo approaches. Throughout the book, authors have shown the value of the WEF nexus approach especially pointing out the synergies and trade-offs that can only be realized when the stakeholders consider WEF together. As pointed out throughout the book, achieving all of the 2030 SDGs is a big challenge that can only be overcome through partnerships and intersectoral understanding and working. The WEF nexus is an approach that can be used to monitor the performance and delivery of the SDGs, particularly Goals 2 (zero hunger), 6 (clean water and sanitation), and 7 (affordable and clean energy).

References

Albrecht, T.R., Crootof, A., Scott, C.A., 2018. The Water-Energy-Food Nexus: a systematic review of methods for nexus assessment. Environ. Res. Lett. 13, 043002.

Bizikova, L., Roy, D., Swanson, D., Venema, H.D., McCandless, M., 2013. The Water-Energy-Food Security Nexus: Towards a Practical Planning and Decision-Support Framework for Landscape Investment and Risk Management. International Development Research Centre (IDRC), Winnipeg, Manitoba, Canada.

Daher, B.T., Mohtar, R.H., 2015. Water–energy–food (WEF) Nexus Tool 2.0: guiding integrative resource planning and decision-making. Water Int. 40, 748–771.

FAO, 2014. The water-energy-food nexus: a new approach in support of food security and sustainable agriculture. In: Dubois, O., Faurès, J.-M., Felix, E., Flammini, A., Hoogeveen, J., Pluschke, L., Puri, M., Ünver, O. (Eds.), Food and Agriculture Organization (FAO), Rome, Italy.

Ferroukhi, R., Nagpal, D., Lopez-Peña, A., Hodges, T., Mohtar, R.H., Daher, B., Mohtar, S., Keulertz, M., 2015. Renewable Energy in the Water, Energy & Food Nexus. The International Renewable Energy Agency (IRENA), IRENA, Abu Dhabi, pp. 1–125.

Giampietro, M., Mayumi, K., Ramos-Martin, J., 2009. Multi-scale integrated analysis of societal and ecosystem metabolism (MuSIASEM): theoretical concepts and basic rationale. Energy 34, 313–322.

Hoff, H., 2011. Understanding the Nexus: Background Paper for the Bonn 2011 Conference: The Water, Energy and Food Security Nexus. Stockholm Environment Institute (SEI), Stockholm, Sweden, p. 52.

Howells, M., Hermann, S., Welsch, M., Bazilian, M., Segerström, R., Alfstad, T., Gielen, D., Rogner, H., Fischer, G., Van Velthuizen, H., 2013. Integrated analysis of climate change, land-use, energy and water strategies. Nat. Clim. Change 3, 621–626.

Liphadzi, S., Mpandeli, S., Mabhaudhi, T., Naidoo, D., Nhamo, L., 2021. The evolution of the water–energy–food nexus as a transformative approach for sustainable development in South Africa. In: Muthu, S. (Ed.), The Water–Energy–Food Nexus: Concept and Assessments. Springer, Kowloon, Hong Kong, pp. 35–67.

Mabhaudhi, T., Nhamo, L., Chibarabada, T.P., Mabaya, G., Mpandeli, S., Liphadzi, S., Senzanje, A., Naidoo, D., Modi, A.T., Chivenge, P.P., 2021. Assessing progress towards sustainable development goals through nexus planning. Water 13, 1321.

Mabhaudhi, T., Nhamo, L., Mpandeli, S., Nhemachena, C., Senzanje, A., Sobratee, N., Chivenge, P.P., Slotow, R., Naidoo, D., Liphadzi, S., 2019. The water–energy–food nexus as a tool to transform rural livelihoods and well-being in southern Africa. Int. J. Environ. Res. Publ. Health 16, 2970.

Mohtar, R.H., Daher, B., 2016. Water-Energy-Food Nexus Framework for facilitating multi-stakeholder dialogue. Water Int. 41, 655–661.

Mpandeli, S., Naidoo, D., Mabhaudhi, T., Nhemachena, C., Nhamo, L., Liphadzi, S., Hlahla, S., Modi, A., 2018. Climate change adaptation through the water-energy-food nexus in southern Africa. Int. J. Environ. Res. Publ. Health 15, 2306.

Naidoo, D., Nhamo, L., Mpandeli, S., Sobratee, N., Senzanje, A., Liphadzi, S., Slotow, R., Jacobson, M., Modi, A., Mabhaudhi, T., 2021. Operationalising the water-energy-food nexus through the theory of change. Renew. Sustain. Energy Rev. 149, 10.

Nhamo, L., Mabhaudhi, T., Mpandeli, S., Dickens, C., Nhemachena, C., Senzanje, A., Naidoo, D., Liphadzi, S., Modi, A.T., 2020. An integrative analytical model for the water-energy-food nexus: South Africa case study. Environ. Sci. Pol. 109, 15–24.

Nhamo, L., Mpandeli, S., Senzanje, A., Liphadzi, S., Naidoo, D., Modi, A.T., Mabhaudhi, T., 2021a. Transitioning toward sustainable development through the water–energy–food nexus. In: Ting, D., Carriveau, R. (Eds.), Sustaining Tomorrow via Innovative Engineering. World Scientific, Singapore, pp. 311–332.

Nhamo, L., Ndlela, B., 2020. Nexus planning as a pathway towards sustainable environmental and human health post Covid-19. Environ. Res. 110376, 7.

Nhamo, L., Ndlela, B., Nhemachena, C., Mabhaudhi, T., Mpandeli, S., Matchaya, G., 2018. The water-energy-food nexus: climate risks and opportunities in southern Africa. Water 10, 567.

Nhamo, L., Rwizi, L., Mpandeli, S., Botai, J., Magidi, J., Tazvinga, H., Sobratee, N., Liphadzi, S., Naidoo, D., Modi, A., Slotow, R., Mabhaudhi, T., 2021b. Urban nexus and transformative pathways towards a resilient Gauteng city-region, South Africa. Cities vol. 116.

Rasul, G., Sharma, B., 2016. The nexus approach to water–energy–food security: an option for adaptation to climate change. Clim. Pol. 16, 682–702.

Roidt, M., de Strasser, L., 2018. Methodology for Assessing the Water-Food-Energy-Ecosystem Nexus in Transboundary Basins and Experiences from its Application: Synthesis. UNECE, New York and Geneva.

Stephan, R.M., Mohtar, R.H., Daher, B., Embid Irujo, A., Hillers, A., Ganter, J.C., Karlberg, L., Martin, L., Nairizi, S., Rodriguez, D.J., 2018. Water–energy–food nexus: a platform for implementing the sustainable development goals. Water Int. 43, 472–479.

UNGA, 2015. Transforming Our World: The 2030 Agenda for Sustainable Development, Resolution Adopted by the General Assembly (UNGA). United Nations General Assembly, New York, USA, p. 35.

Index

Note: 'Page numbers followed by "f" indicate figures and "t" indicate tables.'

A

Africa, transformational change in, 299–301
  capacity building implementation strategy
    individual capacity needs, 313–314
    institutional capacity building plan, 313
    organizational capacity building, 313
  capacity development for upscaling and uptake, 308–314
  capacity needs assessment, 309
  enabling environment for capacity building, 314
  understanding drivers of change, 301–303
  WEF nexus curricula for upscaling, 305–307, 309–312
    governance, 310–311
    multiscaling and interscaling, 311–312
    planning, 303–305
    research in, 301–305
    scenario planning, 312
    tools, 309–310
    upscaling and capacity development, 305–307
Agent-based modeling (ABM), 74–76
Agricultural extension projects, 190
Agricultural sector planning, policy, and legal documents, 154–156, 154t–156t
Agroecology, 287–288
Analytical hierarchical process (AHP), 56–57, 210
Analytical livelihoods framework (ALF), 114, 125–126
Anthropogenic activities, 203–204
AU Malabo Summit Declaration, 281

B

Business-as-usual (BAU) scenario, 260

C

Capacity building implementation strategy
  individual capacity needs, 313–314
  institutional capacity building plan, 313
  organizational capacity building, 313
Capacity needs assessment, 309
Carbon footprints, 22
Catchment level
  analytical livelihoods framework (ALF), 114, 125–126
  application of models, 136–137
  climate, land (food), energy, and water systems approach (CLEWs), 127
  data, 114–117
    sources, 128–135
  energy sector, 131–132
  food sector, 132–135, 133f
  Inkomati-Usuthu Catchment Management Agency (IUCMA), 115, 136
  livelihood, 114
  methodology, 114–117
  model description, 119–128
  model selection, 119–128
  Multi-Scale Integrated Assessment of Society and Ecosystem Metabolism (MuSIASEM), 127–128
  national development plan (NDP), 113
  NexSym, 128
  spatial scale, 135
  Sustainable Development Goals (SDGs), 111
  time/temporal scale, 135–136
  water sector, 129–130, 130f
  WEF nexus available models, 117–119
  WEF Nexus Tool 2.0, 126–127
Causal loop diagrams (CLD), 68–71, 72t, 244
Circular models, 8f
Climate, 103, 103f
Climate change projections, 280–281
Climate Hazards Group Infrared Precipitation with Station data (CHIRPS), 41
Climate, land (food), energy, and water systems approach (CLEWs), 127
Climate, land, energy use, and water strategies (CLEWs), 18–21

327

Common technology interventions and trade-offs, 283t
Complex systems approach and causality, 244–250
Conceptual maps, 68–71, 69f
Conservation agriculture-based sustainable intensification (CASI), 241
Conventional centrifugal pumps, 286
COVID-19, 7

D

Declaration on Research Assessment (DORA), 231
Deregulated retail competition model, 246f
Drivers–pressures–states–impact–response (DPSIR) methodology, 58, 59f

E

Earth Observation-WEF (EO-WEF), 323
  development, 45
  predesign steps of, 35–36
  search table of, 37f
  software design, 36–37
  software development kit (SDK), 36
  Songwe nexus
    climate sector, 41, 42t–43t
    energy sectors, 41–45
    food sectors, 41–45
    land sectors, 41
    socioeconomic sectors, 41–45, 44t
    water sector, 39, 40t
  Songwe River Basin (SRB), 35–36
  user interface of, 38f
  uses, 37–38
Eastern Gangetic Plains (EGP), 241
Economic analysis models, 24–26, 25t
Ecosystem and human health, 104–106, 105f
Ecosystems and anthropocentric perspective, 257–259
  approach, 259–261

Argentina and Brazil, 267–274
comparisons of the WEF nexus in MENA and Latin America, 274–275
MENA and Latin America, 261–274
Energy, 101–102, 102f
  generation operational hours, 249f
  sector, 131–132
Energy sector planning, policy, and legal documents, 152–156, 153t–154t
Energy security, 176, 245
Energy–water link, 248–249
Environmental Sustainability Index (ESI), 82
Epidemic preparedness index (EPI), 215
EUROSTAT data, 57
EXIOBASE, 76–77
Extended conceptual maps, 68–69
Extended conceptual model, 97–106
  climate, 103, 103f
  ecosystem and human health, 104–106, 105f
  energy, 101–102, 102f
  land and food, 99–101, 100f–101f
  socioeconomic system, 103–104, 104f
  water, 97–99, 98f

F

Farmer decision-making, 290
Farmer technology adoption, 286–289
Financing WEF nexus projects, 223–224
  interlinkages within nexus research, 224–225
  role of funding in fostering interdisciplinary dialogue, 226–228
  shared value within multidimensional challenges, 228–229

transboundary systems and the need for interdisciplinary spaces, 225–226
Food and Agriculture Organization (FAO), 18
Food-for-energy relationship, 242
Food-related technology interventions, 285–286
Food sector, 132–135, 133f
Food security, 176, 249–250
Free energy source, 265

G

Geographic information system (GIS), 207
Global Land Data Assimilation System Version 2 (GLDAS 2), 39
Goal-seeking behavior, 69–71
Google Earth Engine (GEE), 35, 41, 323
Greenhouse gas (GHG) emission, 17, 41, 57
Green Revolution, 286–287
Gridded Population of the World Version 4.11 (GPW411), 41–45
Grid electricity and butane, 262
Gross domestic product (GDP), 60–61, 61f

H

Health risk classification categories, 214t
Higher-level nexus studies, 59–61
High level conceptual model, 96–97, 97f
Horizontal upscaling, 305–306
Human Development Index (HDI), 82
Hydroelectric power generation, 190–191
Hydropower, 302–303

I

Independent power producers (IPP), 245
Indices

Environmental Sustainability Index (ESI), 82
Human Development Index (HDI), 82
Sustainable Development Goals (SDGs), 83–84
WEF nexus index, 84
4th Industrial Revolution (4IR), 6–7
Inkomati-Usuthu Catchment Management Agency (IUCMA), 115, 136
Integrated assessment models (IAMs), 79–81
Integrated health indices, 214–215
Integrated natural resource management (INRM), 185
Integrated water resources management (IWRM), 303–304
Integrative analytical WEF nexus (IAWN) model, 192–193
Interdisciplinary research approaches, 223–225
Intergovernmental Panel on Climate Change (IPCC), 280–281
Interjurisdictional Water Basin Authority (AIC), 268–269
International Institute for Applied Systems Analysis (IIASA), 18

J
Jordan nexus challenge, 263f

K
Knowledge generation, 224

L
Lancang-Mekong River Basin, 170
Land and food, 99–101, 100f–101f
Land tenure systems, 285–286
Land use and land change (LULC), 285–286
Land-use conflicts, 285–286
Legal documents, 150–152, 150t–152t
Levelized cost of energy (LCOE), 246
Life cycle analysis (LCA), 26

Life cycle assessment (LCA), 77–79, 78f–79f
Livelihood, 114
Local scale, 51–56, 53f

M
Malaysia, Water-Energy-Food nexus in, 242–251
Mapping risky and vulnerable zoonotic hot spots, 206–208
Millennium Development Goals (MDGs), 186
Model description, 119–128
Modeling multisector and complex systems, 208–210
Modeling vulnerability and resilience, 206–208
Model selection, 119–128
Multicomponent simulation model (WEAP), 269
Multicriteria decision-making (MCDM) process, 210, 289–290
Multiregion input–output (MR)IO, 54, 60, 76–77
Multi-Scale Integrated Assessment of Society and Ecosystem Metabolism (MuSIASEM), 22–24, 127–128
Multi-sector investment, 147–149

N
National development plan (NDP), 113
National Meteorology System (SNM), 273–274
National scale, 56–58, 58f
Natural resources management tool, 192–193
NexSym, 128
Nexus Capacity Score Card, 304
Nexus planning
linking socioecological interactions with, 201–203
simplifying socioecological systems, 203–204
Nexus thinking, 286
Novel infectious diseases, 200

O
Off-grid solar energy project, 190–191
Off-grid solar systems, 191
Options analysis, 260

P
Pairwise comparison matrix (PCM), 210
Palmer Drought Severity Index (PDSI), 41
Paradox of interdisciplinarity, 226
Planning, 3–8
before and after 2015, 7–8
benefits for adopting, 9
circular models, 8f
COVID-19, 7
factors, 5–7
sustainability, 9
4th Industrial Revolution (4IR), 6–7
Poor management of wastewater systems, 186
Precision irrigation technologies, 302

Q
Quantitative nexus research, data challenges of, 28–29

R
Regional development agenda, 147
Regional development context, 149–152
REMap tool, 22
Resource management tools
advantages, 19t–20t
alternative methodologies, 24–28
climate, land, energy use, and water strategies (CLEWs), 18–21
disadvantages, 19t–20t
economic analysis models, 24–26, 25t
environmental impact related, 26
indicators, 28
metrics, 28

Resource management tools (*Continued*)
  multi-scale integrated analysis of societal and ecosystem Metabolism (MuSIASEM) model, 22–24
  quantitative nexus research, data challenges of, 28–29
  REMap tool, 22
  statistics, 27
  systems analysis, 26–27
  Water, Energy, Food Nexus Tool 2.0, 21–22
Robust decision-making framework, 258f
Robust decision support (RDS), 258, 270t
Rural transformation in Africa, 281

S

SADC Centre for Renewable Energy and Energy Efficiency (SACREEE), 146
SADC WEF nexus regional dialogues project, 165
Scales of application
  analytical hierarchical process (AHP) methodology, 56–57
  causal loop diagram (CLD), 52
  cooking sessions, 51–52
  drivers–pressures–states–impact–response (DPSIR) methodology, 58, 59f
  EUROSTAT data, 57
  global-level system behavior modulates, 49
  greenhouse gas (GHG) emissions, 57
  gross domestic product (GDP), 60–61
  higher-level nexus studies, 59–61
  household to subnational, 51–56, 53f
  hyperconnected system, 49
  limits to growth study, 59–60
  local scale, 51–56, 53f
  multiregional input e output (MRIO), 54, 60
  national scale, 56–58, 58f
  spatial interactions in the nexus, 61–62
  system dynamics model (SDM), 55–56
  Urmia Lake Restoration Programme (ULRP), 56
Smallholder agricultural technology adoption in Africa, 279–280
  African context, 280–281
  farmer technology adoption, 286–289
  literature review, 282–286
  research designs for incorporating priori assessment, 289–290
Socioeconomic system, 103–104, 104f
Software development kit (SDK), 36
Songwe River Basin
  case study description, 92–106
  extended conceptual model, 97–106
    climate, 103, 103f
    ecosystem and human health, 104–106, 105f
    energy, 101–102, 102f
    land and food, 99–101, 100f–101f
    socioeconomic system, 103–104, 104f
    water, 97–99, 98f
  high-level conceptual model, 96–97, 97f
Songwe River Basin Development Programme, 93–94
WEF nexus analysis approach, 95–96
WEF nexus, conceptualizing, 96–106
Songwe River Basin (SRB), 35–36
Songwe River Basin Development Programme, 35–36, 93–94
Souss Massa, 262
South and Southeast Asia (SEA), Water-Energy-Food nexus in, 238–241
Southern African Development Community (SADC), 2
  agricultural sector planning, policy, and legal documents, 154–156, 154t–156t
  capacity development, 164
  energy sector planning, policy, and legal documents, 152–156, 153t–154t
  guiding discourse, 164
  legal documents, 150–152, 150t–152t
  multi-sector investment, 147–149
  policy, 150–152
  policy- and decision-making levels, 161, 163
  regional development agenda, 147
  regional development context, 149–152
  regional implementing entities, 161–163
  regional multistakeholder platforms, 162–163
  regional technical level, 161, 163
  SADC Centre for Renewable Energy and Energy Efficiency (SACREEE), 146
  SADC WEF nexus regional dialogues project, 165
  sustainable development, 149–152
  water, energy, and food security, 146–147
  water–energy–food nexus approach, 156–159
    policy- and decision-making levels, 157–159
    regional implementing entities and other partners, 159
    regional technical level, 159
    regional water multistakeholder, 159
    weak coordination of programs, 159
  water sector planning, 150–152, 150t–152t
  WEF nexus conceptualization, 148–149
  WEF nexus framework, 162–163
  WEF nexus governance framework, 160–162, 160f

WEF sectors within, 161
Southern African Development Community (SADC) regional scale, 193
Spatial interactions in the nexus, 61–62
Spatial scale, 135
Stakeholder-driven participatory process, 258
Stakeholder engagement, 260f
Stockholm Energy Institute (SEI), 18
Stockholm International Water Institute (SIWI), 299–300
Successful upscaling, 306
Sustainability, 9
Sustainable Development Goals (SDGs), 16, 83–84, 111, 242–251, 299–300
Sustainable human and environmental health, 199–201
   integrated health indices, 214–215
   linking socioecological interactions with nexus planning, 201–203
   modeling multisector and complex systems, 208–210
   WHEN nexus indices for South Africa, 211–214
   wildlife on human health, 203–208
Sustainable livelihoods approach (SLA), 191
Sustainable resource security
   key messages, 322–325
   WEF nexus, 321–322
System dynamics model (SDM), 55–56, 71–74
Systems analysis, 26–27
Systems dynamics modeling (SDM), 26–27

T

Tethys portal, 36
Three-way interconnectedness, 4f
Timeline for nexus approach in Jordan, 261f
Time/temporal scale, 135–136
Tools
   agent-based modeling (ABM), 74–76
   causal loop diagrams (CLD), 68–71, 72t
   conceptual maps, 68–71, 69f
   EXIOBASE, 76–77
   extended conceptual maps, 68–69
   goal-seeking behavior, 69–71
   integrated assessment models (IAMs), 79–81
   life cycle assessment (LCA), 77–79, 78f–79f
   multiregion input–output (MR)IO, 76–77
   system dynamics modeling, 71–74
Tracing polarities, 69–71
Traditional fossil fuel energy sources, 282
Trans-boundary Basin Nexus Assessment (TBNA) Methodology, 18
Transboundary systems, 226
Tugwi-Mukosi Dam toward water, energy, and food security
   definition, 171–174
   energy security, 176
   food security, 176
   indicators of, 172t
   policy implications, 178–179
   socioeconomic expectations, 175t
   tourism promotion, 176–177
   water security, 174
   WEF linkage conceptual framework, 171

U

United Nations Climate Change Conference, 274
United Nations (UN) Sustainable Development Goals (SDGs), 181–182
Urban water cycle, 247f
Urmia Lake Restoration Programme (ULRP), 56

W

Water, 97–99, 98f
Water and energy links to food sector, 251f
Water–energy–food (WEF) nexus, 235–238
   achievement of, 191
   analysis approach, 95–96
   available models, 117–119
   contents, 3f
   critical findings and key take-home messages, 251–252
   defining, 3–5
   envisaged importance of, 2
   food and nutrition security, 187–188
   index, 84
   in Malaysia, 242–251
   past to present discourse, 181–183
   planning, 3–8
      before and after 2015, 7–8
      benefits for adopting, 9
      circular models, 8f
      COVID-19, 7
      factors, 5–7
      sustainability, 9
      4th Industrial Revolution (4IR), 6–7
   SDGs dimensions and, 185–187
   South and Southeast Asia, 238–241
   Southern African Development Community (SADC), 2
   synergies and trade-offs in, 188–191
   three-way interconnectedness, 4f
   tool for natural resources management, 183–185
   upscaling and outscaling, 192–193
   water–health–environment–nutrition (WHEN), 3–4
Water, Energy, Food Nexus Tool 2.0, 21–22
Water-health-environment-nutrition (WHEN), 3–4
   defining, 202–203
   indices for South Africa, 211–214
   sustainability indicators, 205t

Water-health-environment-nutrition (WHEN) (*Continued*)
and sustainability indicators, 204–206
Water sector, 129–130, 130f

Water security, 246–248
Water system and tariff loops, 248f
WEF Nexus Tool 2.0, 126–127
Wildlife on human health, 203–208

World Economic Forum, 300–301

Z
Zoonotic diseases, 200

Printed in the United States
by Baker & Taylor Publisher Services